FRONTIERS IN MYCOLOGY

ERRATUM
Frontiers in Mycology
Ed. D.L. Hawksworth

Page 122, Figure 6.7
The vertical axis should read:

Yield (mg) in mycorrhizal microcosms

Frontiers in Mycology

Honorary and General Lectures from the Fourth International Mycological Congress, Regensburg, Germany, 1990

Edited by

D.L. Hawksworth

International Mycological Institute

C · A · B International

on behalf of the
International Mycological Association
and the
International Mycological Institute

C · A · B International
Wallingford
Oxon OX10 8DE
UK

Tel: Wallingford (0491) 32111
Telex: 847964 (COMAGG G)
Telecom Gold/Dialcom: 84: CAU001
Fax: (0491) 33508

Published on behalf of the
International Mycological Institute
(an Institute of C · A · B International)
Ferry Lane
Kew
Surrey TW9 3AF
UK
Tel: 081 940 4086
Telex: 9312102252 MI G
Telecom Gold/Dialcom: 84: CAU009
Fax: 081 332 1171

For full details of services available from the International Mycological Institute please contact the Institute Director at this address

and the
International Mycological Association

(Please address all enquiries regarding the Association to the Secretary-General:
Dr C Kurtzman
National Center for Agricultural Utilization Research
1815 N University Street, Peoria, IL 61604, USA
Tel: 309–685–4011
Fax: 309–671–7814)

A CIP catalogue record for this book is available from the British Library

ISBN 0–85198–698–6 √

Typeset by Leaper & Gard Ltd, Bristol
Printed and bound in the UK

Contents

Preface

Mycology, the scientific study of fungi, is relevant to an enormous range of aspects of human endeavour and enquiry. In consequence the subject is rapidly developing in multifarious areas. At the same time published information is also becoming increasingly dispersed, and the need for authoritative reviews consequently more acute for both research scientists and lecturers at colleges and universities.

The series of International Mycological Congresses, held under the auspices of the International Mycological Association (IMA), provide a unique overview of the current state of pure and applied mycology worldwide. The Fourth International Mycological Congress (IMC4), held at the University of Regensburg in Germany from 28 August–3 September 1990, was no exception. This was the largest IMC so far held, attracting nearly 1700 participants drawn from 59 countries. The scientific sessions comprised 33 symposia and 16 workshops grouped under the following themes: Systematics and Evolution; Morphology and Ultrastructure; Ecology; Genetics and Physiology; Biotechnology and Applied Mycology; Pathology; Special Topics. This activity was further supplemented by pre- and post-congress workshops and field excursions.

The IMC4 sessions were complemented by plenary lectures by leading specialists selected by the IMC4 Organizing Committee to report on particularly topical or exciting aspects of mycology. Recognizing that these contributions collectively went some way to fulfilling the current demand for authoritative overviews, these are presented here as snapshots of selected *Frontiers in Mycology.*

These contributions fall into four main categories. First, molecular biology and growth: including molecular aspects of ageing and the potential of fungi as a model system for its study (K. Esser); growth and development as seen from the molecular standpoint (J.G.H. Wessels); and the significance of siderophores in fundamental processes which is only now being fully appreciated (G. Winkelmann). Second, evolution and phylogeny: drawing attention to neoteny, cessation in development at early stages in understanding relationships at higher and lower ranks (H. Kreisel); and the major but often unrecognized problems of analogy and homology in the evolution of lichenized fungi (J. Poelt). Third, the relevance of fungi both in ecosystems and to man: the importance of mycorrhizas in different ecosystems (D.J. Read); the wide range of problems they cause in medicine (O. Male); their particular role as agents of respiratory diseases (J. Lacey); the utilization of lichens by man (D.H.S. Richardson); the potential of modified amatoxins and phallotoxins for research (H. Faulstich); and the array of inputs to biotechnology and importance to that of systematics (J.D. Miller). Fourth, conservation and education: the changes taking place in Europe and the need for active conservation of fungi (E. Arnold); and ways to make mycological teaching more appealing and place it on a firmer base (J. Webster). The text of only one of the General Lectures was not received by the time this volume went to press: New secondary metabolites of basidiomycetes (by J.W. Steglich).

The IMA, founded in 1971, constitutes the Section for General Mycology of the International Union of Biological Sciences (IUBS) and provides the focal point for all aspects of mycology worldwide. While the most visible result of its activities is the series of IMC's including that which led to this volume, it also has a series of active regional committees concerned with Africa, Asia, Europe, and Latin America. The IMA also provides a channel for mycological inputs into broader IUBS initiatives which currently include improvement in the stability of names, and ecosystem function aspects of biodiversity. Further information is included in *IMA News*, distributed to all members of national and regional societies with mycological interests affiliated to the IMA.

Assistance from CAB International and the International Mycological Institute in realizing this present volume is gratefully acknowledged. Further I, and all who participated in IMC4, owe a particular debt to Professor Dr A. Bresinsky (Secretary-General), Professor Dr J. Poelt (Congress President and Chairman of the Organizing Committee), Professor J. Webster (President, IMA 1983–1990), their colleagues, and convenors. The organizers of IMC5, to be held at the University of British Columbia, Vancouver on 14–21 August 1994 will have a hard act to emulate.

D.L. Hawksworth
President, International Mycological Association
13 December 1990

List of Contributors

Dr E. Arnolds, Biological Station, Landbouwuniversiteit Wageningen, Kampsweg 27, 9418 PD Wijster, The Netherlands

Professor Dr K. Esser, Ruhr-Universität Bochum, Postfach 10 21 48, D-4630 Bochum 1, Germany

Professor Dr H. Faulstich, Max-Planck-Institut für medizinische Forschung, Postfach 10 38 20, D-6900 Heidelberg 1, Germany

Professor Dr H. Kreisel, Institut für Allgemeine und Spezielle Mikrobiologie, Ernst-Moritz-Arndt-Universität, Jahnstraße 15, 0-2200 Greifswald, Germany

Dr J. Lacey, AFRC Institute of Arable Crops Research, Rothamsted Experimental Station, Harpenden, Herts AL5 2JQ, UK

Professor Dr O. Male, University of Vienna, Medical School, Alserstrasse 4, A-1090 Vienna, Austria

Dr J.D. Miller, Plant Research Centre, Agriculture Canada, Ottawa, Ontario K1A 0C6, Canada

Professor Dr J. Poelt, Institut für Botanik, Karl-Franzens Universität Graz, Holteigasse 6, A-8010 Graz, Austria

Professor D.J. Read, Department of Animal and Plant Sciences, University of Sheffield, Sheffield S10 2UQ, UK

Professor D.H.S. Richardson, School of Botany, Trinity College, University of Dublin, Ireland

Professor J. Webster, Department of Biological Sciences, University of Exeter, Prince of Wales Road, Exeter EX4 4PS

Professor J.G.H. Wessels, Department of Plant Biology, University of Groningen, Kerklaan 30, 9751 NN Haren, The Netherlands

Professor Dr G. Winkelmann, Mikrobiologie und Biotechnologie, Universität Tübingen, Auf der Morgenstelle 1, D-7400 Tübingen, Germany

Molecular Biology and Growth

1

Molecular Aspects of Ageing: Facts and Perspectives

K. Esser, *Ruhr-Universität Bochum, Postfach 10 21 48, D-4630 Bochum 1, Germany.*

ABSTRACT Ageing, a syndrome which accompanies all forms of life, leads eventually via irreversible alterations in metabolism to cellular death. Senescence is expressed in different phenotypes, considering the wide spectrum from unicellular microbes up to higher organisms (plants, animals and humans). Since the essential metabolic processes do not differ on a molecular level, senescence should also follow a common principle at that level. Research in gerontology has, until quite recently, concentrated on higher organisms, especially animals which are difficult to examine at the molecular level. In order to understand the molecular basis of ageing, a model organism is required which is well suited for both genetical and biochemical analysis. These conditions are fulfilled in an ideal manner by the filamentous fungus *Podospora anserina.* In this ascomycete the onset of senescence occurs after a rather short vegetative growth. A detailed analysis of this syndrome has shown that the onset of senescence depends on nuclear genes, the expression of which is naturally controlled by environmental conditions. Surprisingly, in ageing cells an extrachromosomal genetic element was detected. This genetic trait is identical with a plasmid. In the juvenile cell this plasmid is integrated into the mitochondrial genome as an intron (2539 bp) in a gene which codes for an enzyme of the respiratory chain (subunit 1 of cytochrome oxidase). Once liberated from the mitochondrium, the plasmid is able to propagate autonomously, to "infect" juvenile cells and transform these immediately to senescence. A model is presented which explains the formation of the mitochondrial plasmid and its interdependence to the nuclear genome. Its relevance for ageing in animal systems is discussed. Further, the senescence data obtained with *Podospora* is relevant to cancer research and genetic engineering.

Frontiers in Mycology. Honorary and General Lectures from the Fourth International Mycological Congress, Regensburg 1990. Edited by D.L. Hawksworth.

Introduction

Many complex phenomena in biological systems can be traced back to rather simple basic mechanisms. Ageing, a syndrome which accompanies all forms of life, leading eventually *via* irreversible alterations in metabolism to cellular death, is expressed in different phenotypes. In microorganisms or in single cells of more complex organisms, even in the cell cultures of mammals, ageing may become manifest as a cessation of cellular divisions. In more complex organisms, however, the onset of senescence may become apparent in a decrease of somatic growth, or by failure of asexual or sexual propagation, or a combination of both. As shown in Figure 1.1, the various phases of the life-span of an organism such as a mammal, or a population of microbes such as bacteria, follow the same pattern.

It seems therefore possible, that a general and fundamental event is responsible for the onset of senescence in all its different manifestations, in

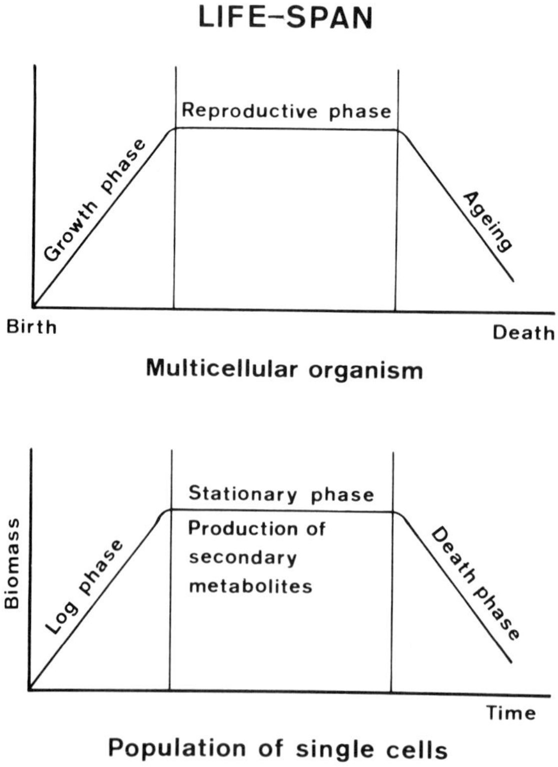

Figure 1.1. Various phases of the life-span of complex organisms as compared to a batch culture of a population of single cells.

just the same way that there is generalized control for the onset of maturity or the productive phase in microbes. Naturally this event is under genetic control, as are all processes of life, and the expression of senescence is expected to depend on environmental conditions according to the general equation:

$$\text{GENOTYPE} + \text{ENVIRONMENTAL CONDITIONS} \rightarrow \text{PHENOTYPE}$$

That formula enables us to draw two main but opposing theories of senescence discussed in the past into a single concept:

1. Senescence is understood as a last and final step of tissue differentiation coded for genetic information, as are all previous steps such as maturation (Comfort, 1974; Orgel, 1963, 1970).
2. Senescence is understood as an "accident" caused by an accumulation of stochastic errors occurring during the whole life-span of an organism at different metabolic levels, such as DNA replication, translation, or membrane synthesis (Hayflick, 1973; Strehler *et al.*, 1971; Strehler, 1976).

It is evident that either of these theories would fulfill the requirements leading to the "phenotype senescence".

In the past, most research activities have been devoted to animal systems. However, due to the difficulties in manipulating them genetically or at the molecular level, most investigations were concerned only with the phenotypic effects of ageing (see for example Danon *et al.*, 1981).

Any endeavour to trace ageing to its molecular basis needs a system exhibited by an organism with a relatively simple organization accessible to both genetic and biochemical investigation at the molecular level. These requirements are fulfilled by the filamentous ascomycete fungus *Podospora anserina* (Sordariales). Some decades ago, Rizet and Marcou observed that this species regularly ages after prolonged vegetative growth. The peripheral cells of the mycelium stop growing and eventually die (Fig. 1.2). These authors showed further that the onset of ageing (25 days in their strains) varies in different races and is therefore dependent on the genotype. A comparable variety could be caused by the alteration of the environmental conditions. The most interesting observation was that in ageing mycelia an infective principle exists, which, when transferred via hyphal fusion to juvenile cells, immediately brings about ageing (Rizet, 1953; Rizet and Marcou, 1954; Marcou, 1961).

This very interesting syndrome of senescence could not be investigated further at that time, because molecular genetics was still in its infancy.

I begin by reviewing our experimental work with *Podospora*, and then integrate these data into a model. Finally perspectives derived from our work on the senescence syndrome of *Podospora* will be discussed with respect to ageing in animal cells and the development of research strategies in cancer and for biotechnology.

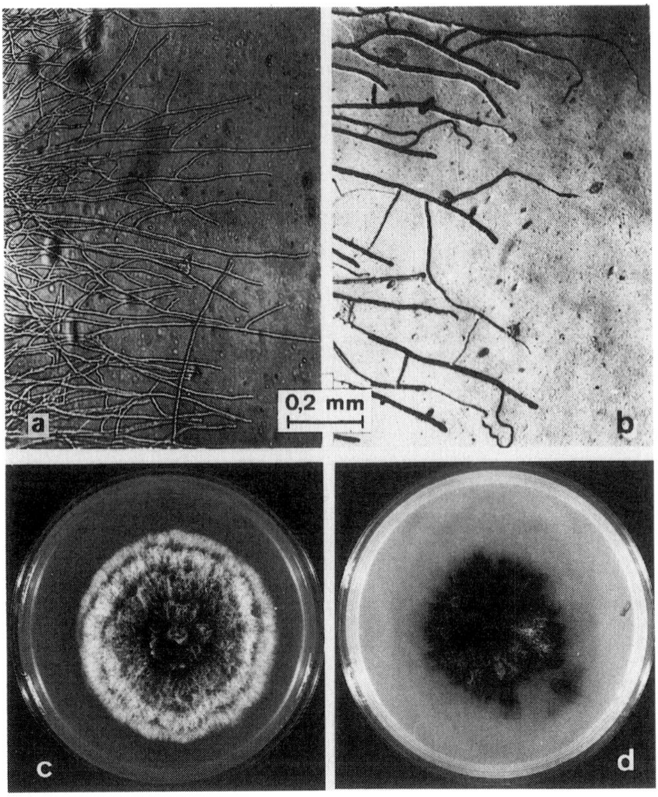

Figure 1.2. *Podospora anserina.* Juvenile and senescent phenotypes; (a) juvenile mycelium; (b) senescent mycelium; (c) juvenile colony; (d) senescent colony. (Adapted from Esser and Tudzynski, 1980.)

Facts

Experimental work with *Podospora*

Chromosomal genetics

In studying a series of morphological mutants of *P. anserina*, the onset of ageing, which in wild strains occurs after about 25 days (corresponding to a mycelial growth length of about 20 cm), was found to be postponed indefinitely by the synergistic action of at least two genes (Esser and Keller, 1976). As may be seen from Figure 1.3, the double mutant *i viv* seems to have "eternal life". It has now been kept in laboratory cultures for almost

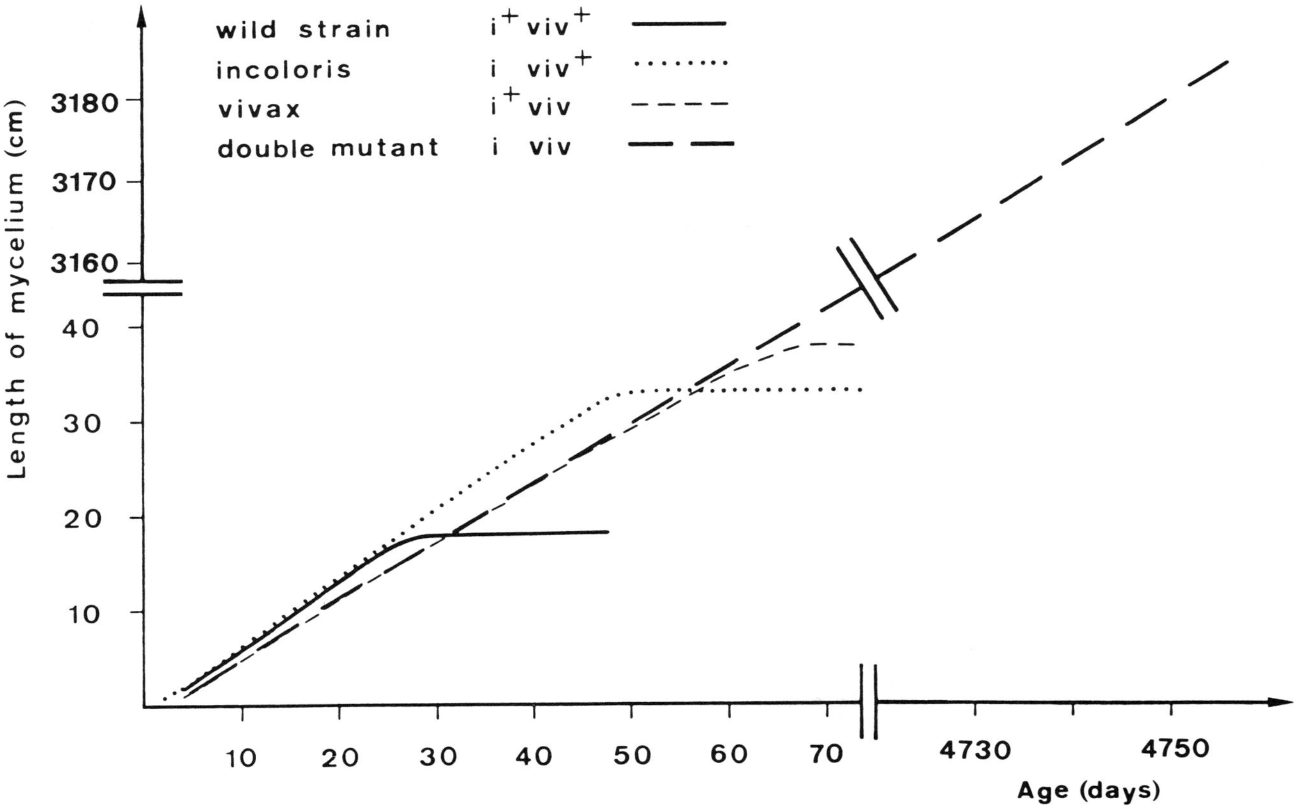

Figure 1.3. *Podospora anserina*. Growth curves of the wild strain i^+viv^+, the mutants *viv* and *i* and the double mutant *i viv*. (Adapted from Esser and Keller, 1976.)

Table 1.1. Characteristics of long-lived double mutants of *P. anserina*. (From Tudzynski *et al.*, 1982.)

		Senescence			pl DNA			
Strain	Life-span	in-ducible	irre-versible	mt DNA	inte-grated	liber-ated	repli-cated	ex-pressed
wild	25 days	+	+	normal	+	+	+	+
i viv	9 years	–		normal	+	+	(+)	–
gr viv	4 years	+	–	normal	+	–	+	+

15 years and its mycelium has grown over a distance of 23 m. It will be noted from the data in Figure 1.3 that the single mutants *i* and *viv* respectively do not show any prolongation of their life-span.

In addition to these two genes there are other morphogenetic genes which have, in a different pairwise combination, a considerable influence on the postponement of ageing. One of these double mutants to be dealt with later (Table 1.1) is *gr viv* which has been growing for almost ten years in consecutive race tubes (Tudzynski and Esser, 1979).

Conclusion: The onset of senescence is under nuclear control.

Physiology

The onset of senescence may also be postponed indefinitely if the fungus is grown on media containing sublethal amounts of inhibitors of the mitochondrial (mt) DNA and of mitochondrial protein synthesis. In contrast to the mutated genes, these substances have no permanent effect since after removal from the medium ageing proceeds. Thus, there is no curing of the infective principle by application of these metabolic inhibitors. These experiments have also revealed the existence of a pre-senescent phase, morphologically indistinguishable from the juvenile phase. During this phase the onset of ageing is only prevented by the mtDNA inhibitors (Esser and Tudzynski, 1977; Tudzynski and Esser, 1977, 1979). Changes in mtDNA in rat liver cells during ageing were also observed by Murray (1982).

Conclusion: The mtDNA must be involved in some way in the onset of senescence.

Extrachromosomal genetics

Since it was shown that protoplasts, when infected with the DNA of senescent strains, become senescent at once, a more detailed analysis of the DNA was performed. According to biochemical and biophysical experiments, ageing cells contain, in addition to nuclear (n) and mtDNA, a third DNA species which was identified as a plasmid (Stahl *et al.*, 1978). This, the first plasmid detected in a filamentous fungus, was termed pDNA (Fig. 1.4).

In order to discover the origin of this plasmid the mtDNA was submitted to a detailed biochemical and biophysical analysis. Having a contour length of 30 μm, corresponding to 95 kb, it was found to be one of the largest chondrioms detected so far in fungi. Its physical map (restriction sites) and genetic map (functional sites) revealed that in juvenile cells the plDNA is an integral part of the mtDNA, whereas in ageing strains it is liberated (Kück *et al.*, 1981; Kück and Esser, 1982; Fig. 1.5).

The *Podospora* plasmid was cloned in a bacterial vector (pBR 322) in *E. coli.* After recovery, it was again found to replicate and express in *Podospora* as earlier postulated (Esser *et al.*, 1980; Stahl *et al.*, 1980).

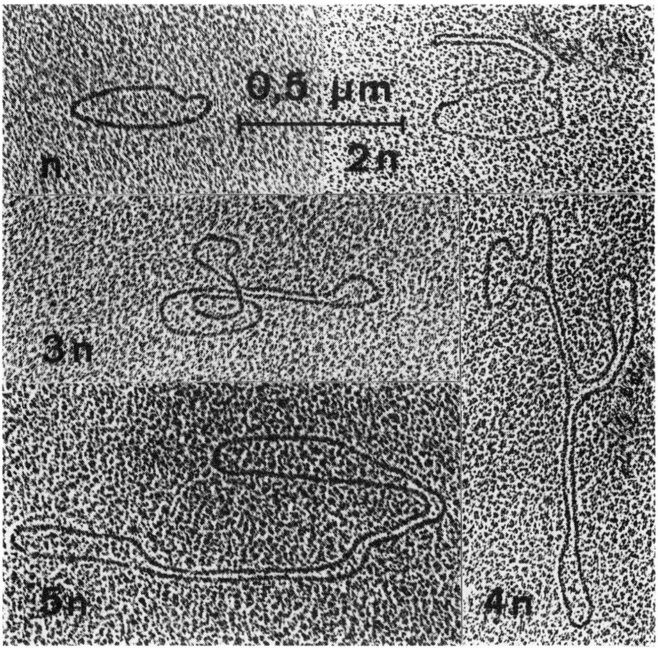

Figure 1.4. *Podospora anserina.* plDNA, the monomer (n) and oligomers (2n–5n) of the open circular form. (From Stahl *et al.*, 1980.)

Conclusion: The senescence plasmid (plDNA) of *Podospora anserina* is part of the mitochondrial DNA. In ageing strains it is excised. Once separated from the mtDNA, the plasmid is able to replicate autonomously and to express its phenotype.

Nuclear mitochondrial correlation

In order to test whether there was a correlation between the two kinds of genetic information which are associated with the senescence phenomenon

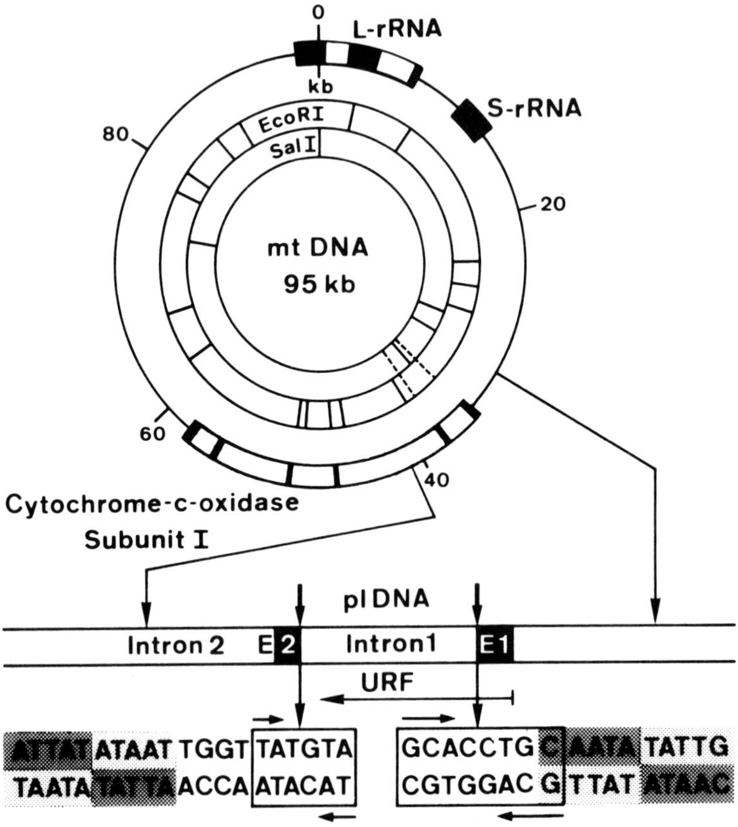

Figure 1.5. *Podospora anserina.* Structure of the mtDNA 95 kb and the mt plasmid. The mtDNA is represented by its physical map (EcoRI- and SalI sites). The genes for the large (L) and small (S) rRNA serve as markers. The integrated plasmid is identical to intron I of the cytochrome-c-oxidase gene, subunit I. It may be seen that from exon I an unidentified reading frame (URF) proceeds into the plasmid. Insertion sites are characterized by palindromes each side of which is followed by inverted repeats. (From Osiewacz and Esser 1984.)

in *Podospora anserina*, the nuclear genes, and the mitochondrial plasmid, the long-lived strains described above were used for transformation experiments (Tudzynski *et al.*, 1980; Stahl *et al.*, 1982). Two types of double mutants were used: the first (*gr viv*) when infected with the plasmid became senescent while the second (*i viv*) did not. Both mutants, however, showed no alterations of their mtDNAs in comparison to the juvenile wild strain. As may be seen from Table 1.1, the mutant *gr viv* is not able to liberate the plasmid, whereas the mutant *i viv*, albeit liberating the plasmid, does not allow its expression. Consequently *gr viv* when infected through a plasmid shows both replication and expression (senescence) of the plasmid. In contrast, in *i viv*, when infected, the plasmid probably replicates, but certainly fails to express. Wright and Cummings (1983) reported that the plasmid entered the nucleus, but this was questioned by Koll (1986). There is some circumstantial evidence from our laboratory that it may also enter the mitochondria.

Conclusion: The liberation and expression of the senescence plasmid depends on the genetic constitution of the nuclear genome.

Molecular approach

In order to obtain information about the function of the plasmid at the molecular level and about its liberation mechanism, a DNA sequence analysis was performed. This revealed first that the *Podospora* plasmid was an intron through a comparison of the DNA sequences at its 5′ and 3′ end with conserved sequences from introns of other eukaryotes, such as *Saccharomyces cerevisiae* and *Zea mays*.

As may be seen from a simplified presentation of the sequencing data (Fig. 1.5), it became evident that the plasmid is flanked on both sides by mtDNA sequences coding for polypeptides. A comparison of the derived amino acid sequence with those of other well-defined mitochondrial genes resulted in the identification of this region. According to these data, the plasmid is deliminated by exon I and II of the cytochrome-c-oxidase (subunit I) gene which is homologous to the corresponding genes of *S. cerevisiae* and *N. crassa*.

Furthermore, it is notable that palindromic sequences at both ends of the plasmid are present which may be involved in the process of its excision. In addition, the plasmid contains an unidentified reading frame (URF) and an origin of replication (ORI). Inverted repeats were observed in both adjacent exons a few base pairs apart (Osiewacz and Esser, 1984).

Conclusion: The *Podospora anserina* plasmid is an intron of a structural mitochondrial gene (subunit I of cytochrome-c-oxidase).

Rearrangements of mtDNA

In addition to the liberation of the plDNA, many other circular DNA sequences were detected in ageing strains of *Podospora* by Cummings and his collaborators (see Kück, 1989). Most of these were found as circular plasmid-like structures and termed beta, gamma, delta, epsilon and theta senDNA. They all exhibited homology to mtDNA and could be mapped at distinct sites of the mitochondrial genome (Fig. 1.6). They differ in one essential way from the plDNA because they are not identical with specific introns or exons. This explains why there are differences in length within each species of senDNA. However, there is one additional interesting feature, that the mitochondrial sites of the senDNAs are "hot spots" of recombination (Schulte *et al.*, 1988) (see Fig. 1.6). From this it follows that in ageing mycelia the mtDNA undergoes numerous recombinations leading to cellular death if essential genetic functions of the mtDNA are eventually lost.

These results are in agreement with ones from *Saccharomyces cerevisiae* where "petite mutants" are known within which all kinds of DNA rearrangements have taken place, even up to a complete loss. However, since this yeast is able to mobilize energy *via* anaerobic glucose degradation, the petite cells do not become senescent.

It is also apparent that the plDNA is involved in the mtDNA rearrangements, because elimination of the plDNA by mutation brings up a certain stability of the mt genome (Kück *et al.*, 1985; Schulte *et al.*, 1988, 1989). These strains grow very slowly but show longevity and never become senescent. Albeit the fact that they are not able to ferment glucose they have overcome the essential loss of the COI gene product by an alternative pathway as known for *Neurospora crassa* (Lambowitz and Slayman, 1971).

Quite recently a very peculiar element was found in the ageing mycelia of race A, a linear plasmid (pAL2-1) of about 8.4 kb with 5′ terminal polypeptides. DNA sequencing has revealed that it contains two URFs with significant homologies to viral RNA or DNA polymerases, respectively. Furthermore, an 85% homology to the third fragment of BglII was found (Fig. 1.6). Thus, the linear plasmid may become liberated from the mtDNA in a similar way to plDNA and the senDNAs. The most interesting observation was that the presence of pAL2-1 considerably postpones the onset of senescence, probably by an interaction involving the liberation of plDNA (Osiewacz *et al.*, 1989; Osiewacz, 1990, and unpubl. data).

Conclusion: In addition to the plDNA in senescent mycelia, other genetic elements are liberated by rearrangements of the mtDNA at specific sites. One of these elements, a linear plasmid, postpones the onset of senescence considerably by interference with the plDNA.

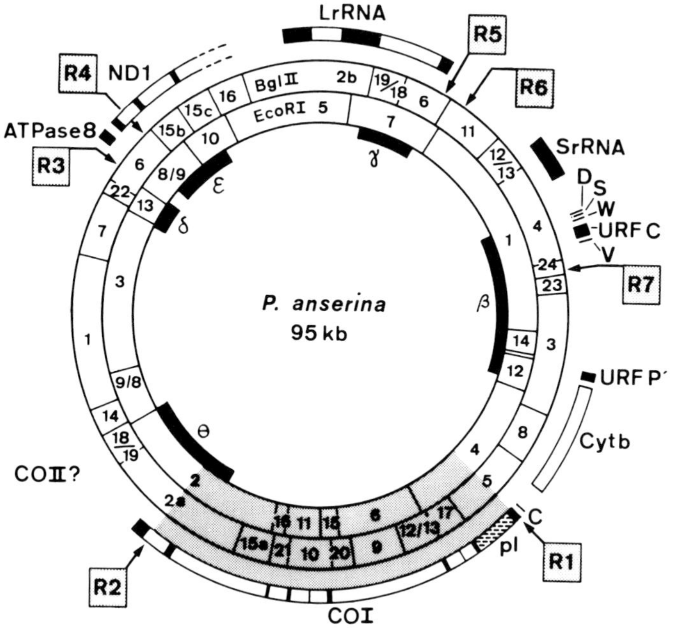

Figure 1.6. Restriction map of wild-type mtDNA, from *Podospora anserina*. The location of various mitochondrial genes is given, as well as amplified variable senDNAs (β, γ, δ, ε, θ) found in senescent cultures. The position of the invariable mobile intron (plDNA) is indicated. The deletion which has occurred in the mutant genomes is *shaded*. In addition, those regions which are marked by *arrows* are involved in rearrangements of mtDNA in mutant strains. The sites of recombinations (R) have been numbered from 1 to 7. Recombinant plasmids containing different fragments used as probes in transcript analysis are shown in the inner circle. (From Schulte *et al.*, 1988.)

The mobile intron model of ageing in Podospora

The combined experimental data allow us to propose a molecular model explaining the onset of senescence in *Podospora*. This model, shown diagramatically in Figure 1.7, is derived from the fact that the plasmid is identical with mitochondrial intron I of the gene for subunit I of the cytochrome-c-oxidase (COI). A further essential point is that this intron (= plasmid) contains an origin of replication (ORI). In juvenile cells the mosaic COI gene is processed and all introns become lost during the maturation of mRNA. Further, the plasmid (= intron) is found as a distinct cccDNA molecule in the cytoplasm of ageing mycelia, and in this state is able to replicate and to transform juvenile into senescent cells. These phenomena may be explained as follows:

1. As a rare event (accident?) DNA splicing may occur. This has been demonstrated for an mtDNA gene (cob) of *Saccharomyces cerevisiae* (Carignani *et al.*, 1983). The base sequence of the unidentified reading frame (URF) of the plDNA probably codes for a DNA maturase (Karsch *et al.*, 1987).

2. The liberated intron I becomes ligated and is thereby converted into a cccDNA plasmid. Ligation of linear DNA has been observed in yeast transformation experiments (Suzuki *et al.*, 1983).

3. Due to the presence of an ORI the newly established circular plasmid should be able to replicate. At this point the fate of the other introns which also might be subject to DNA splicing has to be considered. It is possible that they are unable to ligate and (or), due to the absence of ORI, not able to replicate and thus become lost.

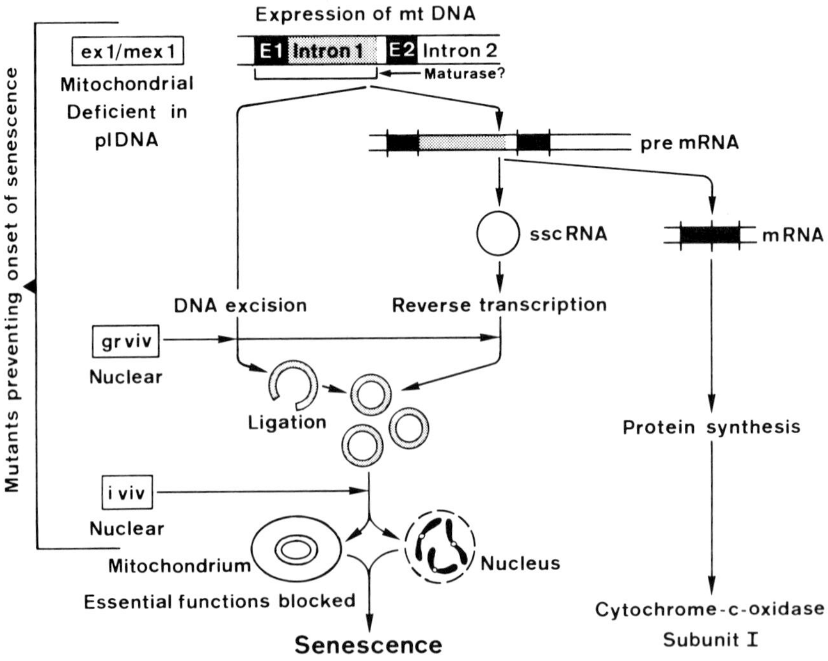

Figure 1.7. *Podospora anserina*. The mobile intron model to explain onset and genetic control of senescence. For details see text. (From Kück *et al.*, 1985; modified from Esser 1985.)

4. Since there is a possibility that the plasmid DNA is able to enter both mitochondria (Stahl *et al.*, 1982) and nuclei (Wright and Cummings, 1983), it is not unreasonable to postulate that the plasmid may act as a transposon. This view is supported by the intact mtDNA intron I being flanked by inverted repeats. Structures of this kind occur at both ends of prokaryotic transposons. Once having entered the mitochondrion, it is possible that the plasmid becomes transposed to mtDNA, probably at its site of homology (= intron I), and thereby blocks the processing of the COI gene. The lack of cytochrome-c-oxidase naturally inhibits the life functions of the cell and promotes senescence. This view is supported both by the possibility of recombination between the plasmid and mtDNA (Stahl *et al.*, 1982), and the absence of cytochrome-c-oxidase in senescent mycelia (Belcour and Begel, 1978).

5. The intervention of the nuclear genome into the process of ageing also needs to be explained. According to the present data, there are three different sites involved. At first the onset of senescence depends on the genome, because, as shown some decades ago, the various wild strains, albeit showing no morphological difference, vary widely in the onset of ageing (e.g. 7 days in race A and 106 days in race E: Marcou, 1961). An attribution to specific genes was not possible at that time. In contrast, the second and the third points of genomic intervention are caused by single genes, as already summarized in Table 1.1. The gene *vivax* plays a central role. When combined with *grisea* one may postulate that DNA splicing is blocked and in combination with the *incoloris* plasmid integration is inhibited.

6. It must not be overlooked that the formation of the plasmid could be explained differently, if one wanted to avoid the rather unusual first step of DNA splicing. This would require normal RNA splicing, but one would need to postulate a reverse transcriptase function at the RNA level to synthesize the double stranded plasmid molecule as has been well established from retrotransposable elements (Michel and Lang, 1985). This view is supported by DNA sequencing: the ssRNA derived from intron I may easily form a secondary structure bringing both ends into close vicinity (Osiewacz and Esser, 1984). In this modification of the model, the genes *vivax* and *grisea* affect RNA. Further support comes first from a comparison of the amino acid sequence of the plDNA with reverse transcriptases from retroviruses and retrotransposons (Michel and Lang, 1985); and secondly Steinhilber and Cummings' (1986) discovery that nuclear-mitochondrial protein extracts DNA polymerase activity with the characteristics of a reverse transcriptase-like enzyme.

7. Finally, another function of the plDNA should be considered. The mobile intron might code for a recombinase, as has been postulated for an intron encoded polypeptide in yeast (Kotylak *et al.*, 1985). This assumption was derived from the data obtained from the analysis of the extrachromosomal longevity mutants (Koll *et al.*, 1985; Schulte *et al.*, 1988, 1989). As briefly mentioned above, mitochondrial rearrangements by recombination are characteristic features of senescence, but if there is no recombinase these events are prevented.

8. From the data discussed here, there are circumstantial indications that the mobile intron identical with the plDNA may code for a multifunctional protein, as found for the gene products of introns in *Saccharomyces cerevisiae* (Lazowska *et al.*, 1980; Macreadie *et al.*, 1985; Colleaux *et al.*, 1986). The functions of a protein of this kind (maturase, reverse transcriptase, recombinase) can only be determined if the gene product of the plDNA is identified. Corresponding experiments are hampered by the enzymes of this type being present only in low amounts.

Conclusion: The mobile intron model certainly has some areas of uncertainty, and is also not the only way to explain the onset of senescence in *Podospora anserina* by an interaction of nuclear and extranuclear elements. However, at present it seems to be the model of least implausibility.

Experimental work with other fungi

Subsequent to or parallel with the studies on *Podospora*, some other fungi became subjects of ageing research. In some cases (*Didymium iridis, Physarum polycephalum*) alterations of nuclear DNA were found. In others, such as *Saccharomyces cerevisiae, Aspergillus amstelodamii, Neurospora crassa, N. intermedia*, and *Podospora curvicolla*, extranuclear elements were detected in ageing cells. In the most thoroughly studied *N. crassa*, Bertrand and Griffith (see Osiewacz, 1990) identified both circular and linear plasmids of mitochondrial origin. However, in none of these cases did the data permit the establishment of a model concerning nuclear/extranuclear interaction with respect to the onset and control of senescence (see Osiewacz, 1990). But there is also no obvious contradiction or gross divergency to the *Podospora* model.

In the same way as in *Podospora*, a complete understanding of the onset of senescence in other fungi will only be obtained if the gene products of the nuclear genes and the extranuclear elements are identified and if the fate and action of the mobile plasmids becomes evident.

Perspectives

In evaluating the general significance of the experimental data obtained during investigations of the senescence syndrome in fungi, especially in *Podospora*, there are several topics which can be considered.

Animal cells

The establishment of a molecular model to explain senescence in a filamentous fungus brings us to the central question: is the *Podospora* case an exception or is ageing in this fungus at least caused by the same basic mechanism as in animal cells? If there is any correspondence, it can be expected only at a molecular level as there are so many differences at later stages (e.g. immune systems in animal cells) that the final mechanism of phenotypic realization may be completely different.

Developing this line of thought leads first to the most thoroughly studied subjects, the fibroblasts. Their pattern of ageing is quite similar to that of *Podospora*: they become senescent rather early after vegetative propagation and they may be easily modified to "eternal life" by genetic alterations (see Hayflick, 1973; Danon *et al.*, 1981). The application of the model involves the participation of mtDNA, but as far as I know the experimental data has become available on DNA studies on young and ageing mitochondria (Shmookler Reis and Goldstein, 1983; Shmookler Reis *et al.*, 1983*a,b*). These authors found that extrachromosomal copies of circular DNA sequences located in the genome of human fibroblasts between Alu repeat clusters were amplified during *in vitro* and *in vivo* ageing. Inter-Alu circles were also found to amplify in lymphocytes. There were no statistically significant differences correlated with growth rate and life-span in the mtDNA of normal and mutant fibroblasts and in permanent cell lines respectively. However, there is some experimental evidence that mtDNA sequences smaller than the mitochondrial genome do exist, the most prominent of which has a size of 0.65 kb. Although the significance of these mtDNA molecular species is not yet understod, these data should encourage further research along these lines. This view is supported by the detection of extrachromosomal cccDNA in the nuclei of monkey and human cells (Krolewski *et al.*, 1982; Calabretta *et al.*, 1982). More knowledge of the origin and interaction of these different DNA circles will prove the general applicability of the Mobile Intron Model.

However, there is another point on which the *Podospora* data might contribute, namely an experimental confirmation of the postulate made in the Introduction, that both widely discussed opposing mechanisms of ageing – differentiation and stochastic errors – are involved in the realization of senescence in this fungus. It could be an accident that the plasmid is liberated from the mtDNA. Certainly the time of its liberation, replication, and integration is coded by genetic information. Further discussion of these two theories should await more experimental data.

Cancer research

It becomes more and more evident that biological phenomena showing rather diverse phenotypes, such as cancer, may be traced back to a common molecular mechanism (Weinberg, 1984). It is therefore very tempting to speculate as to a possible relationship between the senescence phenomenon of *Podospora anserina* and cancer. The "eternal life" of some mutants may be compared with the capacity of a cancer cell (e.g. HeLa cells) to propagate indefinitely. The limited growth of the wild strain, however, is equivalent to the limited growth capacity of normal cells (e.g. fibroblasts). In recent years it has been shown that cancer is both under nuclear and extranuclear control.

As an example of the first, the comprehensive work of Anders *et al.* (1984) may be cited. These authors unequivocally proved that the onset of melanomic tumours in *Xiphoporus* depended on the presence of specific nuclear genes. In *Podospora anserina* permanent growth also depends on the action of specific nuclear genes.

The *Podospora* genes act only synergistically because each mutation alone does not induce longevity (= cancer). These observations are in accordance with a model according to which cancer in mammalian cells is brought about by at least a two-step process, due to the subsequent spontaneously occurring mutations at specific loci (Weinburg, 1984). This theory is supported by experimental results:

1. The *ras*-oncogene, isolated from human bladder carcinoma, is only able to transform normal fibroblasts if these are concomitantly transformed by a second oncogene, such as the viral or cellular *myc* gene (Land *et al.*, 1983).
2. Primary rat kidney cells can only be transformed to cancer cells *via* a double infection with the T24 Harvey *ras*-1 gene (or polyoma virus middle T) and the adenovirus early region 1A (Ruley, 1983).
3. Hamster fibroblasts can only be transformed to cancer cells with a *ras*-gene when previously immortalized by other genetic elements (Newbold and Overell, 1983).

Apart from the fact that as in *Podospora* no single genetic alteration may provoke cancer, the homology goes further in that the synergistic action is not restricted to two specific genes. In contrast, there may be a replacement of one or the other genetic element by a different one, as is also the case in *Podospora* where longevity is obtained by many different gene combinations, such as *i viv* and *gr viv.*

An example of the influence of extrachromosomal elements on cancer is the postulate that oncogenes are activated by the insertion of mt sequences into the genome (Galloway and McDougal, 1983). In *P. anserina* the liberation of mitochondrial plasmids stops unlimited growth. From this it follows that senescence in *P. anserina* is the inverse of the phenomenon of cancer.

Another interesting homology between *Podospora* and mammalian cancer is apparent from the model of Conway (1983) who postulated that cancer depends on the inappropriate expression of non-body introns. In *Podospora* the appropriate expression of the plasmid (= intron) avoids longevity, the equivalent of cancer.

One may object that speculations of this kind, which postulate a molecular mechanism inhibiting the decay of mtDNA as responsible for cancer growth, contradict the discussed models which consider damage in mtDNA as a first step to the onset of cancer (e.g. Wilkie and Evans, 1982). The latter hypothesis is probably an extension of Warburg's theory (Warburg *et al.*, 1924; Warburg, 1956) that cancer may originate from disorders in aerobiosis, and is based mainly on observations that carcinogenic substances cause the degeneration of mtDNA in the yeast *Saccharomyces cerevisiae* (Egilson *et al.*, 1976) or in liver cells (Niranjan *et al.*, 1982). However, an experimental proof would require a study of mtDNA from cancer and non-cancer cells. Detailed data along these lines were presented by Shmookler Reis and Goldstein (1983); they found in a comparative study of mtDNA of normal and mutant fibroblasts and of transformed cells, no evidence of petite-type deletions from human DNA, either at late passage or in individual clones of fibroblasts. These data seem to support our ideas and place speculations on a stronger basis for the development of future research strategies.

Genetic engineering

The discovery of a mitochondrial plasmid has opened up new perspectives for genetic engineering. As shown elsewhere (Esser *et al.*, 1982), based on the postulated homology bacteria/mitochondria (Gray and Doolittle, 1982), plasmids of mitochondrial origin and even parts of mtDNA were found to be valuable vehicles for eukaryotic cloning (Tudzynski and Esser, 1982; Esser *et al.*, 1983). A general possibility for eukaryotic molecular cloning was consequently created. That each eukaryotic cell contains in its mitochondria genetic material which may be used to construct vectors and allow cloning into the donor organism, makes the eukaryote accessible as a host. This may avoid the many complications found in the biotechnological application of gene cloning in *E. coli* strains (MacLeod, 1980).

This concept has the advantage that, due to the presence of an eukaryotic origin of replications, the vectors replicate autonomously and can also be present in numerous copies in the eukaryotic cell (see Esser, 1986). However, during industrial scaling up, integrative vectors, albeit present mostly in one copy, were found to be more stable whereas autonomously replicating vectors may be lost during the many cell divisions which take place in large fermentors (Esser and Mohr, 1986).

Summary

The study of senescence in the ascomycete *Podospora anserina* has shown that the onset of senescence occurs after a rather short period of vegetative growth. A detailed analysis of this syndrome has shown that the onset of senescence depends on nuclear genes, the expression of which is naturally controlled by environmental conditions. It was rather surprising that in ageing cells an extrachromosomal genetic element was also detected.

This genetic trait is identical with a plasmid. In the juvenile cell this plasmid is integrated into the mitochondrial genome as an intron (2539 bp), in a gene which codes for an enzyme of the respiratory chain (subunit 1 of cytochrome oxidase). Once liberated from the mitochondrium, the plasmid is able to propagate autonomously, to "infect" juvenile cells, and transform these immediately to senescence.

The model presented explains the formation of the mitochondrial plasmid and its interdependence with the nuclear genome. This is the first model to explain the onset of senescence on a molecular level. Its relevance for ageing in animal systems is discussed. Further, the senescence data obtained with *Podospora* is relevant to cancer research and genetic engineering.

Last, but not least, is another consequence of this research. For many years plasmids were considered to be a peculiarity of prokaryotes. When the first plasmid (the 2 μm DNA) was found in *Saccharomyces* by Hollenberg and others in the early 1970s, this was considered as an exception. However, the discovery of the plDNA in *Podospora*, and at the same time plasmids in *Zea mays* by Pring evidently encouraged many groups to look for extrachromosomal elements in eukaryotes. This has resulted in mobile or integrated extrachromosomal DNA being detected in many eukaryotes (fungi, plants, and animals) and becoming a normal tool for genetic investigations for both fundamental research and biotechnology (see Esser *et al.*, 1986; Mohr and Esser, 1990; Meinhardt *et al.*, 1990).

Acknowledgements

The experimental work described in this contribution was supported by grants from the Deutsche Forschungsgemeinschaft (Bonn). For reading the manuscript I am indebted to Drs U. Kück (Bochum), P.A. Lemke (Auburn/USA), and U. Stahl (Berlin).

Note

This chapter is a revised and extended version of Esser, K. (1985) Genetic control of ageing, the mobile intron model. In: Bergener, M., Ermini, M. and Stähelin, H.B. (eds) *The 1984 Sandoz Lectures in Gerontology. Thresholds in Ageing.* Academic Press, London, pp. 1–20.

References

Anders, F., Schartl, M., Barnekow, A. and Anders, A. (1984) *Xiphophorus* as an *in vivo* model for studies on oncogenes. *Advances in Cancer Research* 42, 191–275.

Belcour, L. and Begel, O. (1978) Lethal mitochondrial genotypes in *Podospora anserina*, a model for senescence. *Molecular and General Genetics* 163, 113–123.

Calabretta, B., Robberson, D.L., Barrera-Soldana, H.A., Lambrou, T.P. and Saunders, G.F. (1982) Genome instability in a region of human DNA enriched in Alu repeat sequence. *Nature, London* 296, 219–225.

Carignani, G., Groudinski, O., Frezza, D., Schiavon, E., Bergantino, E. and Slonimsky, P.P. (1983) A mRNA maturase is encoded by the first intron of the mitochondrial gene for the subunit I of cytochrome oxidase in *Saccharomyces cerevisiae. Cell* 35, 733–741.

Colleaux, L., D'Auriol, L., Betermier, M., Cottarel, G., Jacquier, A., Galibert, F. and Dujon, B. (1986) Universal code equivalent of a yeast mitochondrial intron reading frame is expressed into *Escherichia coli* as a specific double strand endonuclease. *Cell* 44, 521–533.

Comfort, A. (1974) The position of ageing studies. *Mechanisms of Ageing and Development* 3, 1–10.

Conway, F.J. (1983) A proposed model of cancer as the inappropriate expression of non-body introns. *Theoretical Biology* 100, 1–24.

Danon, D., Shock, N.W. and Marois, M. (1981) *Aging: a Challenge to Science and Society.* Oxford University Press, Oxford.

Egilson, V., Evans, I.H. and Wilkie, D. (1976) Primary antimitochondrial activity of carcinogens in *Saccharomyces cerevisiae.* In: Bucher, T. *et al.* (eds), *Genetics and Biogenesis of Chloroplasts and Mitochondria.* Elsevier/North-Holland Biomedical Press, Amsterdam, pp. 885–895.

Esser, K. (1985) Genetic control of ageing, the mobile intron model. In: Bergener, M., Ermini, M. and Stähelin, H.B. (eds), *The 1984 Sandoz Lectures in Gerontology. Thresholds of Ageing.* Academic Press, London, pp. 1–20.

Esser, K. (1986) Genmanipulation mittels Mitochondrien-DNA. *Spektrum der Wissenschaft* 2, 40–49.

Esser, K. and Keller, W. (1976) Genes inhibiting senescence in the ascomycete *Podospora anserina. Molecular and General Genetics* 144, 107–110.

Esser, K. and Mohr, G. (1986) Integrative transformation of filamentous fungi with respect to biotechnological application. *Process Biochemistry* 21, 153–159.

Esser, K. and Tudzynski, P. (1977) The prevention of senescence in the ascomycete *Podospora anserina* by the antibiotic tiamulin. *Nature, London* 265, 454.

Esser, K. and Tudzynski, P. (1980) Senescence in fungi. In: Thimann, K.V. (ed.),

Senescence in Plants. CRC Press, Boca Raton, vol. 2, pp. 67–83.

Esser, K., Kück, U., Stahl, U. and Tudzynski, P. (1983) Senescence in *Podospora anserina* and its implication for genetic engineering. In: Jennings, D.H. and Rayner, A.D.M. (eds), *The Ecology and Physiology of the Fungal Mycelium.* Cambridge University Press, Cambridge, pp. 342–352.

Esser, K., Stahl, U., Tudzynski, P. and Kück, U. (1982) *Hybridvektor, Verfahren zur Verbesserung der Amplifikation und Expression von Hybridvektoren durch die Verwendung von mitochondrialer DNA.* EP 82104582.0 HOE 81/F 134.

Esser, K., Tudzynski, P., Stahl, U. and Kück, U. (1980) A model to explain senescence in the filamentous fungus *Podospora anserina* by the action of plasmid-like DNA. *Molecular and General Genetics* 178, 213–216.

Esser, K., Kück, U., Lang-Hinrichs, C., Lemke, P., Osiewacz, H.D., Stahl, U. and Tudzynski, P. (1986) *Plasmids of Eukaryotes.* Springer, Berlin.

Galloway, D.A. and McDougal, J.K. (1983) The organic potential of herpes simplex viruses: evidence for a hit-and-run mechanism. *Nature, London* 302, 21–24.

Gray, M.W. and Doolittle, W.F. (1982) Has the endosymbiont hypothesis been proven? *Microbiological Reviews* 46, 1–42.

Hayflick, L. (1973) The biology of human ageing. *American Journal of Medical Sciences* 265, 433–445.

Karsch, T., Kück, U. and Esser, K. (1987) Mitochondrial group I introns from the filamentous fungus *Podospora anserina* code for polypeptides related to maturases. *Nucleic Acids Research* 15, 6743–6744.

Koll, F. (1986) Does nuclear integration of mitochondrial sequences occur during senescence in *Podospora*? *Nature, London* 324, 597–598.

Koll, F., Belcour, L. and Vierny, C. (1985) A 1100-bp sequence of mitochondrial DNA is involved in senescence process in *Podospora*: Study of senescent and mutant cultures. *Plasmid* 14, 106–117.

Kotylak, Z., Lazowska, J., Hawthorne, D.C. and Slonimski, P.P. (1985) Intron encoded proteins of mitochondria: key elements of gene expression and genomic evolution. In: Quagliariello, E., Slater, E.E., Plamieri, F., Saccone, C. and Kroon, A.M. (eds), *Achievements and Perspectives of Mitochondrial Research.* Elsevier, Amsterdam, vol. 2, pp. 1–20.

Krolewski, J.J., Bertelsen, A.H., Humayun, M.U. and Rush, M.G. (1982) Members of the Alu family of interspersed repetitive DNA sequences are in the small circular DNA population of monkey cells grown in culture. *Journal of Molecular Biology* 154, 399–415.

Kück, U. (1989) Mitochondrial DNA rearrangements in *Podospora anserina. Experimental Mycology* 13, 111–120.

Kück, U. and Esser, K. (1982) Genetic map of mitochondrial DNA in *Podospora anserina. Current Genetics* 5, 143–147.

Kück, U., Stahl, U. and Esser, K. (1981) Plasmid-like DNA is part of mitochondrial DNA in *Podospora anserina. Current Genetics* 3, 151–156.

Kück, U., Osiewacz, H.D., Schmidt, U., Kappelhoff, B., Schulte, E., Stahl, U. and Esser, J. (1985) The onset of senescence is affected by DNA rearrangements of a discontinuous mitochondrial gene in *Podospora anserina. Current Genetics* 9, 373–382.

Lambowitz, A.M. and Slayman, C.W. (1971) Cyanide-resistant respiration in *Neurospora crassa. Journal of Bacteriology* 108, 1087–1096.

Land, H., Parada, L.F. and Weinberg, R.A. (1983) Tumorigenic conversion of primary embryo fibroblasts requires at least two cooperating oncogenes. *Nature, London* 304, 596–602.

Lazowska, J., Jacq, C. and Slonimski, P.P. (1980) Sequence of introns and flanking exons in wildtype and box 3 mutants of cytochrome b reveals an interlaced splicing protein coded by an intron. *Cell* 22, 333–348.

MacLeod, A.J. (1980) Biotechnology and the production of proteins. *Nature, London* 285, 136.

Macreadie, I.G., Scott, R.M., Zinn, A.R. and Butow, R.A. (1985) Transposition of an intron in yeast mitochondria requires a protein encoded by that intron. *Cell* 41, 395–402.

Marcou, D. (1961) Notion de longévité et nature cytoplasmique du déterminant de la sénescence chez quelques champignons. *Annales des Sciences Naturelles, Botanique* 12, 653–764.

Meinhardt, F., Kempken, F., Kämper, J. and Esser, K. (1990) Linear plasmids among eukaryotes: fundamentals and application. *Current Genetics* 17, 89–95.

Michel, F. and Lang, B.F. (1985) Mitochondrial class II introns encode proteins related to the reverse transcriptase of retroviruses. *Nature, London* 316, 641–643.

Mohr, S. and Esser, K. (1990) Mobile genetische Elemente bei Eukaryoten – Grundlagen und Anwendung für die Biotechnologie. In: Präve, P., Schlingmann, M., Crueger, W., Esser, K., Thauer, R. and Wagner, F. (eds), *Jahrbuch für Biotechnologie.* Carl Hanser Verlag, München, vol. 3, pp. 5–27.

Murray, M.A. (1982) Changes in mitrochondrial DNA during aging. *Mechanisms of Aging and Development* 20, 233–241.

Newbold, R.F. and Overell, R.W. (1983) Fibroblast immortality is a prerequisite for transformation by EJ c-Ha-ras oncogene. *Nature, London* 304, 648–651.

Niranjan, B.G., Bhat, N.K. and Avadhani, N.G. (1982) Preferential attack of mitochondrial DNA by aflatoxin B1 during hepatocarcinogenesis. *Science, New York* 215, 73–75.

Orgel, L.E. (1963) The maintenance of the accuracy of protein synthesis and its relevance to aging. *Proceedings of the National Academy of Sciences, USA* 49, 517–521.

Orgel, L.E. (1970) The maintenance of accuracy of protein synthesis and its relevance to aging: a correction. *Proceedings of the National Academy of Sciences, USA* 67, 1476–1480.

Osiewacz, H.D. (1990) Molecular analysis of aging processes in fungi. *Mutation Research* 237, 1–8.

Osiewacz, H.D. and Esser, K. (1984) The mitochondrial plasmid of *Podospora anserina*: a mobile intron of a mitochondrial gene. *Current Genetics* 8, 299–305.

Osiewacz, H.D., Hermanns, J., Marcou, D., Triffi, M. and Esser, K. (1989) Mitochondrial DNA rearrangements are correlated with a delayed amplification of the mobile intron (plDNA) in a long-lived mutant of *Podospora anserina. Mutation Research* 219, 9–15.

Rizet, G. (1953) Sur la longévité des souches de *Podospora anserina. Comptes Rendu hebdomadaire des Séances de l'Académie des Sciences, Paris* 237, 1106–1109.

Rizet, G. and Marcou, D. (1954) Longévité et sénescence chez l'ascomycète *Podospora anserina. Comptes Rendus VII. Congress International, Section*

Botanique, Paris 10, 121–128.
Ruley, H.E. (1983) Adenovirus early region 1A enables viral and cellular transforming genes to transform primary cells in culture. *Nature, London* 304, 602–606.
Schulte, E., Kück, U. and Esser, K. (1988) Extrachromosomal mutants from *Podospora anserina*: permanent vegetative growth in spite of multiple recombination events in the mitochondrial genome. *Molecular and General Genetics* 211, 342–350.
Schulte, E., Kück, U. and Esser, K. (1989) Multipartite structure of mitochondrial DNA in a fungal longlife mutant plasmid. *Plasmid* 21, 79–84.
Shmookler Reis, R.J. and Goldstein, S. (1983) Mitochondrial DNA in mortal and immortal human cells. *Journal of Biological Chemisty* 258, 9078–9085.
Schmookler Reis, R.J., Lumpkin, C.K., McGill, J.R., Riabowol, K.T. and Goldstein, S. (1983*a*) Genome alteration during *in vitro* and *in vivo* aging: amplification of extrachromosomal circular DNA molecules containing a chromosomal sequence of variable repeat frequency. *Cold Spring Harbor Symposia on Quantitative Biology* 47, 1135–1139.
Shmookler Reis, R.J., Lumpkin, C.K., McGill, J.R., Riabowol, K.T. and Goldstein, S. (1983*b*) Extrachromosomal circular copies of an 'inter-Alu' unstable sequence in human DNA are amplified during *in vitro* and *in vivo* aging. *Nature, London* 301, 394–398.
Stahl, U., Kück, U., Tudzynski, P. and Esser, K. (1980) Characterization and cloning of plasmid-like DNA of the ascomycete *Podospora anserina. Molecular and General Genetics* 178, 639–646.
Stahl, U., Tudzynski, P., Kück, U. and Esser, K. (1982) Replication and expression of a bacterial-mitochondrial hybrid plasmid in the fungus *Podospora anserina. Proceedings of the National Academy of Sciences, USA* 79, 3641–3645.
Stahl, U., Lemke, P.A., Tudzynski, P., Kück, U. and Esser, K. (1978) Evidence for plasmid-like DNA in a filamentous fungus, the ascomycete *Podospora anserina. Molecular and General Genetics* 162, 341–343.
Steinhilber, W. and Cummings, D.J. (1986) A DNA polymerase activity with characteristics of a reverse transcriptase in *Podospora anserina. Current Genetics* 10, 389–392.
Strehler, B.L. (1976) Elements of a unified theory of ageing, integration of alternative models. In: Platt, D. (ed.), *Alternstheorien.* Schattauer, Stuttgart.
Strehler, B.L., Hirsch, G., Gussek, D., Johnson, R. and Bick, M. (1971) Codon-restriction theory of ageing and development. *Journal of Theoretical Biology* 33, 429–443.
Suzuki, K., Imai, Y., Yamashita, I. and Fukui, S. (1983) *In vivo* ligation of linear DNA molecules to circular forms in the yeast *Saccharomyces cerevisiae. Journal of Bacteriology* 155, 747–754.
Tudzynski, P. and Esser, K. (1977) Inhibitors of mitochondrial function prevent senescence in the ascomycete *Podospora anserina. Molecular and General Genetics* 153, 111–113.
Tudzynski, P. and Esser, K. (1979) Chromosomal and extrachromosomal control of senescence in the ascomycete *Podospora anserina. Molecular and General Genetics* 173, 71–84.
Tudzynski, P. and Esser, K. (1982) Extrachromosomal genetics of *Cephalosporium*

acremonium. II. Development of a mitrochondrial DNA hybrid vector replicating in *Saccharomyces cerevisiae*. *Current Genetics* 6, 153–158.

Tudzynski, P., Stahl, U. and Esser, K. (1980) Transformation to senescence with plasmid-like DNA in the ascomycete *Podospora anserina*. *Current Genetics* 2, 181–184.

Warburg, O. (1956) On the origin of cancer cells. *Science, New York* 123, 309–314.

Warburg, O., Posener, K. and Gegelein, E. (1924) Über den Stoffwechsel der Carcinomzelle. *Biochemische Zeitschrift* 152, 309–344.

Weinburg, R.A. (1984) Molekulare Grundlagen von Krebs. *Spektrum der Wissenschaft*, 1984 (January), 58–71.

Wilkie, D. and Evans, I. (1982) Mitochondria and the yeast cell surface: implications for carcinogenesis. *Trends in Biochemical Science* 7, 147–151.

Wright, R.M. and Cummings, D.J. (1983) Integration of mitochondrial gene sequences within the nuclear genome during senescence in a fungus. *Nature, London* 302, 86–88.

2

Fungal Growth and Development: a Molecular Perspective

J.G.H. Wessels, *Department of Plant Biology, University of Groningen, Kerklaan 30, 9751 NN Haren, The Netherlands.*

ABSTRACT Fungi do not only provide good model systems to study the molecular processes common to all forms of eukaryotic life. Molecular studies starting with the ultimate product of fungal genes, such as the cell wall, or with the activities of genes themselves (reverse genetics) can yield information that is highly relevant to understanding biological features that are unique to fungal life. Such studies generate novel ideas about the way hyphae colonize their substrates by apical extension, and how they may become involved in the formation of aerial structures, such as fruit-bodies.

Fungi as model organisms

Fungi have often been recommended as model organisms for research on general biological problems in eukaryotes. As stated in a well-known textbook on molecular biology (Watson *et al.*, 1987) in the prelude to chapters on eukaryotic cells: "If we can solve the same problem equally well with yeast or with human cells, common sense tells us to stick with the simpler and less expensive system." Indeed, the yeasts *Saccharomyces cerevisiae* and *Schizosaccharomyces pombe* are contributing impressively to our understanding of general biological processes such as the secretion of proteins (Schekman, 1985), and mitosis and meiosis (Nurse, 1985; Kassir *et al.*, 1988, Herskowitz, 1988, 1989), even to the point of analysing human cell cycle genes in yeast (Lee and Nurse, 1987). In addition, genetic and molecular studies on the regulation of transcription in yeast, for example of the *GAL* genes (Johnston, 1987), have greatly contributed to our general understanding of transcriptional regulation in eukaryotes

Frontiers in Mycology. Honorary and General Lectures from the Fourth International Mycological Congress, Regensburg 1990. Edited by D.L. Hawksworth.

(Ptashne, 1988). The success of these studies undoubtedly derives from the ease of genetic manipulations with these unicellular fungi (Struhl, 1983), in addition to their well-known genetic structure as analysed by classical methods.

Some filamentous fungi, notably those that have a good background of classical genetics such as *Neurospora crassa* and *Aspergillus nidulans*, have also yielded to some of the new techniques of molecular genetics so successfully developed for yeasts (Timberlake and Marchall, 1989) and thus may also serve as model organisms. For instance, *A. nidulans* is being used to analyse mitosis (Morris *et al.*, 1989) and the structure of activators of gene expression (Andrianopoulos and Hynes, 1990). These haploid systems will continue to provide important information, particularly because of the availability of mutants. However, the new genetics has also revolutionized the work with animals and plants. Because many of the problems in cell biology have actually been defined in animals and plants, and because the study of these organisms is of direct relevance to medicine and agriculture, it is only natural that they attract most attention from molecular biologists. Some of the animal (e.g. *Drosophila*) and plant systems (e.g. *Arabidopsis thaliana*) also have a long history of classical genetics and a genome size just a few times larger than that of fungi. Therefore, in the future, a glorious role for filamentous fungi, as once bestowed on *Neurospora crassa* in the discovery of the one gene:one enzyme relationship, seems less likely.

Considering the emphasis now being put on studies with animal and plant systems, it is therefore even more regrettable that relatively few molecular studies are being devoted to fungi in their own right. In studies using fungi as model systems, they are often referred to as lower eukaryotes, a misnomer also used once myself (see Zantinge *et al.*, 1979). However, there is nothing lower about fungi when compared to animals and plants. They deserve their own place among the kingdoms of organisms (Whittaker, 1969) on the basis of their ancestry, uniqueness, and importance.

Fungi in their own right

The presence of cell walls sets fungi and plants apart from animals. The importance of cell walls in determining the evolution and life-style of plants has been well described (Bonner, 1974; Alberts *et al.*, 1989). In brief, the division of walled cells in meristems generates the immobile multicellular plant body that depends on light for energy but cannot exploit organic resources for energy and building materials. Although light is a high-quality energy source, the amount of energy acquired is nevertheless relatively small because of the limited surface area exposed to light. However, the absence of mobility and energy-consuming homeostatic mechanisms guarantees that the absorbed energy suffices for the build-up and propagation of the plant. The scarcity of homeostatic mechanisms, compared to animals, is compensated for by a high degree of developmental flexibility in response to environmental conditions.

Perhaps the single most important difference between (filamentous) fungi and (vascular) plants is that fungal cells do not originate by divisions in multicellular entities (meristems) but always grow individually as long filaments by apical extension. The individually growing hyphae form branches and laterally fuse with each other to form a loose network, the mycelium. While the absorption of organic nutrients allows for a high rate of duplication of cytoplasm (Trinci, 1978), the cytoplasm also appears to move constantly towards the growing hyphal apices, leaving highly vacuolated or moribund hyphae behind. A vivid description of this mode of growth, largely based on the observations and concepts of A.H.R. Buller and M. Langeron, has been given by Gregory (1984). The fungus is considered to consist of a cytoplasmic mass which moves inside tubes (cell walls) which it builds gradually as it moves forwards and ramifies into branches while quitting the old tubes. This allows for a high rate of movement and spreading of the active cytoplasm which is essential for acquiring the organic resources necessary for rapid growth. This, of course, is a life-style very different from that of plants. The differences become even more manifest when certain hyphae cease to participate in the acquisition of nutrients and water from the substrate and emerge into the air. Here they may grow as undifferentiated individual hyphae, giving a mouldy appearance to the colony, or they may differentiate into various structures bearing asexual spores. The emerged hyphae may also develop some kind of affinity to each other and, while continuing apical growth, participate in the formation of highly organized multicellular structures, such as fruit bodies, in which certain cells perform meiosis and produce sexual spores. Importantly, all these aerial structures, some very large such as "mushrooms" and "brackets", can only grow because of a massive reallocation of materials from the substrate mycelium into these structures.

While all this has been noticed by traditional studies, molecular analyses can open new vistas into the better understanding of these phenomena, which are so characteristic for fungi. After all, certain molecular "inventions", pertinent to the existence of these phenomena, must have enabled fungi to arise as a separate kingdom and to successfully maintain themselves, apart from plants and animals, for about a billion years. Moreover, they have evolved as an indispensable part of life on earth. Most noticeable is the dependence of plants on mycorrhizas, since the origin of land plants in the Devonian (Nicholson, 1975). Little is yet known of the molecular interactions in these symbiotic associations (Kendrick and Berch, 1985). It is also not surprising that the tendency of fungi to invade living plants makes some of them the most common and severe plant pathogens. Molecular studies on phytopathogenic fungi have just begun to yield interesting results (Leong and Holden, 1989). Finally, their invading growth makes fungi the main decomposers of plants, particularly of their solid woody parts. The molecular biology of wood-degrading fungi is now under intensive investigation because of its biotechnological interest (Broda *et al.*, 1989; Gold *et al.*, 1989).

Schizophyllum commune as a model fungus

Rather than attempting a general review of the molecular biology of fungi related to their development, I will concentrate here on our work with *Schizophyllum commune* because over the years this work has addressed various aspects of the typical fungal way of life. Among the larger basidiomycetes, *S. commune* is by far the easiest to grow and fruit in the laboratory. Consequently, it has served to delineate the mating system of basidiomycetes (Kniep, 1920), a line of research continued in great detail by J.R. Raper and his students (Raper, 1966, 1988). Although genetically less well-known than the ascomycetes *Aspergillus nidulans*, *Neurospora crassa*, and certainly *Saccharomyces cerevisiae*, a reasonable number of genes has been mapped in six linkage groups, while the number of synaptonemal complexes in meiosis indicates eleven chromosomes (see Raper, 1988). The nuclear genome has a size of 36–37 $\times$ 10^6 base pairs, that is about eight times the size of the genome of *Escherichia coli.* Apart from the ribosomal-RNA genes, most DNA sequences are present only once in the genome and 33% of these can be found in the 10000–13000 different mRNAs present in the cytoplasm (see Wessels, 1985).

The salient features of the fungal mode of growth and development can best be seen when growing *S. commune* from a single inoculum (Fig. 2.1). Initially the fungus grows only submerged, colonizing the nutrient agar. Then, a few millimetres behind the colony margin, hyphae escape from the medium and grow into the air. This is a clear example of hyphal differentiation because these hyphae must acquire a hydrophobic surface and, for their growth, become completely dependent on resources retrieved from the submerged mycelium. In other fungi some aerial hyphae may further differentiate into spore-forming structures. The process of conidiophore formation is the subject of intense molecular-genetic studies in *A. nidulans*, and several genes instrumental in this process have been cloned and analysed (Boylan *et al.*, 1987; Adams *et al.*, 1988). In *S. commune* no asexual spores borne on aerial hyphae are formed and the mycelium continues to grow with only the formation of unbranched aerial hyphae. In the light and at low carbon dioxide concentrations, however, the aerial hyphae can embark on a developmental route leading to the formation of fruit-bodies, ultimately measuring several centimetres. When a mycelium grown in the dark is exposed to light (Fig. 2.1) aerial hyphae 2–3 mm behind the advancing colony front can become engaged in a branching and aggregation process leading to a ring of fruit-body primordia, while radial extension of the colony continues but at a slower rate (Raudaskoski and Yli-Mattila, 1985; Yli-Mattila *et al.*, 1989*a*). Normally this only occurs in dikaryons, leading to the fusion of unlike nuclei and the formation of recombinant basidiospores in the mature fruit bodies; the whole process is under the control of the mating-type genes

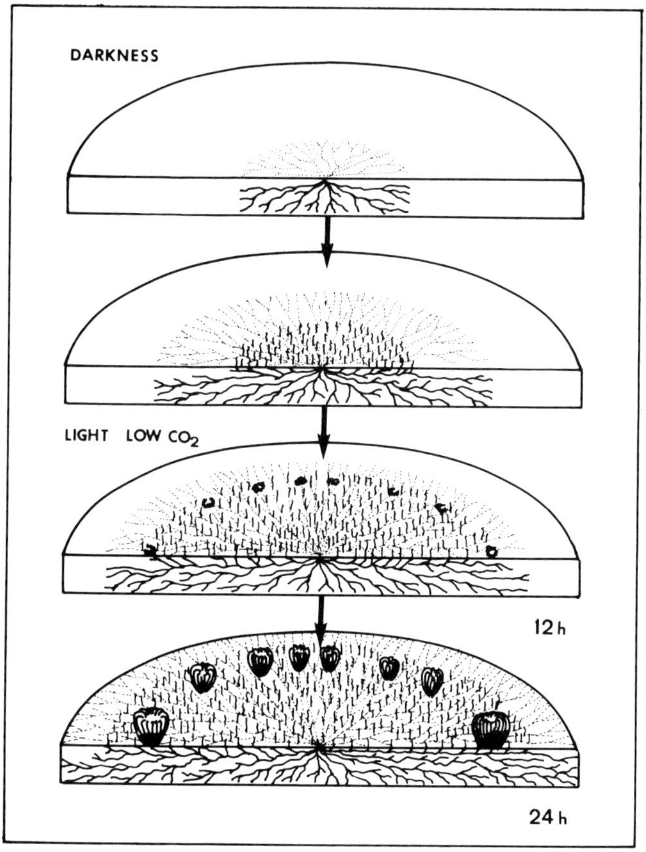

Figure 2.1. Diagram of colony development in *Schizophyllum commune.*

In darkness, only vegetative growth occurs with aerial hyphae differentiating a few millimetres behind the advancing margin of the colony. When illuminated, and at sufficiently low carbon dioxide concentration, such aerial hyphae have a capacity to branch and aggregate, embarking on a morphogenetic route leading to the formation of fruit-bodies. Once formed the fruit-body primordia inhibit formation of new primordia in their neighbourhood. Note that, for clarity, the fruit-body primordia are drawn too large and their numbers too low. (For details see Raudaskoski and Vauras, 1982; van der Valk and Marchant, 1978.)

(Wessels, 1987). However, fruit-bodies (without meiotic cells) may also form in monokaryons carrying so-called haploid fruit-body alleles which bypass the control of the mating-type genes (Yli-Mattila *et al.*, 1989*b*).

The distribution of active cytoplasm in developing colonies of *S. commune* can be seen in autoradiographs in which the ribosomal RNA (rRNA) is hybridized *in situ* with a [^{32}P]-labelled clone of the gene for 18S rRNA (Fig. 2.2). In the dark-grown colony (0 h), the highest concentration of rRNA is in the outermost growing zone matching the region of highest protein synthesis, as measured by incorporation of [^{35}S]sulfate (Yli-Mattila *et al.*, 1989*a*). After 12 h in the light an accumulation of rRNA can already be seen in the ring of developing fruit-body primordia, but after 24 h rRNA is clearly becoming concentrated at the sites where the primordia develop and at the still advancing colony front. Between 24 and 96 h the activity of this front gradually decreases and nearly all rRNA, and thus cytoplasm, apparently moves into the developing primordia. At the same time there is competition between the primordia themselves because at the 96 h state (Fig. 2.2) only about 60% of the primordia still show an 18S rRNA signal (Ruiters and Wessels, 1989*a*). Apparently, there was also translocation of rRNA from abortive to developing primordia.

The above is a modern presentation of a phenomenon discovered earlier using classical analytical methods (Wessels, 1965; Wessels and Sietsma, 1979). These early studies also showed that in the absence of external nutrients, cell wall components from the substrate mycelium and from the abortive primordia are re-utilized to provide for the needs of the developing fruit-bodies. The alkali-insoluble wall complex, containing (1-3)-β/(1-6)-β-glucan and chitin, is degraded while the alkali-soluble (1-3)-α-glucan is not; all these wall components are synthesized at the same time in the developing fruit-bodies. The end result is that the developed fruit-bodies are attached to emptied substrate hyphae and abortive primordia which essentially consist of wall skeletons of (1-3)-α-glucan. This process of cell turnover in the colony may partly depend on the transmigration of cytoplasm, as defined by Gregory (1984), but certainly also involves the degradation of polymers, translocation of the breakdown products, and resynthesis of the polymers, as evident in the case of wall polymers. These processes may be repeated during the formation of basidiospores because detached mature fruit-bodies can shed spores for a long time.

Because the degradation of wall components is such a conspicuous process during fruit-body formation [but also in septal dissolution during nuclear migration in dikaryon formation (see Wessels and Sietsma, 1979)], we made an in-depth study of the structure of the wall of *S. commune*. From this work we have obtained a molecular view of the wall that unexpectedly turned out to be highly relevant for our understanding of its biosynthesis during the apical extension of hyphae, the basic process on which all fungal growth and development depends.

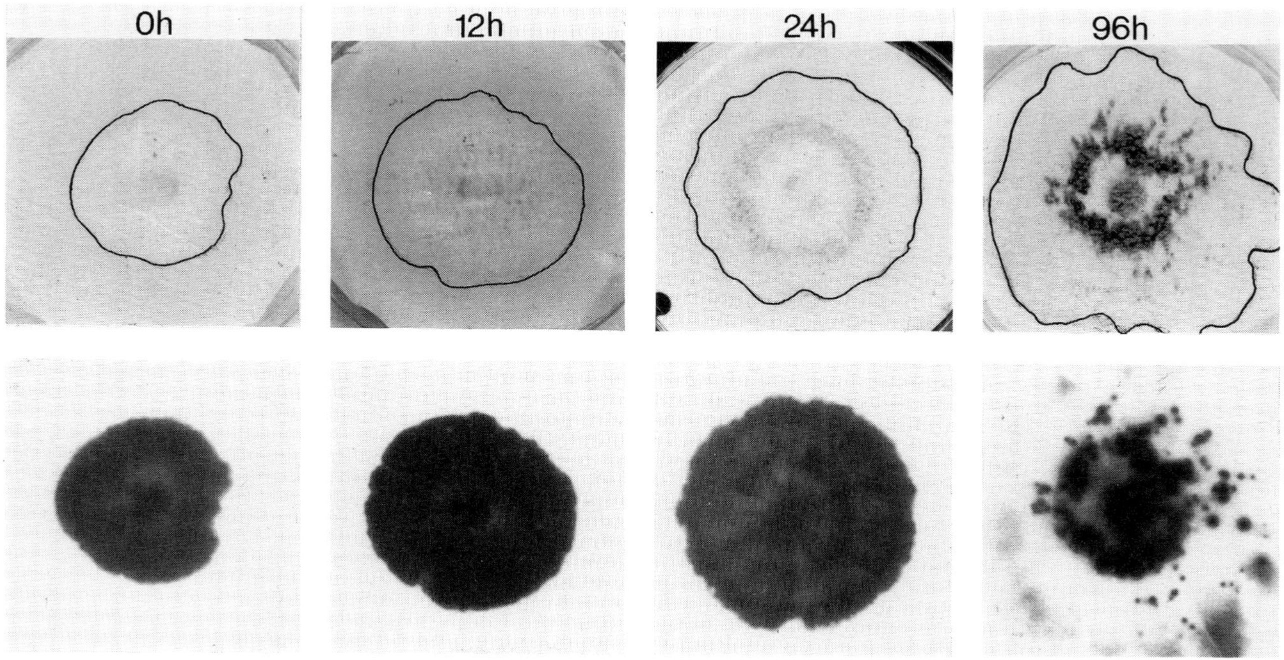

Figure 2.2. Distribution of ribosomal RNA in developing colonies of a *Schizophyllum commune* dikaryon.

Colonies were grown on Gene-screen membranes overlying agar medium, fixed with formaldehyde, and treated with wall-lytic enzymes to permeabilize the walls. The membranes with adhering colonies were then hybridized to a [^{32}P]-labelled clone of 18S rRNA. Upper row: morphological appearance of a 5-d-old dark-grown colony (0 h) at intervals after transfer to light. Because of the white membranes the vegetative mycelium is barely visible and the margins of the colonies have been marked. Lower row: autoradiograms of the same colonies, showing the distribution of the hybridized rRNA clone. (Adapted from Ruiters and Wessels, 1989*a*.)

The hyphal mode of growth

Our work on apical wall growth in fungal hyphae has been the subject of recent reviews (Wessels, 1986, 1990; Wessels *et al.*, 1990) and only the essential findings and some far-reaching speculations are therefore discussed here. This section also serves to underline that a molecular perspective cannot only be based on molecular genetics. The playground of the interacting macromolecules that are responsible for wall growth is so far removed from the genes that studies at this level are also necessary to explain the observed morphological and physiological phenomena.

The basic observation made on the cell wall of *Schizophyllum commune* was that at least two of its major components, the (1-3)-β/(1-6)-β-linked glucan and chitin, are present in the mature wall not only as individual polymers but also covalently linked to each other. This results in a composite in which chitin fibrils, consisting of hydrogen-bonded chitin chains, may be crosslinked by β-glucan chains which are also hydrogen-bonded to each other. Although the β-glucan itself is water/alkali-soluble, in the complex it is highly insoluble because of linkage to the chitin chains. It is suggested that this chitin/β-glucan complex is mainly responsible for the rigidity of the walls in both basidiomycetes and ascomycetes.

If the occurrence of such a molecular structure in the wall is accepted, then it is clear that the hydrogen bonds and covalent bonds between the polymers, responsible for rigidity, can only arise by interactions of the individual polymers within the domain of the wall. Consequently, during biogenesis of the wall there must be a transition in time from an aqueous mixture of individual polymers, extruded by the cytoplasm into the wall, to the cross-linked complex. This probably also represents a transition from a plastic deformable material to the rigid fabric that resists turgor pressure. It should be stressed that this is a general theory of wall biogenesis, not limited to the polymers and crosslinks encountered in the *S. commune* wall.

Our studies with *S. commune* hyphae, using various techniques to probe the chemical and physical nature of the wall polymers *in situ* (see Wessels *et al.*, 1990), have shown that the aforementioned transition indeed occurs at the growing hyphal apex where the individual polymers are extruded into the wall in a steep gradient. At the extreme tip the extrusion of wall polymers is maximal but the cross-linking is minimal. In the posterior direction extrusion of the polymers quickly decreases but more and more cross-links are present. At the point where cross-linking is sufficient to resist the turgor pressure, the wall assumes its cylindrical shape but cross-linking may continue.

These observations led us to formulate the steady-state theory of apical wall growth (Wessels *et al.*, 1983; Wessels, 1986). The theory assumes that during the growth of a hypha there is continuous secretion at the apex of an expandable mixture of wall polymers that is continuously removed at the base of the extension zone as a rigid complex arising by the formation of bonds between and among the polymer chains. In hyphae growing at a constant rate there is thus a steady-state amount of expandable wall material at the apex. This is only possible because of the steep gradient in wall apposition and the continuous forward movement of the apex, away from the zone of rigidification. If the forward movement of the apex is interrupted for a short time, the wall rigidifies over the whole apex and growth is irreversibly blocked. The wall over the apex then assumes a structure as present in the cylindrical wall. These predictions have been verified.

Returning to the view so vividly expressed by Gregory (1984), we can now see that hyphal growth can indeed be likened to the forward

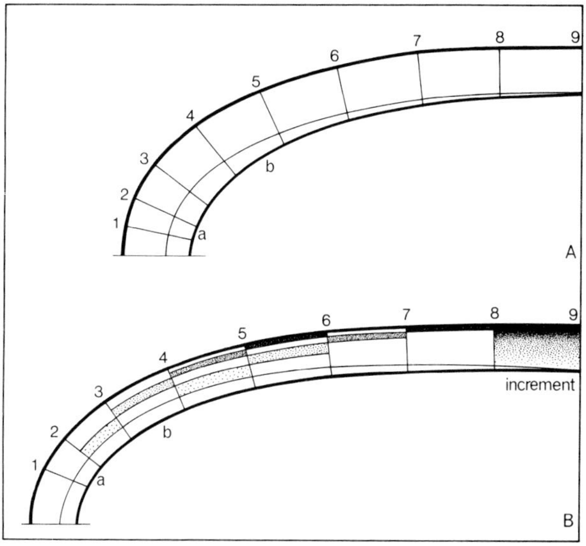

Figure 2.3. Diagram of the extension zone of the hyphal wall to illustrate the steady-state growth theory.

The numbers indicate the displacement of an imaginary point on the surface of the wall at fixed intervals of time. In reality, the addition of viscous wall material according to a gradient as shown in (A), and the expansion of the wall according to the same gradient (B), are simultaneous but here they are separated in time. During wall expansion the added wall material is drawn into the wall to maintain wall thickness. At successive rounds of wall addition and expansion a wall volume (a or b) added to the wall by apposition migrates through the wall, apparently in a posterior direction, from the inside to the outside of the wall while being stretched and cross-linked at the same time (increased stippling). Note that the most stretched and cross-linked wall volumes, which display increased resistance to stretch, accumulate at the outside of the wall. This decelerates expansion until it becomes zero, which marks the base of the extension zone, although at this point the inner portion of the wall has just been deposited and is still assumed to be plastic. (Reproduced with permission from Wessels, 1990.)

movement of a myxomycete, with the important exception that the cytoplasm secretes at its tip a plastic material that gradually hardens and forms a rigid tube. Consequently, an internal hydrostatic pressure can be built up which drives the tip forward at a fast rate.

An important point is that the wall polymers are added to the wall at the hyphal tip by apposition from the inside. They are either synthesized by integral plasma membrane proteins (presumably chitin) or secreted by vesicles that fuse with the expanding plasma membrane. As depicted in Figure 2.3, there must therefore be a flow of wall polymers through the wall. A wall volume added at the extreme tip is pushed to the outside of the wall by newly added wall material, being stretched and crosslinked at the same time. Because the tip moves continuously forward, the stretched and crosslinked wall volume will arrive at the outer surface at some distance from the moving tip. Wall polymers secreted at some distance from the tip will be lodged at a more internal position in the wall.

These considerations lead to certain speculations on tip-associated processes which may serve as working hypotheses for future investigations (Wessels, 1990):

1. Diverse wall polymers are not necessarily all secreted in the same gradient at the apex. Those secreted in a very steep gradient (highest at the very tip) will end up in the outer wall regions, those secreted in a more shallow gradient will be located more internally in the mature wall. At least some of the layering in fungal walls (Burnett, 1979) may thus arise because of a spatial difference, rather than a temporal difference, in their secretion.
2. It is generally assumed that the growing apex is the site of most active protein secretion. Such proteins, some of which (hydrolytic enzymes) may have an important role in invasive growth of the hypha, are probably delivered to the apical surface in vesicles that fuse with the plasma membrane. If the gradient for vesicle fusion is the same as that for wall secretion, the steady-state growth theory would predict that such enzymes are distributed evenly through the wall. Those enzymes delivered in vesicles that fuse at the extreme tip would be carried to the outer regions of the wall which is most stretched and rigidified. Possibly this is the most porous part of the wall because of the mechanical stretch it has undergone. If the gradient of vesicle fusion were steeper than that of wall secretion, then most enzymes would end up in this outer region of the wall and readily diffuse into the medium. This "bulk-flow" hypothesis for enzyme secretion through the wall (Wessels, 1990) solves the paradox that the measured pore size of the wall is generally too small to permit the diffusion of proteins; according to this hypothesis proteins do not pass through the wall by pores but are pushed through the wall by the wall mass added from the inside. The hypothesis couples secretion to extension of the hyphae. This may seem incompatible with the secretion of many enzymes in the ideofase

when net growth of the mycelium is minimal, however, during that phase the mycelium may be redirecting its resources and maintain hyphal growth even though the biomass of the whole mycelium does not increase and may even decrease (*cf.* Fig. 2.2).

3. There is increasing evidence for the general occurrence of mechanosensitive ion channels in the plasma membrane (Gustin *et al.*, 1988; Morris and Sigurdson, 1989). Presumably, such ion channels – some of which may be activated and others inactivated by stretching the membrane – also occur in the plasma membrane of filamentous fungi and are responsible for, for instance, the regulation of turgor. It is conceivable that the gradient in wall plasticity in growing hyphae, in combination with turgor pressure, creates a gradient of stretch in the plasma membrane over the apex and which is maximal at the very tip. A manifestation of this differential activity of mechanosensitive ion channels could be the transcellular ion currents in fungal hyphae (Gow, 1984; Harold *et al.*, 1985; Harold and Caldwell, 1990). The activity of such ion channels, for example Ca^{2+}, could indirectly regulate vesicle fusion and the activity of membrane-bound wall synthases, such as chitin synthase. Alternatively such synthases could be regulated directly by stress in the membrane. By either of these mechanisms, the fluidity of the wall and its tendency to yield to turgor pressure would be coupled to wall-synthetic activity, ensuring that the expanding wall maintains a constant thickness. Also, a temporary local softening of the mature wall by lytic activity would be sufficient to initiate ion currents and wall-synthetic activities and to create a new apical growth centre, as in branch formation.

The emergence of aerial structures

Perhaps the most dramatic event in fungal morphogenesis is the emergence of hyphae from the moist substrate into the air where the water potential can be very low. The function of "undifferentiated" aerial hyphae which form a woolly or felty mat at the substrate/air interface is not clear. Possibly this aerial hyphal mat protects the substrate hyphae from desiccation. Aerial hyphae may also branch and differentiate into spore-bearing structures, such as conidiophores and sporangiophores. Finally they may become engaged in branching and interactions leading to the morphogenesis of multicellular fruit-bodies such as mushrooms. These aerial structures are apparently all designed to lift reproductive hyphae into the air for spore dispersal.

With the objective of understanding these processes at the molecular level, we are studying the emergence of aerial structures in *Schizophyllum commune*, particularly the emergence of fruit-bodies (Figs 2.1 and 2.2). The approach we follow is that of reverse genetics. First, genes of unknown

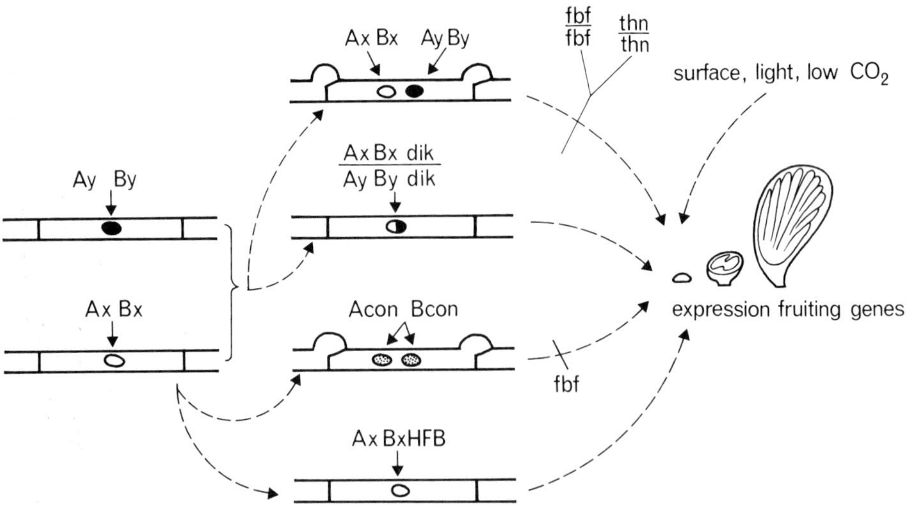

Figure 2.4. Effects of mating-type genes and modifying genes on mycelial morphology and fruiting of *Schizophyllum commune*.

Mating of two monokaryons with different alleles for the *A* and *B* mating-type genes generates the dikaryon which fruits unless both nuclear types contain the recessive mutant alleles *fbf* or *thn*. If both nuclear types contain the recessive *dik* allele, a diploid monokaryon is formed which also fruits. The presence of constitutive mutations in both the *A* and *B* mating-type genes (*Acon Bcon*), converts the monokaryon to a dikaryon (with two identical nuclei in each cell) which normally fruits. Fruiting is now prevented by the presence of one *fbf* allele. In the presence of *HBF* alleles, monokaryons with wild-type mating-type alleles also fruit. (Adapted from Wessels, 1987.)

function are isolated on the basis of their abundant expression during fruit-body formation (or the formation of aerial mycelium). The expression of these genes is then studied in relation to morphogenesis as regulated by various genetic elements. Genes of interest are then sequenced and the derived protein structure is used to establish the location and function of the proteins *in vivo*.

The genetic elements known to control fruiting in *S. commune* are indicated in Figure 2.4. The control of dikaryosis and fruiting by the mating-type genes has long been known and teleologically makes sense because this control ensures that only genetically different mycelia will mate to produce the fruit bodies in which recombination can occur. Although depicted in Figure 2.4 as single genes, the mating-type genes *A* and *B* are actually complex and each contains two genes, called α and β, for which many alleles exist (Raper, 1966, 1988). In the world-wide population of *S. commune*, the estimate is of 9 alleles for *A*α, 32 for *A*β, 9 for *B*α, and 9 for *B*β. When nuclei in a fusion cell carry an allelic difference in at least one of the genes of both the *A* and the *B* complex, a dikaryon is formed and fruiting ensues under permissive environmental conditions. A number of different alleles of the $A\alpha$ gene have now been cloned and characterized (Giasson *et al.*, 1989), but the apparent absence of sequence homology between the alleles makes it difficult to suggest how their interaction leads to a regulatory product unique to the dikaryon, as has been possible for the alleles of the *b* mating-type gene of *Ustilago maydis* (Kronstadt and Leong, 1989; Schulz *et al.*, 1990). Whatever the products of the interactions in *S. commune* will prove to be, similar products are apparently formed in homokaryons carrying single mating-type alleles but with rare mutations that switch on sexual morphogenesis (Raper *et al.*, 1965). If such constitutive mutations are present in alleles of both the *A* and *B* mating-type complex (*Acon Bcon*), the monokaryon is converted into a dikaryon which fruits and forms basidiospores (Fig. 2.4). However, as this homokaryotic dikaryon contains only one type of nucleus, all the basidiospores are then identical.

Although it is the dikaryon that usually forms fruit-bodies, the dikaryotic hyphal morphology is not necessary for fruit-body formation. For instance, a double dose of the recessive mutation *dik* causes precocious fusion of the nuclei in a mating involving different *A* and *B* genes (Koltin and Raper, 1968). The diploid monokaryon that is formed fruits (Fig. 2.4) and forms normal meiospores. In addition, a number of so-called haploid fruit-body alleles (*HFB*, Fig. 2.4) are known that cause wild-type monokaryons to fruit, although meiosis and normal basidiospore formation does not occur (Yli-Mattila *et al.*, 1989*b*). The presence of these alleles, also called *hap*$^+$ genes (Leslie and Leonard, 1979) or *fi*$^+$ genes (Esser *et al.*, 1979) apparently bypasses the need for the presence of different mating-type genes with respect to fruiting. Their presence also enhances fruiting in the dikaryon.

Some other genes are known that do affect the formation of emerged structures but have no effect on the formation of the dikaryon. One of these, identified by the allele *thin* (*thn*), is a frequently occurring spontaneous mutation that converts a monokaryon with woolly aerial hyphae into a mycelium with a wet-looking surface and a few aerial hyphae. In a mating, *thn* behaves as a recessive allele but in double dose in the dikaryon it prevents the formation of both aerial hyphae and fruit-bodies (Schwalb and Miles, 1967). Another recessive mutation, *fbf* (fruit-body formation), has no phenotype in monokaryons because it does not affect the formation of aerial hyphae but only prevents the formation of fruit-bodies. This was discovered as a frequently occurring spontaneous mutation in the fruiting *Acon Bcon* dikaryon which expresses this recessive mutation because of its homokaryotic nature (Springer and Wessels, 1989). In double dose it also prevents fruiting, but not the formation of aerial hyphae, in a normal heterokaryotic dikaryon.

Genes specifically expressed in the fruiting dikaryon of *S. commune*

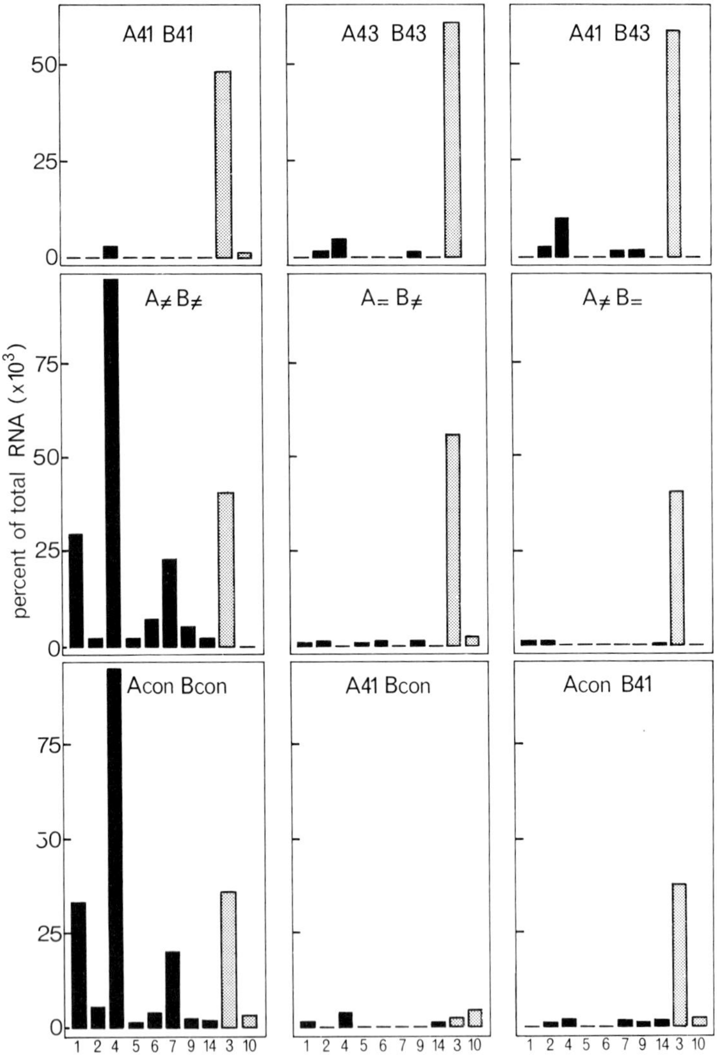

Figure 2.5. Accumulation of specific mRNAs belonging to the class of abundant mRNAs in homokaryons and heterokaryons of *Schizophyllum commune*.

Three wild-type homokaryons (monokaryons) are shown in the upper row. The middle row shows the different heterokaryons that can be formed by mating these monokaryons. The lower row represents homokaryons with constitutive mutations in the mating-type genes. Mycelia were grown as a lawn from mycelial fragments for four days in the light. Fruit-bodies were only formed in the *A≠ B≠* and *Acon Bcon* mycelia. Black bars: dikaryon-specific mRNAs. Shaded bars: mRNAs not specific to the dikaryon. Numbers refer to specific *Sc* genes (*Sc1*, *Sc2*, etc.). (Data from Ruiters *et al.*, 1988.)

were isolated by comparing the messenger RNAs of a monokaryon and a fruiting dikaryon that differed genetically only in the presence of one or two alleles of the mating-type genes (Dons *et al.*, 1984*a*; Mulder and Wessels, 1986). Complementary DNA (cDNA) was prepared on the poly(A)-RNA of the dikaryon and cDNA clones were selected that hybridized to the RNA of the dikaryon only (e.g. cDNA clones for the *Sc1* and *Sc4* genes) or to RNA present in both the monokaryon and dikaryon (e.g. the cDNA for the *Sc3* gene). The isolated cDNA clones were used to clone the genes and to titrate the concentration of the specific mRNAs during development. Figure 2.5 shows the abundance of various specific mRNAs in coisogenic mycelia differing only in the presence of specific mating-type alleles. The dikaryon-specific mRNAs (black bars) are only abundant if two different alleles for the *A* and *B* mating-type genes are present ($A \neq B \neq$) or if there are constitutive mutations in these genes (*Acon Bcon*), conditions also necessary for fruiting. Apparently both mating-type genes are involved because hemi-compatible heterokaryons and homokaryons with a constitutive mutation in only one of the mating-type genes do not support accumulation of these mRNAs or fruiting. These mRNAs were also very low in young cultures of the dikaryons before the advent of fruit-bodies. At the time of initiation of the fruit-bodies they reach the abundances shown in Figure 2.5, for the mRNAs of the *Sc1* and *Sc4* genes up to 0.5% and 3.5% of the total mRNA mass, respectively. Of

Table 2.1. Occurence of fruit-bodies, aerial mycelium, and high concentrations of mRNAs from the *Sc1*, *Sc4*, and *Sc3* genes.

Coisogenic strains (except *Ax Bx HFB*) differed only with respect to mating-type genes and indicated modifying mutations. Mycelia were grown as a lawn from hyphal fragments in the light. (Summarized from Ruiters *et al.*, 1988, Springer and Wessels, 1989, and unpublished results for the *thn* mutation.)

			mRNA[2]		
Genotype[1]	fruit-bodies	aerial hyphae	*Sc1*	*Sc4*	*Sc3*
Ax Bx	–	+	–	–	+
Ax Bx HFB	+	+	+	+	+
Ax Bx thn	–	–	–	–	–
Ax/Ay Bx/By	+	+	+	+	+
Ax/Ay Bx/By thn/thn	–	–	–	–	–
Ax Bcon	–	–	–	–	–
Acon Bx	–	+	–	–	+
Acon Bcon	+	+	+	+	+
Acon Bcon fbf	–	+	–	–	+

[1]*x* and *y* denote wild-type but different alleles of the *A* and *B* mating-type genes
[2]+ = high abundance, – = low abundance of mRNA (*cf.* Fig. 2.5)

the mRNAs present in most mycelia (shaded bars), one (*Sc3* mRNA) was also low in young cultures but rose to a high concentration (up to 1% of the mRNA mass) in all mycelia forming aerial hyphae. The data in Figure 2.5 were obtained by growing cultures as a lawn from mycelial fragments. However, *in situ* hybridizations with colonies grown from a single inoculum showed that the dikaryon-specific mRNAs were only formed at the time and place of formation of the ring of primordia (*cf.* Fig. 2.2), most abundantly in the growing hyphae that were engaged in fruit-body morphogenesis (Ruiters and Wessels, 1989*a,b*). Because of their abundance and similarity in coding sequence (see below) only the genes *Sc1*, *Sc4*, *Sc3*, and their products, will be discussed further.

As shown in Table 2.1, the accumulation of the mRNAs from the *Sc1* and *Sc4* genes is not only controlled by the mating-type genes *A* and *B* but also by the genes *THN* and *FBF* and by the *HFB* alleles. On the other hand, the accumulation of the mRNA for the *Sc3* gene seems to be controlled by the *THN* gene and the *B* mating-type gene only. It is too early to suggest the hierarchy of these multiple genetic control elements. However, in relation to the expressed morphologies, Table 2.1 shows a correlation between expression of the *Sc3* gene and the formation of aerial hyphae on the one hand and between the expression of the *Sc1* and *Sc4* genes and the formation of fruit-bodies on the other. Yet, the formation of these mRNAs is not sufficient to elicit the morphological effects; these mRNAs are also formed under conditions in which no aerial structures can emerge, that is in shaken submerged cultures (Wessels *et al.*, 1987).

Sequencing of the cDNA and genomic clones revealed a remarkable homology between the general structures and the coding sequences of the *Sc1*, *Sc4*, and *Sc3* genes (Dons *et al.*, 1984*b*; Schuren and Wessels, 1990). As shown in Figure 2.6, these three genes clearly belong to a family of evolutionarily-related genes coding for small homologous proteins, each with eight cysteines conserved in the same positions; these proteins we call hydrophobins because they are quite hydrophobic with an average hydrophobicity (Kyte and Doolittle, 1982) of +0.54, +0.59, and +0.90 for pSc1, pSc4, and pSc3, respectively (+0.28, +0.35, and +0.68 if the putative signal peptide is omitted).

The putative signal peptides present in the hydrophobins suggest that these proteins are secreted. Indeed, antibodies raised against synthetic peptides, representing the non-homologous amino acid sequences around the first intron in the *Sc1* and *Sc4* genes (Fig. 2.6), do react with proteins in the medium and in the cell wall of cultures that express these genes. Proteins with properties corresponding to those derived from the cloned sequences have now also been shown to be major components of the medium and the cell walls. In the medium these proteins form a variety of aggregates. In the cell walls they appear in a form that resists extraction with hot sodium dodecylsulfate. They can be extracted by cold formic acid

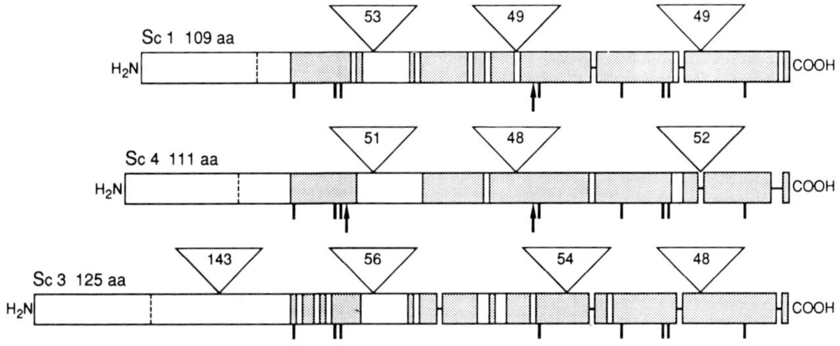

Figure 2.6. Comparison of the coding sequences of the *Sc1, Sc4,* and *Sc3* genes of *Schizophyllum commune.*

The shaded blocks indicate areas of identity of amino acids or conservative substitutions (small interruptions of the sequences are introduced to obtain optimal alignment). Vertical lines under the blocks indicate the positions of the cysteines in the sequence. Arrows indicate the positions of possible *N*– glycosylation sites. The positions of the introns in the genes is shown by triangles within which the lengths of the introns is given in base pairs. The dashed line in the amino-terminal part of the sequence signifies the approximate end of the presumed signal peptide for secretion. (Data from Schuren and Wessels, 1990.)

but then are still of high-molecular weight. Dissociation into a few proteins with apparent molecular weights between 12000 and 30000 occurs after oxidative dissociation of all disulfide bridges with performic acid. The presence of individual members of these proteins is correlated with the presence of the mRNAs for the three hydrophobins (Table 2.1), for example none of them are present in a monokaryon with a *thn* mutation. In the walls of the submerged mycelium they are only found in low concentrations, but they are conspicuously present in the walls of emerged hyphae, including fruit-bodies. For the most abundantly excreted hydrophobins we have established, by *N*-terminal amino acid sequencing, that they are the products of the *Sc3* and *Sc4* genes (Wessels *et al.*, to be published).

Apparently, the hydrophobins can be secreted by the submerged hyphae through the wall into the medium but in hyphae emerging into the air they are anchored in the wall by disulfide bonds as high-molecular-weight complexes that may provide the wall with a hydrophobic surface. The most water-repellent hydrophobin, encoded by the *Sc3* gene, could be necessary for formation of the individually growing aerial hyphae that make up the woolly surface mat. The two other hydrophobins, encoded by the homologous *Sc1* and *Sc4* genes, which are under the control of the mating-type genes, might have the additional role of causing hyphae to adhere to each other during fruit-body formation, e.g. by hydrophobic interactions or formation of disulfide bridges between hyphal surfaces.

It is unknown whether a specific signal is required for converting the medium-secreted hydrophobins into wall-bound hydrophobins. The simplest hypothesis would be that the association of hydrophobins with the wall faces all hyphae that secrete these proteins and encounter a high redox potential. The hydrophobins secreted at their apices would now be prevented from diffusing into the medium but form a hydrophobic coating on the surface of the hyphae, determining their destiny for life in the air.

Acknowledgements

I wish to thank my colleagues Drs J.H. Sietsma, C.A. Vermeulen, and O.M.H. de Vries for comments on the manuscript, and Dr O.M.H. de Vries for permission to quote unpublished results.

References

Adams, T.H., Boylan, M.T. and Timberlake, W.E. (1988) *brlA* is necessary and sufficient to direct conidiophore development in *Aspergillus nidulans*. *Cell* 54, 353–362.

Alberts, B., Bray, D., Lewis, J., Raff, M., Roberts, K. and Watson, J.D. (1989) *Molecular Biology of the Cell.* Garland, New York–London.

Andrianopoulos, A. and Hynes, M.J. (1990) Sequence and functional analysis of the positively acting regulatory gene *amdR* from *Aspergillus nidulans*. *Molecular and Cellular Biology* 10, 3194–3203.

Bonner, J.T. (1974) *On Development.* Harvard University Press, Cambridge, Mass.

Boylan, M.T., Mirabito, P.M., Willets, C.E., Zimmerman, C.R. and Timberlake, W.E. (1987) Isolation and physical characterization of three essential conidiation genes from *Aspergillus nidulans*. *Molecular and Cellular Biology* 7, 3113–3118.

Broda, P., Sims, P.F.G. and Mason, J.C. (1989) Lignin biodegradation: A molecular approach. *Essays in Biochemistry* 24, 82–114.

Burnett, J.H. (1979) Aspects of the structure and growth of hyphal walls. In: Burnett, J.H. and Trinci, A.P.J. (eds), *Fungal Walls and Hyphal Growth.* Cambridge University Press, London, pp. 1–25.

Dons, J.J.M., Springer, J., De Vries, S.C. and Wessels, J.G.H. (1984*a*) Molecular cloning of a gene abundantly expressed during fruiting body initiation in *Schizophyllum commune*. *Journal of Bacteriology* 157, 802–808.

Dons, J.J.M., Mulder, G.H., Rouwendal, G.J.A., Springer, J., Bremer, W. and Wessels, J.G.H. (1984*b*) Sequence analysis of a split gene involved in fruiting from the fungus *Schizophyllum commune*. *EMBO Journal* 3, 2101–2106.

Esser, K., Saleh, F., and Meinhardt, F. (1979) Genetics of fruit-body production in

higher Basidiomycetes. II. Monokaryotic and dikaryotic fruiting in *Schizophyllum commune. Current Genetics* 1, 85–88.
Giasson, L., Specht, C.A., Milgrim, C., Novotny, C.P. and Ullrich, R.C. (1989) Cloning and comparison of *A*α mating-type alleles of the Basidiomycete *Schizophyllum commune. Molecular and General Genetics* 218, 72–77.
Gold, M.H., Wariishi, H. and Valli, K. (1989) Extracellular peroxydase involved in lignin degradation by the white rot fungus *Phanerochaete chrysosporium.* In: Whitaker, J.R. and Sonnet, P.E. (eds), *Biocatalysis in Agricultural Biotechnology.* American Chemical Society, pp. 127–140.
Gow, N.A.R. (1984) Transhyphal electrical currents in fungi. *Journal of General Microbiology* 130, 3313–3318.
Gregory, P.H. (1984) The fungal mycelium: an historical perspective. *Transactions of the British Mycological Society* 82, 1–11.
Gustin, M.C., Zhou, X.-L., Martinac, B. and Kung, C. (1988) A mechanosensitive ion channel in the yeast plasma membrane. *Science, New York* 242, 762–765.
Harold, F.M. and Caldwell, J.H. (1990) Tips and currents: electrobiology for apical growth. In: Heath, I.B. (ed.), *Tip Growth of Plant and Fungal Cells.* Academic Press, San Diego, pp. 59–90.
Harold, F.M., Kropt, D.L. and Caldwell, J.H. (1985) Why do fungi drive electric currents through themselves? *Experimental Mycology* 9, 183–186.
Herskowitz, I. (1988) Life cycle of the budding yeast *Saccharomyces cerevisiae. Microbiological Reviews* 52, 536–553.
Herskowitz, I. (1989) A regulatory hierarchy for cell specialization in yeast. *Nature, London* 342, 749–757.
Johnston, M. (1987) A model fungal gene regulatory mechanism: the *GAL* genes of *Saccharomyces cerevisiae. Microbiological Reviews* 51, 458–476.
Kassir, Y., Granot, D. and Simchen, G. (1988) *IME*1, a positive regulatory gene of meiosis in *S. cerevisiae. Cell* 52, 853–862.
Kendrick, B. and Berch, S. (1985) Mycorrhiza: applications in agriculture and forestry. In: Moo-Young, M. (ed.), *Comprehensive Biotechnology.* Vol. 4. Pergamon Press, Oxford, pp. 110–152.
Kniep, H. (1920) Über morphologische und physiologische Geschlechtsdifferenzierung (Untersuchungen an Basidiomyzeten). *Verhandlungen der Physisch-Medische Gesellschaft Würzburg* 46, 1–18.
Koltin, Y. and Raper, J.R. (1968) Dikaryosis: Genetic determination in *Schizophyllum. Science, New York* 60, 85–86.
Kronstadt, J.W. and Leong, S.A. (1989) Isolation of two alleles of the *b* locus of *Ustilago maydis. Proceedings of the National Academy of Sciences, USA* 86, 978–982.
Kyte, J. and Doolittle, R.F. (1982) A simple method for displaying the hydropathy character of a protein. *Journal of Molecular Biology* 157, 105–132.
Lee, M.G. and Nurse, P. (1987) Complementation to clone a human homologue of the fission yeast cell cycle control gene *cdc* 2. *Nature, London* 327, 31–35.
Leong, S.A. and Holden, D.W. (1989) Molecular genetic approaches to the study of fungal pathogenesis. *Annual Review of Phytopathology* 27, 463–481.
Leslie, J.F. and Leonard, T.J. (1979) Monokaryotic fruiting in *Schizophyllum commune*: genetic control of the response to injury. *Molecular and General Genetics* 175, 5–12.

Morris, C.E. and Sigurdson, W.J. (1989) Stretch-inactivated ion channels coexist with stretch-activated ion channels. *Science, New York* 243, 807–809.

Morris, N.R., Osmani, S.A., Engle, D.B. and Doonan, H. (1989) The genetic analysis of mitosis in *Aspergillus nidulans. Bioassays* 10, 196–201.

Mulder, G.H. and Wessels, J.G.H. (1986) Molecular cloning of RNAs differentially expressed in monokaryons and dikaryons of *Schizophyllum commune* in relation to fruiting. *Experimental Mycology* 10, 214–227.

Nicholson, T.H. (1975) Evolution of vesicular-arbuscular mycorrhizas. In: Sanders, F.E., Mosse, B. and Tinker, P.B. (eds), *Endomycorrhizas.* Academic Press, London, pp. 25–34.

Nurse, P. (1985) Cell cycle controls in yeast. *Trends in Genetics* 1, 51–55.

Ptashne, M. (1988) How eukaryotic transcriptional activators work. *Nature, London* 335, 683–689.

Raper, C.A. (1988) *Schizophyllum commune*, a model for genetic studies of the Basidiomycetes. In: Sidhu, G.S. (ed.), *Genetics of Plant Pathogenic Fungi.* Academic Press, London, pp. 511–522.

Raper, J.R. (1966) *Genetics of Sexuality in Higher Fungi.* Ronald Press, New York.

Raper, J.R., Boyd, D.H., and Raper, C.A. (1965) Primary and secondary mutations at the incompatibility loci in *Schizophyllum. Proceedings of the National Academy of Sciences, USA* 53, 1324–1332.

Raudaskoski, M. and Vauras, R. (1982) Scanning electron microscope study of fruit-body differentiation in *Schizophyllum commune. Transactions of the British Mycological Society* 78, 475–481.

Raudaskoski, M. and Yli-Mattila, T. (1985) Capacity for photoinduced fruiting in a dikaryon of *Schizophyllum commune. Transactions of the British Mycological Society* 85, 145–151.

Ruiters, M.H.J. and Wessels, J.G.H. (1989*a*) *In situ* localization of specific RNAs in whole fruiting colonies of *Schizophyllum commune. Journal of General Microbiology* 135, 1747–1754.

Ruiters, M.H.J. and Wessels J.G.H. (1989*b*) *In situ* localization of specific RNAs in developing fruit bodies of the basidiomycete *Schizophyllum commune. Experimental Mycology* 13, 212–222.

Ruiters, M.H.J., Sietsma, J.H. and Wessels, J.G.H. (1988) Expression of dikaryon-specific mRNAs of *Schizophyllum commune* in relation to incompatibility genes, light, and fruiting. *Experimental Mycology* 12, 60–69.

Schekman, R. (1985) Protein localization and membrane traffic in yeast. *Annual Review of Cell Biology* 1, 115–143.

Schulz, B., Bannuett, F., Dahl, M., Schlessinger, R., Schäfer, W., Martin, T., Herskowitz, I. and Kahmann, R. (1990) The *b* alleles of *U. maydis*, whose combinations programme pathogenic development, code for polypeptides containing a homeodomain. *Cell* 60, 295–306.

Schuren, F.H.J. and Wessels, J.G.H. (1990) Two genes specifically expressed in fruiting dikaryons of *Schizophyllum commune*: homologies with a gene not regulated by the mating-type genes. *Gene* 90, 199–205.

Schwalb, M.N. and Miles, P.G. (1967) Morphogenesis of *Schizophyllum commune.* I. Morphological variation and mating behavior of the thin mutation. *American Journal of Botany* 54, 440–446.

Springer, J., and Wessels, J.G.H. (1989) A frequently occurring mutation that blocks

the expression of fruiting genes in *Schizophyllum commune. Molecular and General Genetics* 219, 486–488.
Struhl, K. (1983) The new yeast genetics. *Nature, London* 305, 391–397.
Timberlake, W.E. and Marchall, A. (1989) Genetic engineering of filamentous fungi. *Science, New York* 244, 1313–1317.
Trinci, A.P.J. (1978) The duplication cycle and vegetative development in moulds. In: Smith, J.E. and Berry, D. (eds), *The Filamentous Fungi.* Vol. 3. Edward Arnold, London, pp. 132–163.
van der Valk, P. and Marchant, R. (1978) Hyphal ultrastructure in fruit-body primordia of the basidiomycetes *Schizophyllum commune* and *Coprinus cinereus. Protoplasma* 95, 57–72.
Watson, J.D., Hopkins, N.H., Roberts, J.W., Steitz, J.A. and Weiner, A.M. (1987) *Molecular Biology of the Gene.* Vol. 1. Benjamin/Cummings Publishing, Menlo Park, California.
Wessels, J.G.H. (1965) *Morphogenesis and Biochemical Processes in Schizophyllum commune Fr.* North-Holland Publishing, Amsterdam.
Wessels, J.G.H. (1985) Gene expression during basidiocarp formation in *Schizophyllum commune.* In: Timberlake, W.E. (ed.), *Molecular Genetics of Filamentous Fungi.* A.R. Liss, New York, pp. 193–206.
Wessels, J.G.H. (1986) Cell wall synthesis in apical hyphal growth. *International Review of Cytology* 104, 37–79.
Wessels, J.G.H. (1987) Mating-type genes and the control of fruiting in basidiomycetes, *Antonie van Leeuwenhoek* 53, 307–317.
Wessels, J.G.H. (1990) Role of cell wall architecture in fungal tip growth generation. In: Heath, B. (ed.), *Tip Growth of Plant and Fungal Cells.* Academic Press, San Diego, pp. 1–29.
Wessels, J.G.H. and Sietsma, J.H. (1979) Wall structure and growth in *Schizophyllum commune.* In: Burnett, J.H. and Trinci, A.P.J. (eds), *Fungal Walls and Hyphal Growth.* Cambridge University Press, London, pp. 27–48.
Wessels, J.G.H., Mulder, G.H., and Springer, J. (1987) Expression of dikaryon-specific and non-specific mRNAs of *Schizophyllum commune* in relation to environmental conditions and fruiting. *Journal of General Microbiology* 133, 2557–2561.
Wessels, J.G.H., Sietsma, J.H. and Sonnenberg, A.S.M. (1983) Wall synthesis and assembly during hyphal morphogenesis in *Schizophyllum commune. Journal of General Microbiology* 129, 1599–1605.
Wessels, J.G.H., Mol, P.C., Sietsma, J.H. and Vermeulen, C.A. (1990) Wall structure, wall growth, and fungal cell morphogenesis. In: Kuhn, P.J., Trinci, A.P.J., Jung, M.J., Goosey, M.W. and Copping, L.G. (eds), *Biochemistry of Cell Walls and Membranes in Fungi.* Springer Verlag, Berlin, pp. 81–95.
Whittaker, R.H. (1969) New concepts of kingdoms of organisms. *Science, New York* 163, 150–159.
Yli-Mattila, T., Ruiters, M.H.J., and Wessels, J.G.H. (1989*a*) Photoregulation of dikaryon-specific mRNAs and proteins by UV-A light in *Schizophyllum commune. Current Microbiology* 18, 289–295.
Yli-Mattila, T., Ruiters, M.H.J., Wessels, J.G.H., and Raudaskoski, M. (1989*b*) Effect of inbreeding and light on monokaryotic and dikaryotic fruiting in the homobasidiomycete *Schizophyllum commune. Mycological Research* 93, 535–542.

Zantinge, B., Dons, H., and Wessels, J.G.H. (1979) Comparison of poly(A)-containing RNAs in different cell types of the lower eukaryote *Schizophyllum commune*. *European Journal of Biochemistry* 101, 251–260.

3

Importance of Siderophores in Fungal Growth, Sporulation and Spore Germination

G. Winkelmann, *Mikrobiologie und Biotechnologie, Universität Tübingen, Auf der Morgenstelle 1, D-7400 Tübingen, Germany.*

Abstract During iron limitation most fungi excrete siderophores which function in solubilization, transport and storage of iron. Three main siderophore classes have so far been described, the fusarinines, coprogens, rhodotorulic acid, and ferrichromes, all of which show considerable structural variation. Transport of siderophores in fungi is energy-dependent and requires specific recognition of the siderophore structure, the iron coordination center and its surrounding N-acyl groups. Different mechanisms of uptake have been described which involve transport of the intact chelate across the membrane and internal release of the iron, or uptake of iron from the chelate without uptake of the ligand. Some fungi are unable to produce any siderophores but are still able to utilize iron from various other sources. Fungi which do not depend on siderophores possess specific ferric reductase and seem to transport ferrous iron across the plasma membrane. Experiments with siderophore-free mutants suggest that siderophores are also required for the formation of conidia. Moreover, germination is dependent on the presence of certain cellular siderophores which are released and metabolized during germination of spores.

Introduction

Although iron is essential to virtually all living organisms, its solubility and bioavailability is generally low. Therefore, specific mechanisms for iron uptake have evolved. A comprehensive treatise on iron transport in

Frontiers in Mycology. Honorary and General Lectures from the Fourth International Mycological Congress, Regensburg 1990. Edited by D.L. Hawksworth.

microbes, plants and animals has appeared recently (Winkelmann *et al.*, 1987). In fungi different iron uptake mechanisms exist, of which the siderophore-mediated iron transport system has been studied in most detail. Siderophores are small molecular (500–1000 Da), ferric specific, iron complexing microbial compounds, which function in the solubilization, transport and storage of iron. Moreover, regulation of siderophore biosynthesis is an integral part of its definition. Otherwise, every cellular product possessing iron-binding properties would be regarded as a siderophore. Siderophore biosynthesis in fungi is regulated by the prevailing iron content of the medium in which they are growing. Since the intracellular iron concentration in fungi is metabolically linked to the extracellular iron concentration in the medium, any significant decrease in available iron induces the biosynthesis and excretion of siderophores. This rescue mechanism helps to overcome iron stress in many fungi. Moreover, the ability to transport iron into the cells of the fungus has to be proved before a compound can be regarded as a cell-specific siderophore. Constitutive mutants of siderophore biosynthesis have been obtained recently (S.A. Leong, pers. comm.), which continue to produce siderophores even in the presence of high concentrations of the medium, confirming the view that siderophore biosynthesis is regulated at the transcriptional level, as it was found earlier in bacteria (Ernst *et al.*, 1978; Hantke, 1984).

Siderophore structures

Ornithine is an essential constituent of all fungal hydroxamate-type siderophores. It is present as N^{δ}-acyl-N^{δ}-hydroxyornithine, which constitutes the iron-binding hydroxamic acid residues of fungal siderophores. Mutants defective in the biosynthesis of ornithine are unable to biosynthesize hydroxamate-type siderophores, as has been shown in the triple mutant *Neurospora crassa arg-5 ota aga* (Winkelmann and Zähner, 1973). Another mutation which affects siderophore biosynthesis has recently been described in *Ustilago maydis* (Wang *et al.*, 1989). This mutation is a defect in the oxygenase gene which is regarded as the first step in hydroxamate-type siderophore biosynthesis. Mutants defective in siderophore biosynthesis are valuable tools for transport studies because ligand exchange events are excluded. The main fungal siderophore classes are: fusarinines, coprogens, rhodotorulic acid, and ferrichromes.

Fusarinines

The fusarinines (Fig. 3.1) are a group of simple hydroxamic acid derivatives of ornithine containing anhydromevalonic acid (5-hydroxy-3-methyl-pent-

2-enoic acid). The anhydromevalonic acid may be present either in a *cis*- or *trans*-configuration. The hydroxamic acid *cis*-fusarinine (Emery, 1965), or N^{δ}-(*cis*-5-hydroxy-3-methyl-pent-2-enoyl)-N^{δ}-hydroxy-L-ornithine may be polymerized into linear dimers (fusarinine A), linear trimers (fusarinine B), or cyclic trimers (fusarinine C), via an ester between the carbonyl group of the orthinine and the 5-hydroxy group of the anhydromevalonic acid residue (Diekmann and Zähner, 1967; Sayer and Emery, 1968; Jalal *et al.*, 1986, 1987). The cyclic trimer of *cis*-fusarinine has also been named fusigen (Diekmann and Zähner, 1967). The *cis*-fusarinine (mono-, di-, timer) are widespread among fungi in the general *Fusarium*, *Gibberella*, *Nectria*, *Trichotecium*, *Gliocladium*, and *Penicillium* (Diekmann, 1968). The ester bonds of the fusarinines are extremely labile, which renders these compounds unsuitable for a role as extracellular iron-sequestering agents in alkaline environments. However, acetylation of the α-amino groups of the

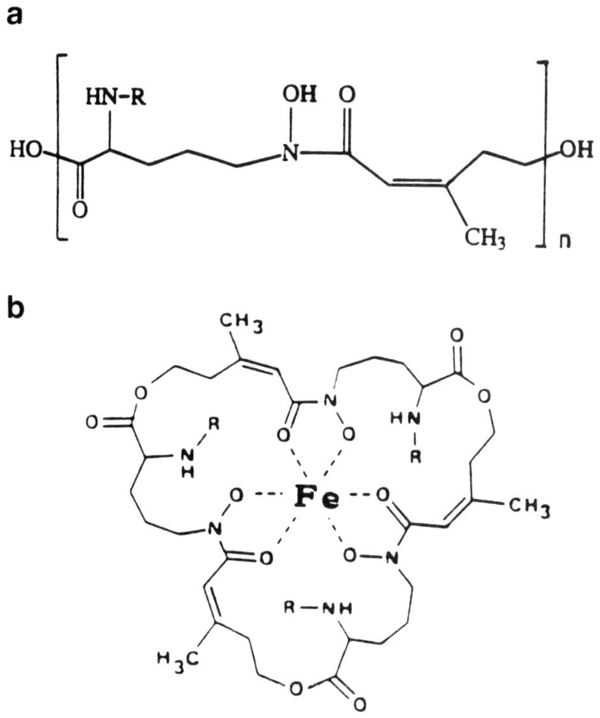

Figure 3.1. (a) Structural formulae of linear fusarinines (R = H): fusarinine, n = 1; fusarinine A, n = 2; fusarinine B, n = 3; and acetyl fusarinines (R = acetyl) correspondingly. (b) Cyclic ferric fusarinines (R = H) and cyclic ferric acetyl fusarinines (R = acetyl): ferric fusarinine C (= fusigen); ferric *N, N′, N″*-triacetyl fusarinine C (= triacetylfusigen).

fusarinines leads to the N^{α}-acetyl-fusarinines, among which the *N*, *N′*, *N″*-triacetyl fusarinine C (triacetylfusigen) revealed the highest chemical stability. Thus, the acetylated trimer of fusarinine seems to be the actual siderophore in many fungi of the genus *Aspergillus* (Diekmann and Krezdorn, 1975; Anke, 1977) and also Agonomycetales (Moore and Emery, 1976; Adjimani and Emery, 1987, 1988). Although the crystal structure of the ferric complex of *N*, *N′*, *N″*-triacetyl-fusarinine C showed a Λ-*cis* absolute configuration when crystallized from ethanol/benzene, the molecule exists predominantly as a Δ-*cis* diastereomer in aqueous solution, as determined by circular dichroism (Hossain *et al.*, 1980).

Coprogens

The first indication that siderophores of the coprogen family existed came from Hesseltine and his coworkers (Hesseltine *et al.*, 1952; Pidacks *et al.*, 1953) who observed that the dung extracts usually employed for the culture of *Pilobolus* species could be replaced by fermentation liquors of a number of species of bacteria and fungi. The addition of 5×10^{-9}g ml^{-1} of coprogen isolated from a culture of a *Penicillium* species facilitated the growth and fruiting of *Pilobolus kleinii.* That compound turned out to be equally effective in all species of *Pilobolus* studied. The structure of

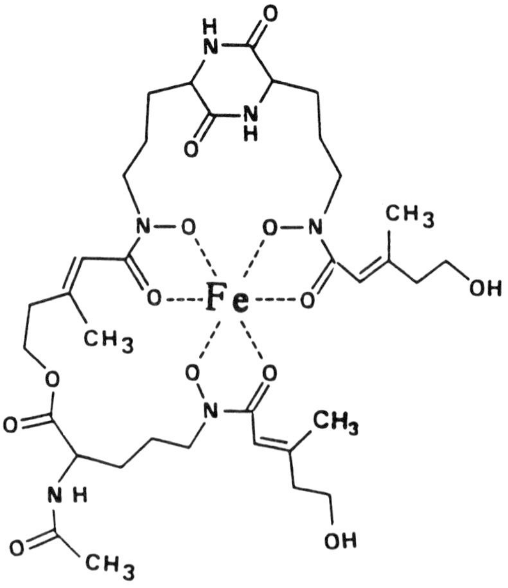

Figure 3.2. Structural formula of coprogen.

coprogen, elucidated by Keller-Schierlein and Diekmann (1970), proved to be a linear trihydroxamate siderophore composed of three transfusarinine molecules, of which two were linked head-to-head in a diketopiperazine ring and the third was linked by an ester group (Fig. 3.2). Coprogen was found in low-iron cultures of *Neurospora crassa* and some *Penicillium* strains (Winkelmann and Huschka, 1987). Several derivatives of coprogen have since been described, for example neocoprogen I and neocoprogen II from cultures of *Curvularia lunata* (Hossain *et al.*, 1987), triornicin and isotriornicine (= neocoprogen I) from *Epicoccum nigrum* (Frederick *et al.*, 1981, 1982), as well as N^{α}-dimethylcoprogen and hydroxycoprogens isolated from *Alternaria alternata* and *A. longipes* respectively (Jalal *et al.*, 1988, 1989). N^{α}-desacetyl coprogen (= coprogen B) has been isolated from *Neurospora crassa* (Diekmann, 1970), *Histoplasma capsulatum* (Burt, 1982), *Verticillium dahliae* (Harrington and Neilands, 1982), and *Stemphylium botyrosum* (Manulis *et al.*, 1987). As shown by Diekmann (1970), dimerum acid is a constituent of coprogen which may originate either as a biosynthetic precursor or as a breakdown product in coprogen-producing cultures. Dimerum acid is composed of two molecules of *trans*-fusarinine joined head-to-head forming a diketopiperazine ring.

Rhodotorulic acid

The structure of rhodotorulic acid (Fig. 3.3) is very similar to that of dimerum acid. The two outer anhydromevalonic acid residues of dimerum acid are replaced by acetyl residues in rhodotorulic acid (Atkin and Neilands, 1968). Thus, rhodotorulic acid may be regarded as a member of the coprogen family. It is a typical siderophore of heterobasidiomycete yeasts, such as *Rhodotorula, Rhodosporidium, Leucosporidium, Sporidiobolus,* and *Sporobolomyces* (Atkin *et al.*, 1970). As *Ustilago violacea* also produced rhodotorulic acid instead of the commonly found ferrichromes, it has recently been transferred to a separate genus as *Microbotryum violaceum* (Deml and Oberwinkler, 1982). Like dimerum acid, rhodotorulic acid (RA) is a dihydroxamate-type siderophore which forms a 3 : 2 (Fe_2RA_3) chelate with ferric iron in aqueous solution. Below pH 3 the complex dissociates into a simple 1 : 1 complex which is cationic and stable to below pH 0. Circular dichroism studies showed a Δ configuration about the two iron atoms in Fe_2RA_3 (Carrano and Raymond, 1978*a*).

Ferrichromes

Ferrichrome was originally detected in the culture medium of *Ustilago sphaerogena* (Neilands, 1952), and has subsequently been found in various *Aspergillus* and *Penicillium* strains (Winkelmann and Huschka, 1987).

a

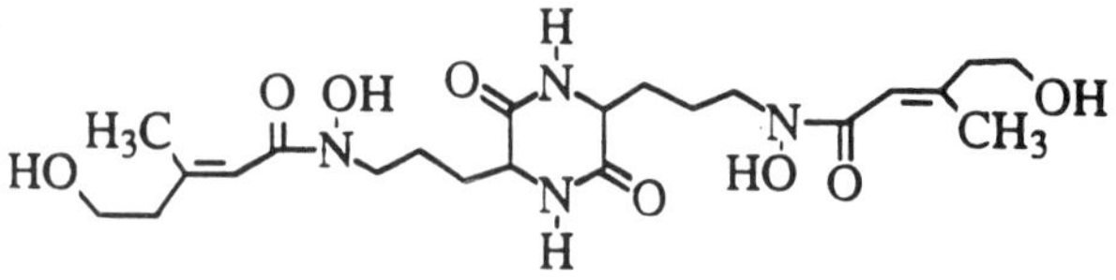

b

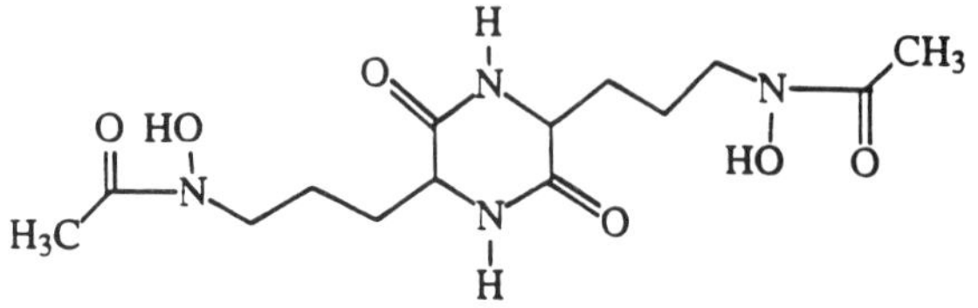

Figure 3.3. Structural formulae of (a) dimerum acid, and (b) rhodotorulic acid.

Ferrichrome is a cyclic hexapeptide containing three glycyl residues and three ^5N-acetyl-^5N-hydroxyornithine residues, designed to form a hexacoordinate ferric iron complex with a Λ-*cis*, *cis* absolute configuration around the iron atom (Fig. 3.4). The very high complex stability with ferric iron ($K_f = 10^{29}$) renders the ferrichromes efficient iron transport molecules. In hydroxamate siderophores ferrous iron is only weakly bound allowing easy reductive removal of iron by the fungal cells. During complex formation with Fe^{3+}, the desferri-ferrichrome releases three protons resulting in an uncharged complex. Moreover, complex formation is associated with characteristic configurational changes which have been followed by ^{13}NMR spectroscopy (Llinas *et al.*, 1970, 1976). Although the hexapeptide structure predominates in the ferrichrome family, a heptapeptide ferrichrome has also been detected which, because of the four glycyl residues, was named tetraglycylferrichrome (Deml *et al.*, 1984). Further ferrichrome-type siderophores have since been isolated in a variety of other fungal genera (Winkelmann and Huschka, 1987). Ferricrocin is the principal siderophore of the *Aspergillus fumigatus* group, possessing a -gly-ser-gly- sequence instead of the -gly-gly-gly- in ferrichrome. (Diekmann and Krezdorn, 1975). Ferricrocin is also produced by *Neurospora crassa* (Horowitz *et al.*, 1976), and by the dermatophyte *Microsporum gypseum* (Bentley *et al.*, 1986). A sequence -gly-ala-gly- was detected in ferrichrome C, isolated from a culture of the species now known as *Myxozyma melibiosi* (Atkin *et al.*, 1970) and placed in the family Lipomycetaceae (van der Walt *et al.*, 1981). Ferrichrome C has also been isolated from low-iron cultures

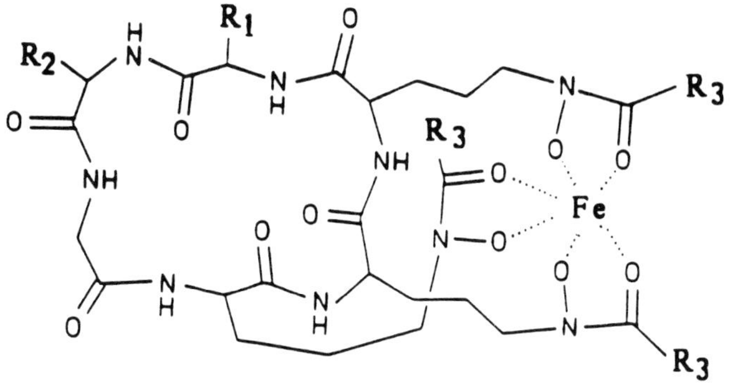

Figure 3.4. Structural formulae of ferrichromes. Ferrichrome: $R_1 = R_2 = H$, $R_3 = CH_3$; ferricrocin: $R_1 = H$, $R_2 = CH_2OH$ $R_3 = CH_3$; ferrichrome C: $R_1 = H$, $R_2 = CH_3$, $R_3 = CH_3$; ferrichrysin: $R_1 = R_2 = CH_2OH$, $R_3 = CH_3$; ferrirubin: R_1 $R_2 = CH_2OH$, $R_3 =$ *trans* $CH{=}C\ (CH_3){-}CH_2{-}CH_2OH$; ferrirhodin: $R_1 = R_2 = CH_2OH$, $R_3 =$ *cis* $CH{=}C(CH_3){-}CH_2{-}CH_2OH$; ferrichrome A: $R_1 = R_2 = CH_2OH$, $R_3 = CH{=}C(CH_3){-}CH_2{-}COOH$, asperchromes: $R_1 = R_2 = CH_2OH$, $R_3 = CH_3$ and *trans* $CH{=}C(CH_3){-}CH_2{-}CH_2OH$ at different ratios.

of *Neovossia indica* (Deml *et al.*, 1984), as well as from species of *Dipodascopsis* and *Zygozyma* (van der Walt *et al.*, 1990). However, species of *Lipomyces* studied by van der Walt *et al.* (1990) did not produce any hydroxamate siderophores.

Ferrichrysin, produced by *Aspergillus melleus* (Keller-Schierlein and Deér, 1963) and *A. ochraceus* (Jalal *et al.*, 1984), contains a -ser-ser-gly- sequence. A ferrichrysin peptide backbone is also present in ferrichrome A isolated from *Ustilago* (Garibaldi and Neilands, 1955; Emery, 1971; van der Helm *et al.*, 1980), as well as in ferrirubin and ferrirhodin isolated from strains of *Penicillium* (Zähner *et al.*, 1963) and *Botrytis cinerea* (Konetschny-Rapp *et al.*, 1988), all of which possess more bulky N-acyl residues compared to ferrichrome, ferricrocin and ferrichrysin (Fig. 3.4).

While all members of the ferrichrome family possess three identical hydroxamic acid residues, asperchromes isolated from *A. ochraceus* contain three or two dissimilar δ-N-acyl groups (Jalal *et al.*, 1984, 1988). All asperchromes are minor products of *A. ochraceus* and seem to result from enzyme non-specificity. Members of the ferrichrome family are often simply referred to as ferrichromes, emphasizing their common structural features. However, their transport properties may differ remarkably depending on the organism studied and the existing siderophore transport system. Thus, ferrichrome and ferrichrome A are transported by different systems in *Ustilago* species (Ecker *et al.*, 1982*a*, *b*; Ecker and Emery, 1983). Furthermore, while ferrichromes possessing δ-N-acetyl residues are transported quite well in *Neurospora crassa*, ferrirubin, possessing δ-N-anhydromevalonic acid residues, is not transported at all and inhibits the uptake of all other siderophores (Winkelmann, 1974; Huschka *et al.*, 1986).

Growth stimulating effect

The requirement of siderophore iron for fungal growth has been documented in a variety of investigations. It was the growth stimulating property of coprogen which allowed its detection by dung-inhabiting *Pilobolus kleinii* species. Siderophore-auxotrophic organisms respond particularly well by enhanced growth when siderophores are added to the growth medium. The bacterium *Arthrobacter flavescens* JG9, which is auxotrophic for hydroxamate siderophores, is now widely used to detect minimal amounts of fungal siderophores. However, not only auxotrophic fungi depend on siderophore iron. Nearly all known fungi, irrespective of their ability to produce siderophores, are able to utilize siderophores for growth. The normal development of fungi starts with germination from spores and the subsequent growth of mycelia, which after a while begins to biosynthesize and excrete siderophores in order to collect environmental iron. At this stage mycelia have already developed to a certain extent at the expense of iron stores, but then seem to stop until iron from siderophores has been accumulated. Growth then continues until further iron is needed for growth or the development of fruiting bodies and spores. This kind of discontinuous iron demand and siderophore biosynthesis is observed under laboratory conditions using defined artificial media. There is every reason to assume that a similar alternating process of siderophore overproduction and growth occurs in the natural environment. It is therefore questionable whether siderophores are continuously produced under natural conditions. As a result of limited carbon and energy sources, fungal siderophore production is probably normally switched off and their biosynthesis only resumes under iron-stress conditions.

Certain fungi, such as species of *Saccharomyces*, *Mucor*, *Phycomyces*, *Geotrichum*, and *Candida*, are unable to synthesize hydroxamate siderophores. In *S. cerevisiae* iron (III) is reduced and transported as ferrous iron across the plasma membrane (Lesuisse *et al.*, 1987; Lesuisse and Labbé, 1989; Dancis *et al.*, 1990). However, even in fungi which do not produce siderophores, the utilization of various siderophores has been reported which suggests that siderophore transport systems still exist in them or that membrane-located reductases are also able to reduce iron from hydroxamates. That siderophores from other fungi can be utilized seems to be advantageous to certain ecological groups. The ecological significance of siderophores has been recognized by Bossier *et al.* (1988).

Transport mechanisms

The basis for growth is the transport of nutrients. The actual transport mechanisms of siderophores have been discussed in detail by Winkelmann (1986). There are obviously different routes and mechanisms, some of which require transport of the intact chelate molecule into the interior and others which operate by delivering iron to membrane-located acceptors with or without the reduction of ferric to ferrous iron. Gallium incorporated into siderophores has been found to be the metal of choice when reductive removal during siderophore transport has been studied as gallium cannot be reduced. The so-called "taxicab" mechanism in *Rhodotorula pilimanae* has been shown to transport not only Fe(III)-rhodotorulate but also Ga(III)-rhodotorulate, suggesting that a reduction step was not involved (Carrano and Raymond, 1987*b*). This is in contrast to the taxicab mechanism in *Ustilago sphaerogena*, where ferrichrome A is transported with a concomitant reduction (Ecker *et al.*, 1982*a*). While all these mechanisms deliver iron to the fungal cells in the form of siderophores, the place of reductive removal of ferric iron seems to be variable. In most cases reductive removal occurs inside the cells, but in others there is a reductive removal at the outside of the cell membrane. Moreover, in certain cases, reduction takes place not only with siderophore-bound iron but also or exclusively with ferric iron. This latter mode of iron uptake has been shown to occur in *Saccharomyces cerevisiae* by Dancis *et al.* (1990). These authors isolated a mutant of *S. cerevisiae* lacking ferric reductase and the ability to take up ferric iron. Furthermore, they presented genetic evidence that the lack of ferric reductase and the uptake deficiency were both due to a single mutation, designated *fre-1.* As pointed out by Emery (1987), reductases play an important role in the iron uptake of aerobic organisms and may be of broader significance then assumed earlier.

Recognition of siderophores

Although iron-sufficient and iron-deficient cultured cells of *Neurospora crassa* had different transport rates during siderophore uptake, the overproduction of cytoplasm membrane proteins could not be detected (Huschka and Winkelmann, 1989). This is in contrast to enteric bacteria, where several outer membrane proteins, functioning as siderophore receptors, are overproduced in low-iron media (Hantke, 1983). Obviously, the transport of siderophores in fungal plasma membranes is not accompanied by an overproduction of plasma membrane receptor or transport proteins. However, kinetic data suggest that the transport of siderophores in fungi is dependent on specific recognition of the various siderophores (Huschka *et al.*, 1985). For example, several strains of *Penicillium* and *Aspergillus* transport ferrichrome-type siderophores, but are unable to transport coprogen. On the other hand, ferrichromes and coprogens are equally well transported by *N. crassa*, indicating the presence of two routes with specific recognition sites for both siderophores in the plasma membrane (Huschka *et al.*, 1986). There is also evidence that not only trihydroxamate siderophores can be recognized by the fungal transport systems, but also mono-di- and trihydroxamates seem to be recognized and transported with considerable efficiency in certain fungi (Jalal *et al.*, 1987). For example, while in *Gliocladium virens* and *Fusarium dimerum* the trihydroxamates, coprogen and ferricrocin, were poorly transported, the dihydroxamates rhodotorulic acid and dimerum acid, and to a lesser extent also the monohydroxamates, *cis* and *trans*-fusarinine, showed high iron uptake rates. These results indicate that the various siderophore structures have their complementary recognition sites within the fungal membrane. It is tempting to speculate that the membrane contains a variety of recognition sites within a siderophore transport system, which reflects the structural diversity of siderophores, and may originate as a result of an evolution of siderophores and their cognate transport systems.

Stereospecificity

Specific recognition of siderophores has also been demonstrated using enantiomeric ferrichromes, such as entantio-ferrichrome and enantio-ferricrocin (Winkelmann, 1979; Winkelmann and Braun, 1981; Huschka *et al.*, 1985). These results clearly showed that recognition of siderophores in fungal membrane transport systems is stereospecific or stereoselective, favouring the natural isomer composed of L-ornithine and excluding the isomer synthesized from D-ornithine. The reversed configuration of the D-ornithyl residues forces the iron centre to adopt a Λ-*cis* absolute configuration which is the mirror image (optical isomer) of the natural Λ-*cis* configuration. The N-acyl residues surrounding the iron atom have also been shown to influence uptake of siderophores in *Neurospora crassa* (Winkelmann *et al.*, 1988). For example, in the ferrichromes an increasing number of *trans*-anhydromevalonic acid residues are inhibitory, but in the coprogens, however, these residues exert a positive effect. These results indicate that the iron centre and its surrounding N-acyl residues are crucial for recognition by fungal siderophore transport systems (Winkelmann, 1990).

Energy requirement

Studies using different siderophores and different fungal strains have confirmed that iron uptake via siderophores is an active transport process which consumes cellular energy (Emery, 1971; Wiebe and Winkelmann, 1975; Huschka *et al.*, 1983; Jalal *et al.*, 1987; Mor *et al.*, 1988; Mor and Barash, 1990). The active transport of siderophores is inhibited by respiratory poisons and uncouplers. For example, cyanide (KCN), azide (NaN_3), 2,4-dinitrophenol (DNP), and carbonyl cyanide phenylhydrazone derivatives (FCCP), are potent inhibitors of siderophore transport. These inhibitors are active irrespective of the actual siderophore transport mechanism involved. Decreasing the temperature has the same effect, confirming the view that iron uptake from siderophores requires metabolic energy. Where does this energy come from? Two principal ways of transport energy have been proposed: (1) a direct involvement of cellular ATP and the phosphorylation of proteins with a subsequent conformation change of transport proteins; or (2) the formation of a membrane potential at the expense of ATP hydrolysis by membrane ATPases which energize the membrane for a variety of transport processes. The fungal plasma membrane H^+-ATPase is known to generate an electrochemical proton gradient across the membrane (200 mV inside negative) which provides energy for nutrient-proton cotransport (Bowman and Bowman, 1986; Harper *et al.*, 1989). Siderophore transport in *Neurospora crassa* has been shown to be sensitive to slight changes of the membrane potential (Huschka *et al.*, 1983). Moreover, ATPase inhibitors have been shown to completely inhibit siderophore transport after membrane depolarization, indicating that the membrane potential is essential for the functioning of siderophore transport across the plasma membrane in fungi. Metabolic inhibitors have also been used to dissect active and non-active uptake in siderophore transport kinetics. For example, uptake curves of Fe-rhodotorulate analogues showed considerable deviation from saturation kinetics, which could be eliminated by substracting residual transport rates in the presence of sodium azide (Müller *et al.*, 1985).

Siderophores in spores

In 1971 Horowitz and his coworkers reported the presence of a germination factor in conidia of *Neurospora crassa* (Charlang and Horowitz, 1971, 1974). When conidia were exposed to media whose water activity had been brought down to 0.95 by the addition of electrolytes or non-electrolytes, they lost a substance that was essential for their germination. The germination factor was identified later as ferricrocin (Horowitz *et al.*, 1976). Ferricrocin and coprogen are both produced by the mycelium of *N. crassa* when grown under conditions of iron limitation. However, while desferricoprogen is excreted into the medium, desferri-ferricrocin remained inside the cells. In addition two other siderophores were found in the mycelium, ferrichrome C and an unidentified siderophore. Ferricrocin and ferrichrome C were also found in conidia, suggesting a direct relationship of mycelial and conidial siderophores. Bioassays showed the ferricrocin content to be 1.43 mg per mg dry weight, 25–30 times higher than that in a 48 h mycelium (Charlang *et al.*, 1981). Later measurements confirmed the exceptional high ferricrocin content of conidia in *N. crassa,* which was approximately 10^{-15}moles spore^{-1} (Huschka and Winkelmann, 1989). Mössbauer spectroscopic measurements revealed that about 47% of the total iron content of conidia (*N. crassa* 74A, wild type) represented ferricrocin bound iron (Matzanke *et al.*, 1987*a*). The residual iron species in conidia, determined after substraction of the ferricrocin resonances. could not be assigned to ferritin, heme, cytochromes, ferredoxin, or to any known intracellular iron protein. Moreover, studies on the metabolic utilization of ^{57}Fe-labelled coprogen in a siderophore-free mutant of *N. crassa* showed a slow metabolization of coprogen and the biosynthesis of ferricrocin (Matzanke *et al.*, 1987*b*). These results were evidence that ferricrocin in *N. crassa* is an intracellular siderophore possessing a function in iron storage (Matzanke *et al.*, 1988). A distinction between extra- and intracellular siderophores has also been made in *Aspergillus nidulans* and *Penicillium chrysogenum* (Charlang *et al.*, 1981). The cellular siderophore of *A. nidulans* is ferricrocin, while that of *P. chrysogenum* is ferrichrome. A characteristic feature of the cellular siderophores is their 'population effect' which allows outgrowth at high densities of conidia but not at low densities. The population effect has been shown to be due to the reversible loss of conidial siderophores (Charlang *et al.*, 1981). Whether ferricrocin is metabolized inside the spores during germination, or released from spores into the medium to be metabolized by the growing mycelia, is still unresolved. Our results using Mössbauer spectroscopy have shown a rapid disappearance of conidial ferricrocin during germination of *N. crassa* suggesting that intracellular siderophores, although metabolically inert during hyphal growth, are metabolized during germination. Furthermore, a mutant of *N. crassa* (*arg-5, ota, aga*), unable to synthesize any siderophore in an ornithine-free medium, did not sporulate on minimal agar. Summarizing these results, it appears that in addition to their well-known function in solubilization, transport, and storage, siderophores play an important role in the regulation of internal processes, such as fungal sporulation and germination.

References

Adjimani, J.P. and Emery, T. (1987) Iron uptake in Mycelia sterilia EP-76. *Journal of Bacteriology* 169, 3664–3668.

Adjimani, J.P. and Emery, T. (1988) Stereochemical aspects of iron transport in Mycelia sterilia EP-76. *Journal of Bacteriology* 170, 1377–1379.

Anke, H. (1977) Metabolic products of microorganisms. 163. Desferritriacetylfusigen, an antibiotic from *Aspergillus deflectus*. *Journal of Antibiotics* 30, 125–128.

Atkin, C.L. and Neilands, J.B. (1968) Rhodotorulic acid, a diketopiperazine dihydroxamic acid with growth factor activity. I. Isolation and characterization. *Biochemistry* 7, 3734–3739.

Atkin, C.L., Neilands, J.B. and Phaff, H.J. (1970) Rhodotorulic acid from species of *Leucosporidium*, *Rhodosporidium*, *Rhodotorula*, *Sporidiobolus* and *Sporobolomyces*, and a new alanine-containing ferrichrome from *Cryptococcus melibiosum*. *Journal of Bacteriology* 103, 722–733.

Bentley, M.D., and Aderegg, R.J., Szaniszlo, P. and Davenport, R.F. (1986) Isolation and identification of the principal siderophore of the dermatophyte *Microsporum gypseum*. *Biochemistry* 25, 1455–1457.

Bossier, P., Hofte, M. and Verstraete, W. (1988) Ecological significance of siderophores in soil. In: Marshall, K.C. (ed.), *Advances in Microbial Ecology*. Plenum Press, New York, pp. 385–413.

Bowman, B.J. and Bowman, E.J. (1986) H^+-ATPases from mitochondria, plasma membranes, and vacuoles of fungal cells. *Journal of Membrane Biology* 94, 83–97.

Burt, W. (1982) Identification of coprogen B and its breakdown products from *Histoplasma capsulatum*. *Infection and Immunity* 35, 990–996.

Carrano, C.J. and Raymond, K.N. (1978*a*) Coordination chemistry of microbial iron transport compounds. 10. Characterization of the complexes of rhodotorulic acid, a dihydroxamate siderophore, *Journal of the American Chemical Society* 100, 5371–5374.

Carrano, C.J. and Raymond, K.N. (1978*b*) Coordination chemistry of microbial iron transport compounds, rhodotorulic acid and iron uptake in *Rhodotorula pilimanae*. *Journal of Bacteriology* 136, 69–74.

Charlang, G., Ng, B., Horowitz, N.H. and Horowitz, R.M. (1981) Cellular and extracellular siderophores of *Aspergillus nidulans* and *Penicillium chrysogenum*. *Molecular and Cellular Biology* 1, 94–100.

Charlang, G.W. and Horowitz, N.H. (1971) Germination and growth of *Neurospora* at low water activities. *Proceedings of the National Academy of Sciences, USA* 68, 260–262.

Charlang, G. and Horowitz, N.H. (1974) Membrane permeability and loss of germination factor from *Neurospora crassa* at low water activities. *Journal of Bacteriology* 177, 261–264.

Dancis, A., Klausner, R.D., Hinnebusch, A.G. and Barriocanal, J.G. (1990) Genetic evidence that ferric reductase is required for iron uptake in *Saccharomyces cerevisiae*. *Molecular and Cellular Biology* 10, 2294–2301.

Deml, G. and Oberwinkler, F. (1982) Studies in heterobasidiomycetes. Part 24. On

Ustilago violacea (Pers.) Rouss. from *Saponaria officinalis* L. *Phytopathologische Zeitschrift* 104, 345–356.

Deml, G., Voges, K., Jung, G. and Winkelmann, G. (1984) Tetraglycylferrichrome – The first heptapeptide ferrichrome. *FEBS Letters* 173, 53–57.

Diekmann, H. (1968) Stoffwechselprodukte von Mikroorganismen. 68. Mitteilung. Die Isolierung und Darstellung von trans-5-hydroxy-3-methylpenten-(2)-säure. *Archives of Microbiology* 62, 322–327.

Diekmann, H. (1970) Stoffwechselprodukte von Mikroorganismen. 81. Mitteilung. Vorkommen und Strukturen von Coprogen B und Dimerumsäure. *Archives of Microbiology* 73, 65–76.

Diekmann, H. and Krezdorn, E. (1975) Stoffwechselprodukte von Mikroorganismen. 150. Ferricrocin, Triacetylfusigen und andere Sideramine aus Pilzen der Gattung *Aspergillus*, Gruppe *Fumigatus*. *Archives of Microbiology* 106, 191–194.

Diekmann, H. and Zähner, H. (1967) Konstitution von Fusigen und dessen Abbau zu Δ^2-Anhydromevalonsäurelacton. *European Journal of Biochemistry* 3, 213–218.

Ecker, D.J. and Emery, T. (1983) Iron uptake from ferrichrome A and iron citrate in *Ustilago sphaerogena*. *Journal of Bacteriology* 155, 616–622.

Ecker, D.J., Lancaster, J.R. and Emery, T. (1982) Siderophore iron transport followed by electron paramagnetic resonance spectroscopy. *Journal of Biological Chemistry* 257, 8623–8626.

Ecker, D.J., Passavant, C.W. and Emery, T. (1982) Role of two siderophores in *Ustilago sphaerogena*. Regulation of biosynthesis and uptake mechanisms. *Biochimica Biophysica Acta* 720, 242–249.

Emery, T. (1965) Isolation, characterization, and properties of fusarinine, a δ-hydroxamic acid derivative of ornithine. *Biochemistry* 4, 1410–1417.

Emery, T. (1971) Role of ferrichrome as a ferric ionophore in *Ustilago sphaerogena*. *Biochemistry* 10, 1483–1488.

Emery, T. (1987) Reductive mechanisms of iron assimilation. In: Winkelmann, G., van der Helm, D., Neilands, J.B. (eds), *Iron Transport in Microbes, Plants and Animals*. VCH Verlagsgesellschaft, Weinheim, pp. 235–250.

Ernst, J.F., Bennet, R.L. and Rothfield, L.I. (1978) Constitutive expression of the iron enterochelin and ferrichrome uptake systems in a mutant strain of *Salmonella typhimurium*. *Journal of Bacteriology* 135, 928–934.

Frederick, C.B., Bentley, M.D. and Shive, W. (1981) Structure of triornicin, a new siderophore. *Biochemistry* 20, 2436–2438.

Frederick, C.B., Bentley, M.D. and Shive, W. (1982) The structure of the fungal siderophore, isotriornicin. *Biochemical and Biophysical Research Communication* 105, 133–138.

Garibaldi, J.A. and Neilands, J.B. (1955) Isolation and properties of ferrichrome A. *Journal of the American Chemical Society* 77, 4846–4847.

Hantke, K. (1983) Identification of an iron uptake system specific for coprogen and rhodotorulic acid in *Escherichia coli* K12. *Molecular and General Genetics* 191, 301–306.

Hantke, K. (1984) Regulation of ferric iron transport in *E. coli*. Isolation of a constitutive mutant. *Molecular and General Genetics* 182, 288–292.

Harper, J.F., Surowy, T.K. and Sussman, M.R. (1989) Molecular cloning and sequence of cDNA encoding the plasma membrane proton pump (H^+-ATPase) of *Arabidopsis thaliana. Proceedings of the National Academy of Sciences, USA* 86, 1234–1238.

Harrington, G.J. and Neilands, J.B. (1982) Isolation and characterization of dimerum acid from *Verticillium dahliae. Journal of Plant Nutrition* 5, 675–682.

van der Helm, D., Baker, J.R., Eng-Wilmot, D.L., Hossain, M.B. and Loghry, R.A. (1980) Crystal structure of ferrichrome and a comparison with the structure of ferrichrome A. *Journal of the American Chemical Society* 102, 4224–4231.

Hesseltine, C.W., Pidacks, C., Whitehill, A.R., Bohonos, N., Hutchings, B.L. and Williams, J.H. (1952) Coprogen, a new growth factor for coprophilic fungi. *Journal of the American Chemical Society* 74, 1362.

Horowitz, N.H., Charlang, G., Horn, G. and Williams, N.P. (1976) Isolation and identification of the conidial germination factor of *Neurospora crassa. Journal of Bacteriology* 127, 135–140.

Hossain, M.B., Eng-Wilmot, D.L., Loghry, R.A. and van der Helm, D. (1980) Circular dichroism, crystal structure, and absolute configuration of the siderophore ferric *N,N'N"*-triacetylfusarinine $FeC_{39}H_{57}N_6O_{15}$. *Journal of the American Chemical Society* 102, 5766–5773.

Hossain, M.B., Jalal, M.A.F., Benson, B.A., Barnes, C.L. and van der Helm, D. (1987) Structure and conformation of two coprogen-type siderophores: neocoprogen I and neocoprogen II, *Journal of the American Chemical Society* 109, 4948–4954.

Huschka, H., Jalal, M.A.F., van der Helm, D. and Winkelmann, G. (1986) Molecular recognition of siderophores in fungi: Role of iron-surrounding N-acyl residues and the peptide backbone during membrane transport in *Neurospora crassa. Journal of Bacteriology* 167, 1020–1024.

Huschka, H., Müller, G. and Winkelmann, G. (1983) The membrane potential is the driving force for siderophore iron transport in fungi. *FEMS Microbiology Letters* 20, 125–129.

Huschka, H., Naegeli, H.U., Leuenberger-Ryf, H., Keller-Schierlein, W. and Winkelmann, G. (1985) Evidence for a common siderophore transport system but different siderophore receptors in *Neurospora crassa. Journal of Bacteriology* 162, 715–721.

Huschka, H. and Winkelmann, G. (1989) Iron limitation and its effect on membrane proteins and siderophore transport in *Neurospora crassa. Biology of Metals* 2, 108–113.

Jalal, M.A.F., Love, S.K. and van der Helm, D. (1986) Siderophore mediated iron(III) uptake in *Gliocladium virens.* 1. Properties of *cis*-fusarinine, *trans*-fusarinine, dimerum acid, and their ferric complexes. *Journal of Inorganic Biochemistry* 28, 417–430.

Jalal, M.A.F., Love, S.K. and van der Helm, D. (1987) Siderophore mediated iron(III) uptake in *Gliocladium virens.* 2. Role of ferric mono- and dihydroxamates as iron transport agents. *Journal of Inorganic Biochemistry* 29, 259–267.

Jalal, M.A.F. and van der Helm, D. (1989) Siderophores of highly phytopathogenic *Alternaria longipes. Biology of Metals* 2, 11–17.

Jalal, M.A.F., Love, S.K. and van der Helm, D. (1988) N^{α}-dimethylcoprogens: three novel trihydroxamate siderophores from pathogenic fungi. *Biology of Metals* 1, 4–8.

Jalal, M.A.F., Mocharla, R., Barnes, C.L., Hossain, M.B., Powell, D.R., Eng-Wilmot, D.L., Grayson, S.L., Benson, B.A. and van der Helm, D. (1984) Extracellular siderophores from *Aspsergillus ochraceus*. *Journal of Bacteriology* 158, 683–688.

Jalal, M.A.F., Hossain, M.B., van der Helm, D. and Barnes, C.L. (1988) Structure of ferrichrome-type siderophores with dissimilar N^{δ}-acyl groups: asperchrome B_1, B_2, B_3, D_1, D_2, and D_3. *Biology of Metals* 1, 77–89.

Keller-Schierlein, W. and Deér (1963) Stoffwechselprodukte von Mikroorganismen. 45. Mitteilung. Zur Konstitution von ferrichrysin und Ferricrocin. *Helvetica Chimica Acta* 46, 1907–1920.

Keller-Schierlein, W. and Diekmann, H. (1970) Stoffwechselprodukte von Mikroorganismen. 85. Mitteilung. Zur Konstitution des Coprogens. *Helvetica Chimica Acta* 53, 2035–2044.

Konetschny-Rapp, S., Jung, G., Huschka, H. and Winkelmann, G. (1988) Isolation and identification of the principal siderophore of the plant pathogenic fungus *Botrytis cinerea*. *Biology of Metals* 1, 90–98.

Lesuisse, E. and Labbé, P. (1989) Reductive and non-reductive mechanisms of iron assimilation by the yeast *Saccharomyces cerevisiae*. *Journal of General Microbiology* 135, 257–263.

Lesuisse, E., Raguzzi, F. and Crichton, R.R. (1987) Iron uptake by the yeast *Saccharomyces cerevisiae*: involvement of a reduction step. *Journal of General Microbiology* 133, 3229–3236.

Llinas, M., Klein, M.P. and Neilands, J.B. (1970) Solution conformation of ferrichrome, a microbial iron transport cyclohexapeptide, as deduced by high resolution proton magnetic resonance. *Journal of Molecular Biology* 52, 399–414.

Llinas, M., Wilson, D.M., Klein, M.P. and Neilands, J.B. (1976) ^{13}C Nuclear magnetic resonance of the ferrichrome peptides: Structural and strain contribution to the conformational state. *Journal of Molecular Biology* 104, 853–864.

Manulis, S., Kashman, Y., and Barash, I. (1987) Identification of siderophores and siderophore-mediated uptake of iron in *Stemphylium botryosum*. *Phytochemistry* 26, 1317–1320.

Matzanke, B.F., Bill, E., Trautwein, A.X., and Winkelmann, G. (1987*a*) Role of siderophores in iron storage in spores of *Neurospora crassa* and *Aspergillus ochraceus*. *Journal of Bacteriology* 169, 5873–5876.

Matzanke, B.F., Bill, E., Müller, G.I., Trautwein, A.X., and Winkelmann, G. (1987*b*) Metabolic utilization of ^{57}Fe-labeled croprogen in *Neurospora crassa*. An in vivo Mössbauer study. *European Journal of Biochemistry* 162, 643–650.

Matzanke, B.F., Bill, E., Trautwein, A.X., and Winkelmann, G. (1988) Ferricrocin functions as the main intracellular iron-storage compound in mycelia of *Neurospora crassa*. *Biology of Metals* 1, 18–25.

Moore, R.E. and Emery, T. (1976) N^{α}-Acetylfusarinines: isolation, characterization and properties. *Biochemistry* 15, 2719–2723.

Mor, H. and Barash, I. (1990) Characterization of siderophore-mediated iron transport in *Geotrichum candidum*, a non-siderophore producer. *Biology of Metals* 2, 209–213.

Mor, H., Pasternak, M., and Barash, I. (1988) Uptake of iron by *Geotrichum candidum*, a non-siderophore producer. *Biology of Metals* 1, 99–105.

Müller, G., Barclay, S.J. and Raymond, K.N. (1985) The mechanism and specificity of iron transport in *Rhodotorula pilimanae* probed by synthetic analogs of rhodotorulic acid. *Journal of Biological Chemistry* 260, 13916–13920.

Neilands, J.B. (1952) A crystalline organo-iron pigment from the smut fungus *Ustilago sphaerogena. Journal of the American Chemical Society* 74, 4846–4847.

Pidacks, C., Whitehill, A.R., Pruess, L.M., Hesseltine, C.W., Hutchings, B.L., Bohonos, N. and Williams, J.H. (1953) Coprogen, the isolation of a new growth factor required by *Pilobolus* species. *Journal of the American Chemical Society* 75, 6064–6065.

Sayer, J.M. and Emery, T.F. (1968) Structures of naturally occurring hydroxamate acids, fusarinines A and B. *Biochemistry* 7, 184–190.

van der Walt, J.P., Botha, A. and Eicker, A. (1990) Ferrichrome production by Lipomycetaceae *Systematic and Applied Microbiology* 13, 131–135.

van der Walt, J.P., Weijman, A.C.M. and von Arx, J.A. (1981) The anamorphic yeast gene *Myxozyma* gen. nov. *Sydowia* 34, 191–198.

Wang, J., Budde, A.D. and Leong, S.A. (1989) Analysis of ferrichrome biosynthesis in the phytopathogenic fungus *Ustilago maydis*: cloning of an ornithine-N^5-oxygenase gene. *Journal of Bacteriology* 171, 2811–2818.

Wiebe, C. and Winkelmann, G. (1975) Kinetic studies on the specificity of chelate iron uptake in *Aspergillus. Journal of Bacteriology* 123, 837–842.

Winkelmann, G. (1974) Metabolic products of microorganisms. 132. Uptake of iron by *Neurospora crassa* III. Iron transport studies with ferrichrome-type compounds. *Archives of Microbiology* 98, 39–50.

Winkelmann, G. (1979) Evidence for stereospecific uptake of iron chelates in fungi. *FEBS Letters* 97, 43–46.

Winkelmann, G., (1986) Iron uptake systems in fungi. In: Swinburne, T.R. (ed.), *Iron, Siderophores and Plant Diseases.* Plenum Press, New York and London, pp. 7–14.

Winkelmann, G. (1990) Structural and stereochemical aspects of iron transport in fungi. *Biotechnology Advances* 8, 207–231.

Winkelmann, G., Berner, I. and Huschka, H. (1988) Structure-activity relationship of siderophores in fungi. *Journal of Plant Nutrition* 11, 883–892.

Winkelmman, G. and Braun, V. (1981) Stereoselective recognition of ferrichrome by fungi and bacteria. *FEMS Microbiology Letters* 11, 237–241.

Winkelmann, G., van der Helm, D. and Neilands, J.B. (eds) (1987) *Iron Transport in Microbes, Plants and Animals.* VCH Verlagsgesellschaft, Weinheim.

Winkelmann, G. and Huschka, H. (1987) Molecular recognition and transport of siderophores in fungi. In: Winkelmann, G., van der Helm, D. and Neilands, J.B. (eds), *Iron Transport in Microbes, Plants and Animals.* VCH Verlagsgesellschaft, Weinheim, pp. 317–336.

Winkelmann, G., and Zähner, H. (1973) Stoffwechselprodukte von Mikroorganismen. 115. Mitteilung. Eisenaufnahme bei *Neurospora crassa.* I. Zur Spezifität des Eisentransportes. *Archives of Microbiology* 88, 49–60.

Zähner, H., Keller-Schierlein, W., Hütter, R., Hess-Leisinger, K. and Deér, A. (1963) Stoffwechselprodukte von Mikroorganismen. 40. Mitteilung. Sideramine aus Aspergillaceen. *Archives of Microbiology* 45, 119–135.

Evolution and Phylogeny

4

Neoteny in the Phylogeny of Eumycota

H. Kreisel, *Institut für Allgemeine und Spezielle Mikrobiologie, Ernst-Moritz Arndt-Universität, Jahnstraße 15, 0-2200 Greifswald, Germany.*

Abstract The role of abbreviation of life-cycles in fungal phylogeny is illustrated with examples from various groups of Eumycota, with particular emphasis on rusts and yeasts and yeast-like fungi. The abbreviation of life-cycles is interpreted as a case of neoteny.

The phylogeny of fungi remains a rather enigmatic theme, although many interesting thoughts and hypotheses on this subject have been expressed by mycologists from the time of Sachs (1874) and Brefeld (1877) to the present. Even if we restrict discussion to the Eumycota, many fundamental questions remain unresolved.

In several cases we can establish certain phylogenetic relations amongst either taxa or life-forms, but the real direction remains doubtful: which of two characters is primary (*plesiomorphic*), and secondary (*apomorphic*)? That applies to morphological as well as biological or ecological characters (Table 4.1).

It is generally assumed that evolution leads from simple to complicated structures, and from monomerous to more complicated life-cycles. Complications result from processes of:

1. Prolongation of certain phases, e.g. the increasing extension of the dikaryophase in Ascomycetes, Teliomycetes, and Basidiomycetes.
2. Summation of structures to form a "composite", e.g. the summation of perithecia or stromata of Hypocreales, Xylariales, etc.; the summation of apothecia in Cyttariales; the summation of conidiophores into acervuli, pycnidia, and coremia of Deuteromycetes.
3. Additional reproductive structures as in the different spore types of rust fungi (Pucciniales).

Frontiers in Mycology. Honorary and General Lectures from the Fourth International Mycological Congress, Regensburg 1990. Edited by D.L. Hawksworth.

Table 4.1. Characters in fungi – which is the earlier state?

Morphological Characters	
Filamentous growth	Yeast-like growth (budding)
Simple septa (no clamps)	Septa with clamps
Spermatiogamy	Angiogamy (cystogamy)
Epigeous fructification	Hypogeous fructification
Gymnocarpy	Angiocarpy
Ecological Characters	
Parasitism	Saprobism
Polyphagous parasitism	Monophagous parasitism
Mycorrhizal symbiosis	Saprophytism
Lichenization	No lichenization
K-strategists	*R*-strategists

Nevertheless, we must admit that the contrary phenomenon is also manifest in the phylogeny of fungi, in particular the abbreviation of certain phases in life-cycles, resulting in a simplification of structures and cycles.

If we consider an "ideal" or "average" life-cycle of a hypothetical "Fungus normalis" (Fig. 4.1), it can be appreciated that there are several phases in which abbreviation has apparently occurred in certain groups of fungi:

1. Loss of sex leads inevitably to the solely anamorphic fungi, the so-called deuteromycetes. There are thousands of species which have apparently lost their ability to form teleomorphs – many descended from Ascomycetes, others from Teliomycetes (rusts), yeasts, etc. Several have been studied quite thoroughly from a genetical point of view.
2. Loss of sexual structures (gametocysts, ascogonia, antheridia, etc.) leading to somatogamy, characteristic of Ustomycetes (smuts), Basidiomycetes, most yeasts, and a few Ascomycetes (e.g. *Morchella*).
3. Loss of the dikaryotic phase (i.e. of conjugated mitoses) leading to secondary haplontic life-cycles which are evidently abbreviated. A "classic" example, already documented by de Bary (1879) and discussed further by Klebahn (1904), Mordvilko (1925), Gäumann (1963) and others, is the descendence of autoecious microcyclic rusts from heteroecious macrocyclic species on the aecial hosts of their relatives. In effect that means an emancipation of the fungus from one of its two alternative hosts (Fig. 4.2, Table 4.2).
4. Loss of fruit-body formation has been postulated by several authors in various ascomycetes: a line of simplification leads from perithecia via cleistothecia to gymnothecia or even naked asci, and simultaneously from

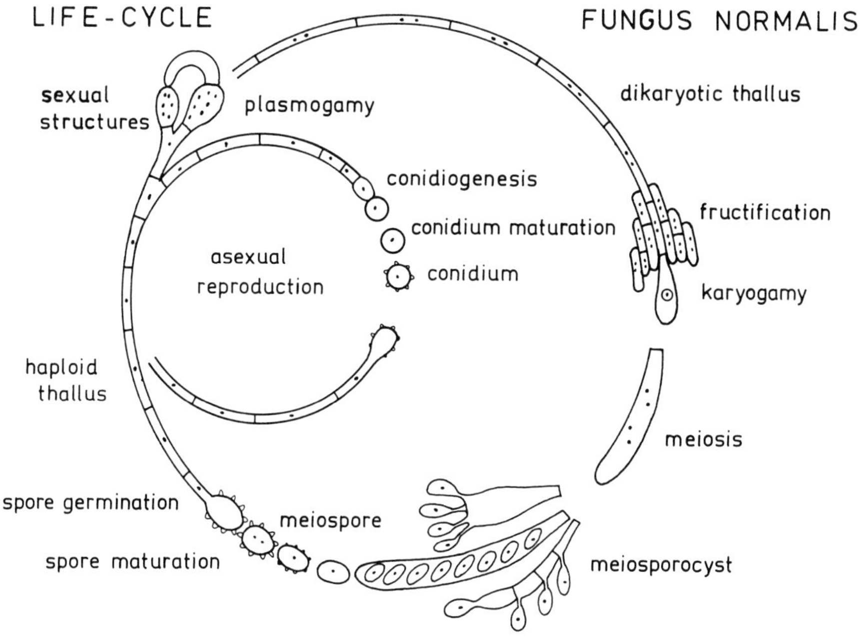

Figure 4.1. Life-cycle of a hypothetical "normal" filamentous fungus.

the inoperculate ascus with apical apparatus to self-autolyzing "prototunicate" asci without any apical structure and without the ability to violently discharge ascospores (i.e. a loss of autochory). Examples have been documented, for example by Cain (1956, 1972) and Malloch (1987).

5. Abbreviation of spore maturation. Hoiland (1983) has drawn attention to the interesting relationship between the agaricoid genera *Cortinarius* (subgenera *Dermocybe, Leprocybe, Phlegmacium*) and *Tricholoma.* Both have tricholomoid fruit-bodies with emarginate gills, a similar pigment spectrum of anthraquinones (biphyscion, endocrocin, flavomannins), and live as ectomycorrhizas, but their basidiospores are different. Basidiospores of *Tricholoma* appear juvenile and immature as compared with those of *Cortinarius* (Table 4.3).

Similar relationships can be found between other genera of dark- and light-spored Hymenomycetes (Table 4.4). In epigeous Gasteromycetes it seems in some cases as if the frequent and widespread "phylogenetically successful" species have relatively small, smooth or weakly ornamented basidiospores, while apparently "ancient" species, with limited areas and (or) restricted to extreme habitats, have rather large and strongly ornamented basidiospores (Table 4.4).

Many moulds, members of the Hyphomycetes, have dry, globose air-

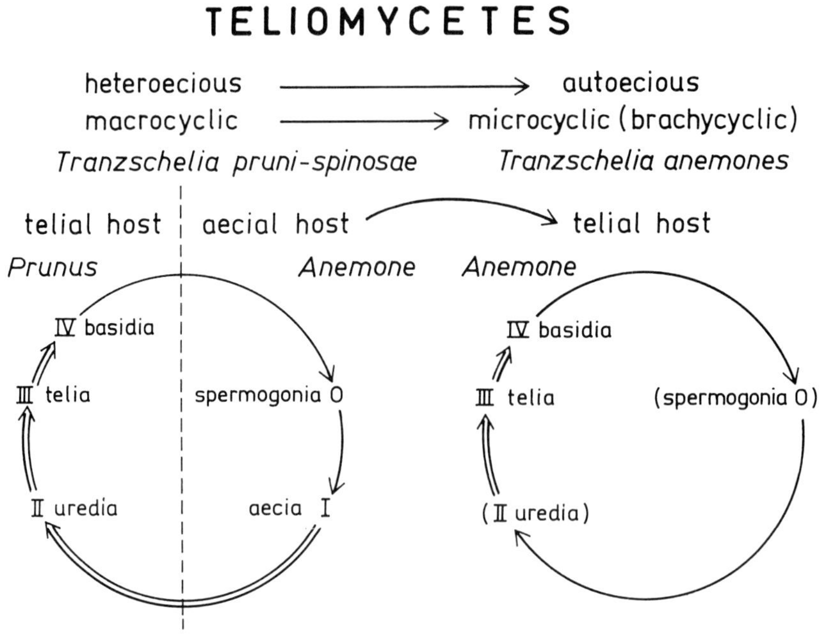

Figure 4.2. Comparison of life-cycles in macrocyclic heteroecious and microcyclic autoecious rust fungi (Pucciniales).

transported conidia which are distributed in a similar manner to the basidiospores of Gastromycetes. Similarly, the widespread and ubiquitous species have small and smooth or nearly smooth conidia as compared with their relatives, which are generally rare or specialized either to soil or to other particular habitats. This can be demonstrated in *Penicillium* (Table 4.5).

6. One of the most intriguing problems in the phylogeny of fungi is the origin of the yeasts. Some authors have regarded these unicellular Eumycota as primitive representatives of the fungi which can be derived directly from other unicellular ancestors (Nadson, 1911; Schussnig, 1954, 1960: 896), but such ideas have mainly been expressed by microbiologists rather than mycologists. This hypothesis is substantiated by yeasts generally having much smaller genomes (5000 to 30000 kb) than filamentous Asco-, Usto-, or Basidiomycetes (40000 to 120000 kb; F. Böttcher, pers. comm.).

Others regard the yeasts as a monophyletic branch, derived from filamentous and dimorphic Endomycetes (e.g. *Dipodascus*, *Eremascus*, *Endomyces fibuliger*) or from primitive Ascomycetes (Guilliermond, 1909, 1940; Gäumann, 1926, 1963).

Table 4.2. Supposed affiliations of microcyclic rusts*.

Heteroecious – Macrocyclic	**Autoecious – Microcyclic**
Chrysomyxa rhododenri	*Chrysomyxa abietis*
0, I *Picea*	III *Picea*
II, III *Rhododendron*	
Puccinia coronata	*Puccinia mesneriana*
0, I *Rhamnus*	III *Rhamnus*
II, III *Poaceae*	
Puccinia longissima	*Puccinia sedi*
0, I *Sedum*	III *Sedum*
II, III *Koeleria*	
Puccinia veratri	*Puccinia epilobii*
0, I *Epilobium*	III *Epilobium*
II, III *Veratrum*	
Tranzschelia pruni-spinosae	*Tranzschelia anemones*
0, I *Anemone*	0, III *Anemone, Hepatica, Thalictrum*
II, III *Prunus*	
Uromyces caricis-sempervirentis	*Uromyces phyteumatum*
0, I *Phyteuma*	III *Phyteuma*
II, III *Carex sempervirens*	
Uromyces rumicis	*Uromyces ficariae*
0, I *Ranunculus ficaria*	II, III *Ranunculus ficaria*
II, III *Rumex*	
Uromyces veratri	*Uromyces cacaliae*
0, I *Adenostyles*	III *Adenostyles, Cacalia*
II, III *Veratrum*	
Uromyces pisi, Uromyces punctatus, Uromyces striatus	*Uromyces laevis, Uromyces scutellatus*, etc.
0, I *Euphorbia*	(II), III *Euphorbia*
II, III Fabaceae	
Heteroecious – Demicyclic	**Autoecious – Demicyclic**
Gymnosporangium clavariiforme, Gymnosporangium juniperinum, etc.	*Gymnosporangium bermudianum*
0, I Rosaceae (tribe Maloideae)	I, III *Juniperus*
III *Juniperus*	

*See Fig. 4.2 for meaning of life-cycle stages 0–IV.

Both hypotheses are questioned by the great diversity in ultrastructural and biochemical characters of yeasts which contrasts greatly with the rather small number of 470 or 500 yeast species (Barnett *et al.*, 1983; Kreger-van Rij, 1984).

Table 4.3. Characters of basidiospores (Hoiland, 1983).

Cortinarius (primitive)	**Tricholoma** (neotenic)
Pigmented (brown)	Hyaline
Ornamented (verrucose)	Smooth
Binucleate	Uninucleate
Volume 13 to 60 μm^3	Volume 3.5 to 20 μm^3

Table 4.4. Basidiospore characters in related genera.

Primitive	**Neotenic**
Hymenomycetidae	
Boletellus	*Gyroporus, Xerocomus*
Paxillus, Tapinella	*Hygrophoropsis, Omphalotus*
Melanomphalia	*Camarophyllopsis, Hygrotrama*
Omphaliaster	*Omphalina*
Cortinarius	*Tricholoma*
Gasteromycetidae	
Bovista trachyspora	*Bovista pusilla*
Disciseda reticulata	*Disciseda candida*
Geasteropsis conrathii	*Geastrum fimbriatum*
Myriostoma coliforme	*Geastrum fimbriatum*
Lycoperdon cokeri	*Lycoperdon pyriforme*
Lycoperdon rimulatum	*Lycoperdon perlatum*
Lycoperdon subvelatum	*Lycoperdon perlatum*
Endemic/Rare	**Widespread/Frequent**

Table 4.5. Conidium characters in *Penicillium*.

Primitive	**Neotenic**
P. asperosporum	*P. aurantiogriseum*
P. fennelliae	*P. chrysogenum*
P. megasporum	*P. griseofulvum*
P. sacculum	*P. funiculosum*
P. verruculosum	*P. viridicatum*
P. lignorum	*P. funiculosum*
More specialized/Rare	**Ubiquitous/Abundant**

Table 4.6. Teleomorphic structures in yeasts and yeast-like fungi.

Asci originating from rudimentary gamocysts	*Dipodascus* *Lipomyces*
Asci originating by somatogamy	*Pichia* *Saccharomyces* *Schizosaccharomyces*
Asci originating from teliospores	*Mixia* *Protomyces* *Taphrina*
Fissitunicate asci enclosed in pseudothecia	Ascogenous black yeasts
Promycelia (septate, forming sporidia)	*Anthracoidea* *Graphiola* *Leucosporidium* *Rhodosporidium* *Ustilago*
Pseudobasidia (aseptate, no sterigmata)	*Aessosporon* *Filobasidiella* *Filobasidium* *Mrakia* *Tilletia*
Phragmobasidia (septate, forming ballistospores)	*Auricularia* *Exidia* *Hirneola* *Tremella*
Holobasidia (aseptate, forming ballistospores)	*Exobasidium* *Tulasnella*

This striking diversity among yeasts concerns:

- Teleomorphic structures (Table 4.6).
- Ultrastructure and the arrangement of budding scars. Streiblová and Beran (1963) and Streiblová *et al.* (1964) distinguished four types of budding; still another has subsequently been found in the basidiomycetous yeasts. Some of the budding types appear comparable to certain modes of conidiogenesis in Asco- and Deuteromycetes: bipolar budding (annelloconidia), lateral budding (solitary blastoconidia), while others do not. The budding type seen in fission yeasts (*Schizosaccharomyces*) seems to be unique amongst the fungi.
- Ultrastructure of cell walls: homogenous in ascogenous and most anascogenous yeasts, but lamellar in basidiomycetous yeasts (Marchant and Smith, 1967).
- Ultrastructure of septa, in as far as true hyphae in promycelia are produced. At least seven types can be distinguished (Kreger-van Rij and Veenhuis, 1973; von Arx and van der Walt, 1986; Moore, 1987; Table 4.7).

- Chemical composition of cell walls, e.g. chitin and mannan content (Miller and Phaff, 1958; Bartnicki-Garcia, 1968), DBB-reaction (positive in basidiomycetous yeasts; van der Walt and Hopsu-Havu, 1976), amyloid sheaths (in *Dipodascopsis*, and *Lipomyces*).
- G + C contents in DNA: low in ascogenous yeasts, and high in basdiomycetous yeasts (Kurtzman, 1985; Jahnke, 1987).
- Ubiquinone systems (types of coenzyme Q), ranging from Q-5 to Q-10 and hydrogenated ubiquinones. In general, the ubiquinone type is genus-specific in yeasts (Table 4.8). No ubiquinone has been found in *Hasegawaea japonica*, a segregate from *Schizosaccharomyces* (Yamada and Banno, 1987).

Table 4.7. Septal structures in yeasts and yeast-like fungi.

Multiple micropores	*Arxula* *Dipodascus* (incl. *Geotrichum*) *Endomyces* *Saccharomycopsis* *Sporothrix* sect. *Farinosa* p.p.
Single central micropores	*Cephaloascus* *Dipodascopsis* *Hyphopichia* *Pichia* *Sporothrix* sect. *Farinosa* p.p. *Yarrowia*
Simple pores with Woronin bodies	*Sporothrix* sect. *Sporothrix* *Taphrina*
Simple pores	*Chionosphaera* *Exobasidium* *Graphiola* *Ustilago* *Rhodosporidium* (incl. *Rhodotorula*)
Dolipores, no parenthesome	*Christiansenia* *Filobasidiella* (incl. *Cryptococcus*)
Dolipores, with imperforate parenthesome	*Auricularia* *Exidia* *Moniliella* *Trichosporonoides* *Tulasnella*
Dolipores, with vesicular parenthesome	*Sirobasidium* *Tremella*

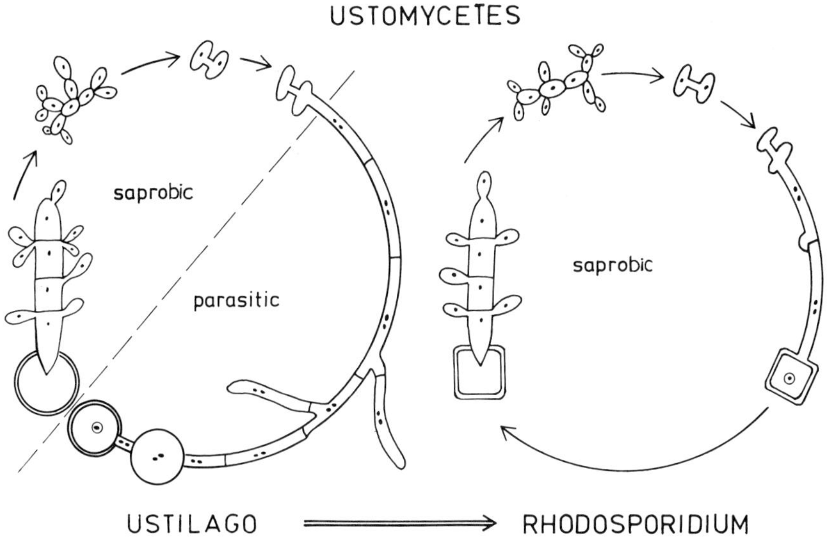

Figure 4.3. Comparison of life-cycles in Ustilaginales: *Ustilago* (metabiotrophic parasite) and *Rhodosporidium* (saprobe).

All these criteria lead to the conclusion that the yeasts have evolved polyphyletically from different groups of filamentous or dimorphic fungi, as has been expressed, amongst others, by Lodder (1934), Cain (1972), Redhead and Malloch (1977), Kreisel (1983, 1988), van der Walt (1987), von Arx and van der Walt (1987), and Guého *et al.* (1989) (Table 4.9). This concept was substantiated further by 5S rRNA sequence analysis (Huysmans *et al.*, 1983) which places, e.g. *Schizosaccharomyces pombe* and *Filobasidium floriform* far away from *Saccharomyces*, *Pichia*, etc.

If yeasts are polyphyletic, then their evolution from dimorphic fungi must be interpreted as an example of *neoteny*, as an extreme reduction of the thallus to a single cell which corresponds to the physiologically most active part of the hypha, the growing hyphal tip. That means a wide variety of complicated morphological structures, and the genes which are coding such structures, have been "dropped" or made "redundant". This concept also explains the poverty of yeasts in secondary metabolites; no antibiotics or mycotoxins have been found and only a few pigments demonstrated in yeasts. Yeasts are an extreme case of the ruderal life strategy (Pugh, 1980).

The transition from the mycelial form to the yeast form may be illustrated by the metabiotrophic dimorphic phytoparasites, which occur in two classes of fungi:

Ascomycetes : *Taphrina, Protomyces, Mixia*
Ustomycetes : *Ustilago, Graphiola, Exobasidium*

Table 4.8. Ubiquinones in yeasts and yeast-like fungi.

Coenzyme	Anamorphic genera (+ = DBB Positive)	Endomycetes and Ascomycetes	Ustomycetes and Basidiomycetes
Q-5		*Nematospora* p.p.	
Q-6	*Candida* p.p. (*C. glabrata*, *C. holmii*, *C. stellata*) *Kloeckera*	*Arxiozyma* *Hanseniaspora* *Nadsonia* *Nematospora* p.p. *Pachytichospora* *Saccharomyces* *Saccharomycodes* *Waltiozyma* *Zygosaccharomyces*	
Q-7	*Candida* p.p. (*C. diversa*, *C. utilis*)	*Ambrosiozyma* *Emmonsiella* *Endomycopsis* *Hormoascus* *Issatchenkia* *Pichia* s.str. *Torulopsis* *Williopsis*	*Kondoa*
Q-8	*Candida* p.p. (*C. ernobii*, *C. mucilaginea*) + "*Rhodotorula*" *infirmominiata*	*Botryoascus* *Hyphopichia* *Pachysolen* *Pichia* p.p. *Saccharomycopsis* *Zygozyma*	*Cystofilobasidium* *Mrakia*
Q-9	*Arxula* + *Bensingtonia* *Brettanomyces* *Candida* s.str. (*C. maltosa*, *C. parapsilosis*, *C. tropicalis*) *Sporothrix* sect. *Farinosae* + *Sterigmatomyces*	*Cephaloascus* *Dekkera* *Dipodascopsis* *Dipodascus* *Holleya* *Lipomyces* *Lodderomyces* *Metschnikowia* *Octosporomyces* *Wickerhamia* *Wingea* *Yarrowia* *Zygoascus*	*Auricularia* *Carcinomyces* *Filobasidium* p.p. *Hirneola* *Leucobasidium* *Sterigmatosporidium*
Q-10	+ *Ballistosporomyces* + *Bullera* *Candida* p.p. (*C. aeria*, *C. marina*)	*Schizosaccharomyces* s.str. *Taphrina* *Waltomyces*	*Aessosporon* *Doassansia* *Fellomyces* *Filobasidiella*

Table 4.8. *Continued.*

Coenzyme	Anamorphic genera (+ = DBB Positive)	Endomycetes and Ascomycetes	Ustomycetes and Basidiomycetes
			Filobasidium p.p.
	+ *Cryptococcus*		*Graphiola*
	+ *Rhodotorula*		*Phaffia*
	Saitoella		*Rhodosporidium*
	+ *Sporobolomyces*		*Tremella*
	+ *Sporothrix*		*Ustilago*
	sect. *Luteoalba*		
	+ *Tilletiopsis*		
Q-10 (H_2)	*Hormonema*		*Erythrobasidium*
	+ *Sporobolomyces*		*Microbotryum*
	elongatus		
	Sporothrix s.str.		
None		*Hasegawaea*	

By "loosing" the obligately parasitic, dikaryotic mycelial phase, the saprobic, monokaryotic yeast phase has become completely free from any host plant and assumed an independent life as a typical yeast (Fig. 4.3).

It remains to be discussed whether the dimorphic fungi have evolved from purely filamentous fungi, or both yeasts and filamentous fungi have evolved from dimorphic fungi but in different directions. It should be remembered that some authors, such as Savile (1955, 1968, 1971) and Shaffer (1975), have stressed the fundamental position of the dimorphic order Taphrinales in the evolution of Ascomycetes as well as of Teliomycetes (rust fungi), although the ultrastructure of *Taphrina* (Syrop and Beckett, 1976; von Arx *et al.*, 1982) corresponds much more to that of the Ascomycetes than to either Teliomycetes or Basidiomycetes.

In this connection it should be noted that several large groups of Eumycota do not include any yeasts or dimorphic taxa. These are the Trichomycetes, Endogonales, most orders of ascomycetes (including all lichenized groups), Teliomycetes, Hymenomycetidae, and Gasteromycetidae. Among these groups, the lichenized ascomycetes, some holobiotrophic Ascomycetes (Laboulbeniales, Meliolales), the holobiotrophic Teliomycetes, and the symbiotic VA mycorrhiza-forming Endogonales can be regarded as phylogenetically rather old, ancestral groups if we accept the hypothesis of Demoulin (1974), Kohlmeyer (1975) and others that biotrophism rather than parasitism is an ancestral character in fungi, prior to saprobism.

More information on this subject must be accumulated before it can be

Table 4.9. Polyphyletic origin of yeasts.

Class	Order	Yeast taxa
Zygomycetes	Mucorales	"*Mucor* yeasts"
Endomycetes	Endomycetales	Ascogenous yeasts p.p.
Ascomycetes	Taphrinales	*Saitoella* yeast phase of *Taphrina*
	Eurotiales	Ascogenous yeasts p.p.
	Ophiostomatales	Ascogenous yeasts p.p.
	Dothideales	Ascogenous "black yeasts"
Ustomycetes	Ustilaginales	*Erythrobasidium*
		Rhodosporidium
		Yeast phase of *Microbotryum*
		Yeast phase of *Ustilago*
	Graphiolales	Yeast phase of *Graphiola*
	Tilletiales	*Aessosporon*
		Mrakia
		Sporobolomyces
	Exobasidiales	Yeast phase of *Exobasidium*
	Cryptobasidiales	Yeast phase of *Microstroma*
Basidiomycetes	Filobasidiales	*Cryptococcus*
		Filobasidiella
		Filobasidium
		Crystofilobasidium
	Auriculariales	Yeast phase of *Auricularia*
		Yeast phase of *Hirneola*
	Tremellales	*Bensingtonia*
		Moniliella
		Trichosporonoides
		Yeast phase of *Tremella*
Uncertain position		*Hasegawaea*
		Octosporomyces
		Schizosaccharomyces

decided whether dimorphic fungi arose prior to filamentous fungi, or vice versa.

Acknowledgements

I thank Professor Dr F. Böttcher for useful information and Dr F. Schauer for stimulating discussion on advances in yeast taxonomy.

References

von Arx, J.A. and van der Walt, J.P. (1986) Are yeast cells of Endomycetales homologues of conidia of Eurotiales? *Persoonia* 13, 161–171.

von Arx, J.A. and van der Walt, J.P. (1987) Ophiostomatales and Endomycetales. *Studies in Mycology, Baarn* 30, 167–176.

von Arx, J.A., van der Walt, J.P. and Liebenberg, N.V.D.M. (1982) The classification of *Taphrina* and other fungi with yeast-like cultural states. *Mycologia* 74, 285–296.

Barnett, J.A., Payne, R.W. and Yarrow, D. (1983) *Yeasts: characteristics and identification.* Cambridge University Press, Cambridge.

Bartnicki-Garcia, S. (1968) Cell wall chemistry, morphogenesis and taxonomy of fungi. *Annual Review of Microbiology* 22, 87–108.

Bary, A. de (1879) *Aecidium abietinum. Botanische Zeitung* 37, 825–830.

Brefeld, I. (1877) *Botanische Untersuchungen über Schimmelpilze.* Vol. 3. Arthur Felix, Leipzig.

Cain, R.F. (1956) Studies of coprophilous ascomycetes. II. *Phaeotrichum,* a new genus in a new family, and its relationships. *Canadian Journal of Botany* 34, 675–687.

Cain, R.F. (1972) Evolution of the fungi. *Mycologia* 64, 1–14.

Demoulin, V. (1974) The origin of ascomycetes and basidiomycetes. The case for a red algal ancestry. *Botanical Review* 40, 315–345.

Gäumann, E. (1926) *Vergleichende Morphologie der Pilze.* Gustav Fischer, Jena.

Gäumann, E. (1963) *Die Pilze.* 2nd edn. Birkhäuser, Basel and Stuttgart.

Guého, E., Kurtzman, C.P. and Peterson, S.W. (1989) Evolutionary affinities of heterobasidiomycetous yeasts estimated from 18S and 25S ribosomal RNA sequence divergence. *Systematic and Applied Microbiology* 12, 230–236.

Guilliermond, A. (1909) Sur la phylogenèse des levures. *Comptes rendus hebdomadaires de la Société Biologique de Paris* 66, 998–1000.

Hoiland, K. (1983) *Cortinarius* subgenus *Dermocybe. Opera Botanica* 71, 1–113.

Huysmans, E., Dams, E., Vandenberge, A. and de Wachter, R. (1983) The nucleotide sequences of the 5S rRNAs of four mushrooms and their use in studying the phylogenetic position of basidiomycetes among the Eukaryota. *Nucleic Acids Research* 11, 2871–2880.

Jahnke, K.-D. (1987) Assessing natural relationships by DNA analysis – techniques and applications. *Studies in Mycology, Baarn* 30, 227–246.

Klebahn, H. (1904) *Die wirtswechselnden Rostpilze.* Borntraeger, Berlin.

Kohlmeyer, J. (1975) New clues to the possible origin of ascomycetes. *BioScience* 25, 86–93.

Kreger-van Rij, N.J.W. [ed.] (1984) *The Yeasts – a taxonomic study.* 3rd edn. Elsevier, Amsterdam.

Kreger-van Rij, N.J.W. and Veenhuis, M. (1973) Electron microscopy of septa in ascomycetous yeasts. *Antonie van Leeuwenhoek* 39, 481–490.

Kreisel, H. (1983) Abstammung und Evolution der Pilze. In: Michael, E., Hennig, B. and Kreisel, H. (eds), *Handbuch für Pilzfreunde.* Vol. 5. 2nd edn. Gustav Fischer, Jena, pp. 9–25, 63–66.

Kreisel, H. (1988) Abstammung und systematische Einordnung der Pilze. *Biologische Rundschau* 26, 65–77.

Kurtzman, C.P. (1985) Molecular taxonomy of the fungi. In: Bennett, J.W. and Lasure, L.L. (eds), *Gene Manipulations in Fungi.* Academic Press, New York, pp. 35–66.

Lodder, J. (1934) Die anaskosporogenen Hefen. 1. Hälfte. *Verhandelingen der Koninklijke Nederlandse Akademie van Wetenschapen, serie 2,* 32, 1–256.

Malloch, D. (1987) Ordinal relationships among reduced ascomycetes. *Studies in Mycology, Baarn* 30, 177–186.

Marchant, R. and Smith, D.G. (1967) Wall structure and bud formation in *Rhodotorula glutinis. Archiv für Mikrobiologie* 58, 248–256.

Marchant, R. and Smith, D.G. (1968) Bud formation in *Saccharomyces cerevisiae* and a comparison with the mechanism of cell division in other yeasts. *Journal of General Microbiology* 53, 163–169.

Miller, M.V. and Phaff, H.J. (1958) On the cell wall composition of apiculate yeasts. *Antonie van Leeuwenhoek* 24, 225–238.

Moore, R.T. (1987) Micromorphology of yeasts and yeast-like fungi and its taxonomic implications. *Studies in Mycology, Baarn* 30, 203–226.

Mordvilko, A. (1925) Die Evolution der Zyklen und die Heterözie bei den Rostpilzen. *Centralblatt für Bakteriologie, serie 2,* 22, 181–204, 505–530.

Nadson, G.A. (1911) [The sexual process in yeasts and bacteria.] (In Russian) *Russkij Vratch* 51, 2093.

Pugh, G.J.P. (1980) Strategies in fungal ecology. *Transactions of the British Mycological Society* 75, 1–14.

Redhead, S.A. and Malloch, D.W. (1977) The Endomycetaceae: new concepts, new taxa. *Canadian Journal of Botany* 55, 1701–1711.

Sachs, J. (1874) *Lehrbuch der Botanik.* 4th edn. Engelmann, Leipzig.

Savile, D.B.O. (1955) A phylogeny of the basidiomycetes. *Canadian Journal of Botany* 33, 60–104.

Savile, D.B.O. (1968) Possible interrelationships between fungal groups. In: Ainsworth, C.G. and Sussman, A.S. (eds), *The Fungi: an advanced treatise.* Vol. 3. Academic Press, New York and London, pp. 649–675.

Savile, D.B.O. (1971) Coevolution of the rust fungi and their hosts. *Quarterly Review of Biology* 46, 211–218.

Schussnig, B. (1954) *Grundriß der Protophytenkunde.* Gustav Fischer, Jena.

Schussnig, B. (1960) *Handbuch der Protophytenkunde,* Vol. 2. Gustav Fischer, Jena.

Shaffer, R.L. (1975) The major groups of basidiomycetes. *Mycologia* 67, 1–18.

Streiblová, E. and Beran, K. (1963) Types of multiplication scars in yeasts, demonstrated by fluorescence microscopy. *Folia Microbiologica, Praha* 8, 221–227.

Streiblová, E., Beran, K. and Pokorny, V. (1964) Multiple scars, a new type of yeast scars in apiculate yeasts. *Journal of Bacteriology* 88, 1104–1111.

Syrop, M. and Beckett, A. (1976) Leaf curl disease of almond caused by *Taphrina deformans.* III Ultrastructural cytology of the pathogen. *Canadian Journal of Botany* 54, 293–305.

van der Walt, J.P. (1987) The yeasts – a conspectus. *Studies in Mycology, Baarn 30,* 19–31.

van der Walt, J.P. and Hopsu-Havu, V.K. (1976) A colour reaction for the differentiation of ascomycetous and hemibasidiomycetous yeasts. *Antonie van Leeuwenhoek* 42, 157–163.

Yamada, Y. and Banno, I. (1987) *Hasegawaea* gen. nov., an ascosporogenous yeast genus for the organisms whose asexual reproduction is by fission and whose ascospores have smooth surfaces without papillae and which are characterized by absence of coenzyme Q and by the presence of linoleic acid in cellular fatty acid compositions. *Journal of General and Applied Microbiology* 33, 295–298.

5

Homologies and Analogies in the Evolution of Lichens

J. Poelt, *Institut für Botanik, Karl-Franzens Universität Graz, Holteigasse 6, A-8010 Graz, Austria.*

Abstract The problems homologous and analogous characters present in understanding the evolution of lichens are discussed with particular reference to lecideine ascomata, thallus anatomy and morphology, asexual diaspores, pycnidia and conidia, secondary metabolites, lichen biology, and growth forms. The need for caution in the use of such characters in the development of taxonomies is stressed. Foliose lichen thalli are interpreted as precocious thalline exciples in which the development of hymenial tissues is delayed.

Introduction

The great German morphologist Karl von Goebel once mentioned that lichens interested him so little during his life-time because they seemed to be boring. In his time it was not possible to recognize and study the exciting aspects of these organisms in an adequate way. And, furthermore, the extremely formalistic taxonomic system used for lichens at that time made one believe that lichenology only consisted of putting material collected in the field into man-made well-defined taxonomic boxes. How does this appear today?

It is well known that lichens, which are fungi living symbiotically with algae, have originated independently within the ascomycetes and, to a minor degree, within the basidiomycetes. That analogy is clear. Within the ascomycetes there are at this time about 46 orders distinguished (Eriksson and Hawksworth, 1990) and out of those no less than 16 contain lichenized taxa to a smaller or larger degree. Out of about 238 families, 81 consist entirely of lichens or include at least some lichenized taxa. This shows

Frontiers in Mycology. Honorary and General Lectures from the Fourth International Mycological Congress, Regensburg 1990. Edited by D.L. Hawksworth.

that lichenization has occurred polyphyletically many times in the taxonomic system of the group and, very likely, originated at very different times. We must therefore be concerned with very many analogies, when we try to understand even the basic trends of evolution of lichens.

A few examples of homologies and analogies are discussed here.

Ascomata and thalli

The lecideine apothecium, coloured dark to completely black by a not yet chemically understood process of carbonization, has over decades given the impression of a convincing character, so that it has become essential in defining genera and even families. All the crustose lichens distinguished by such fruiting bodies, producing single-celled ascospores and associated with green algae, were accommodated in an elephantine genus *Lecidea*; this was for over 100 years considered as a "crux lichenologorum", but nevertheless it was presented as a more or less natural unit. Hertel and his coworkers (e.g. Hertel and Rambold, 1985; Hertel, 1987) have shown that the lichens treated in *Lecidea* have to be separated into no less than 20 families in a more natural system. The lecideine structure of the ascomata must be considered the result of ecologically directed analogous (or partly convergent) evolution (Fig. 5.1). In this connection it should also be mentioned, that the spore-based system, the "holy cow" of former times, has also proved to be as unnatural. In different natural genera, the form and division of spores can vary from one-celled to multiseptate or even strongly muriform. Many examples of this could be cited. Even the seemingly unique dry spore mass, the mazaedium in lichens, united for a long time within a single order Caliciales, has evolved at least twice according to Tibell (1984).

The structure of the fruiting bodies, which are in general considered as very conservative, can therefore vary considerably. If these structures are so variable, how much more must this be true for thalline structures.

Crustose lichens exposed to the harsh conditions of coastal rocks, especially on northern seashores, are distinguished from their inland relatives by major alterations in thallus anatomy, to the extent that their normal basic structure seems to be obscured. Thallus structures in lichens from this habitat but of quite different families are very similar, and therefore analogously evolved (Poelt and Romauch, 1977; Fig. 5.2). Comparable alterations also occur in freshwater lichens, whose adaptations have been studied much less. Such examples demonstrate that a thorough examination of the functional anatomy of lichens still remains to be carried out.

A type of thallus, which was for a long time regarded as sorediate and also essentially primitive, that of the leprarioid lichens, should not be confused with undeveloped states of other lichens, as sometimes suggested

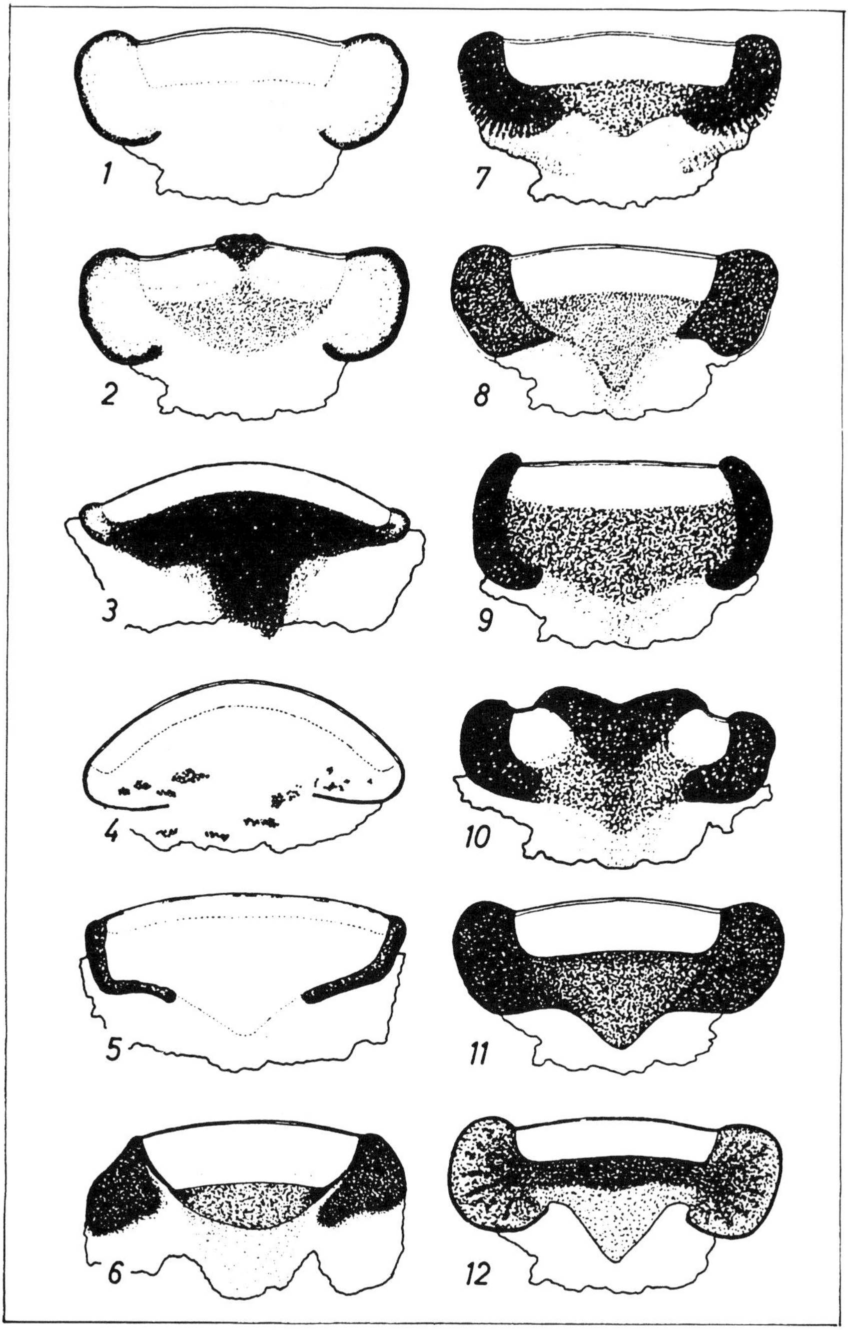

Figure 5.1. Variability of lecideine ascomatal structures in *Lecidea s. ampl.* (after Hertel, 1967).

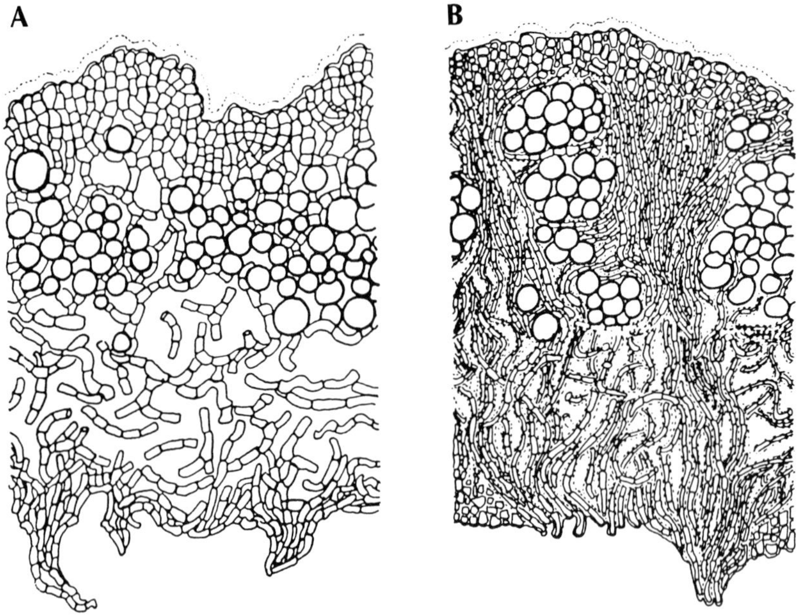

Figure 5.2. Vertical sections of *Caloplaca* thalli from different habitats. A, *C. granulosa*, an inland species with distinct cortical and medullary layers; B, *C. verruculifera* from the seashore where most of the medulla is filled by compact prosoplectenchyma (after Poelt and Romauch, 1977).

in earlier times. These lichens are in reality highly derived specialists adapted to survival in overhangs and bark fissures not directly wetted by rain. They are specialized in being able to take up water vapour. Their cotton wool-like structure and hyphae impregnated with wax-like unwettable compounds, are closely connected with their particular ecology. Leprarioid lichens have evolved in very different orders and families of lichens, but only in those where the photobionts belong to the green algae; physiologists have shown that cyanobacteria are not able to take up water vapour directly. It is nevertheless difficult to assign leprarioid lichens to their natural groups because, for as yet unknown reasons, they only rarely develop ascomata. As a result these lichens have been put into artificial groupings, which are untenable. For example, two leprarioid species containing anthraquinones have been united into the genus *Leproplaca* (Laundon, 1974). In one of these species, *L. xantholyta*, apothecia have subsequently been discovered showing that this is only a leprarioid species of the well-known genus *Caloplaca*.

Asexual diaspores

Isidia, soredia, or other lichenized asexual diaspores occur in very many different groups of lichens; they must also be the result of analogous, or possibly sometimes convergent, evolution. An account of the exact typology of these bodies, which are anything but monomorphous (e.g. the "sorediigenous paraisidia" of *Physcia tribacia*, usually called soredia; Scutari, 1990) as frequently described, needs also to be prepared.

Pycnidia and conidia

Within the lichenized ascomycetes, structures of very different shape and anatomy are collectively described as pycnidia, and give rise to pycnospores (or pycnoconidia) (e.g. Vobis, 1980). Most of these in my view function as male sexual organs, i.e. "spermogonia" producing "spermatia". The cells of the spermatia are very small and low in "plasma" content, and therefore entirely unsuited to vegetative reproduction. However, there is also a growing number of examples of lichens which produce true, often large, conidia rich in plasma content. These conidia may improve the reliability of reproduction especially in short-lived lichens. A recently discovered species in the genus *Cheiromycina*, *C. petri* (Hawksworth and Poelt, 1990) with very unusual dichotomously forked conidia can be cited as an example

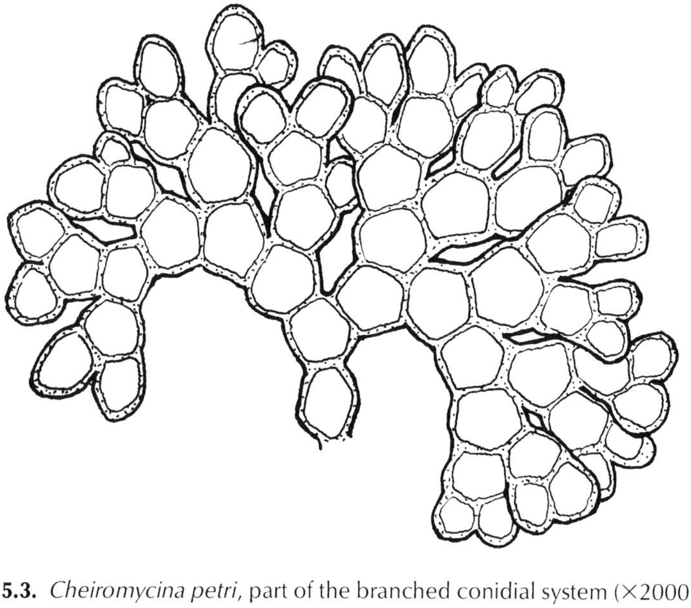

Figure 5.3. *Cheiromycina petri*, part of the branched conidial system (×2000).

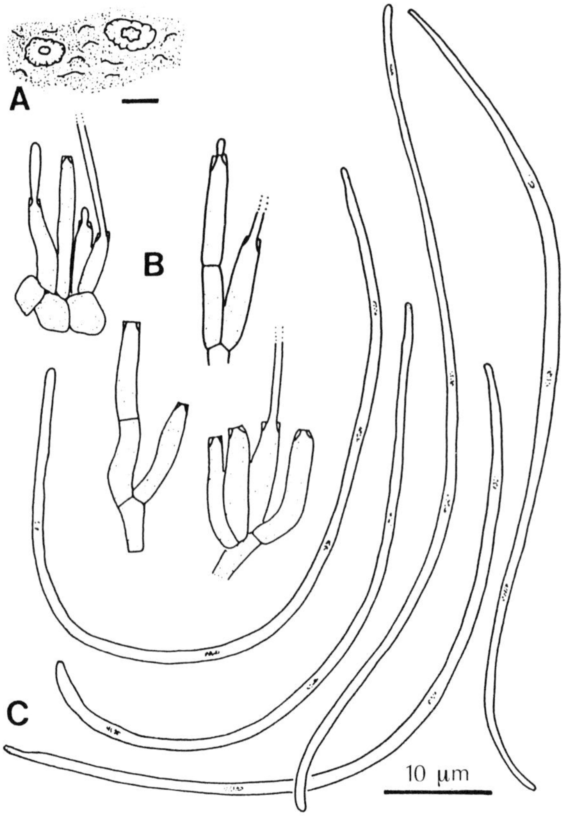

Figure 5.4. *Lichingoldia gyalectiformis*. A, Conidiomata, surface view; B, conidiogenous cells; C, conidia (after Hawksworth and Poelt, 1986).

(Fig. 5.3). Because there was a previous mention of aquatic lichens above, a second example can be selected from this habitat. The "Ingoldian" aquatic hyphomycetes with filiform or tetraradiate conidia have been frequently discussed in recent years (e.g. Webster, 1987). *Lichingoldia gyalectiformis*, described from a Norwegian mountain rivulet (Hawksworth and Poelt, 1986), shows that such more or less filiform "hydroconidia" can also arise from lichens (Fig. 5.4). Even tetraradiate conidia have been found in the foliicolous lichen genus *Lasioloma* (Vězda, in Poelt, 1986; Fig. 5.5). The tetraradiate structure may have a role in anchoring the conidia which are probably dispersed by water drops during heavy tropical rains.

A special type of conidia are the "thallospores" ("thalloconidia") only described from the single genus *Umbilicaria* (Poelt, 1977; Hasenhüttl and Poelt, 1978; Hestmark, 1990). A short survey has shown, however, that thallospores may occur in very similar structures in different genera and

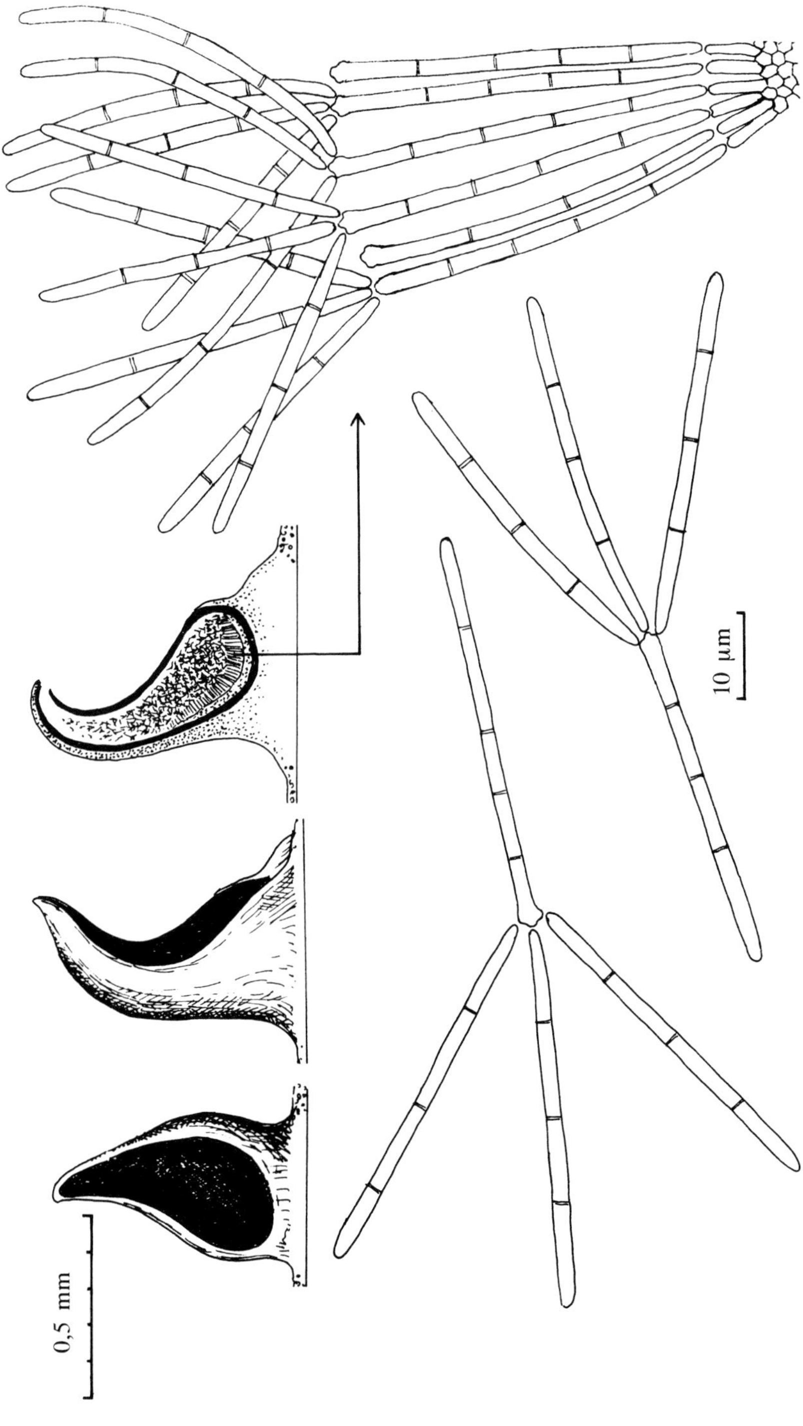

Figure 5.5. *Lasioloma javanicum*, campylidia and tetraradiate conidia. Drawing by A. Vězda.

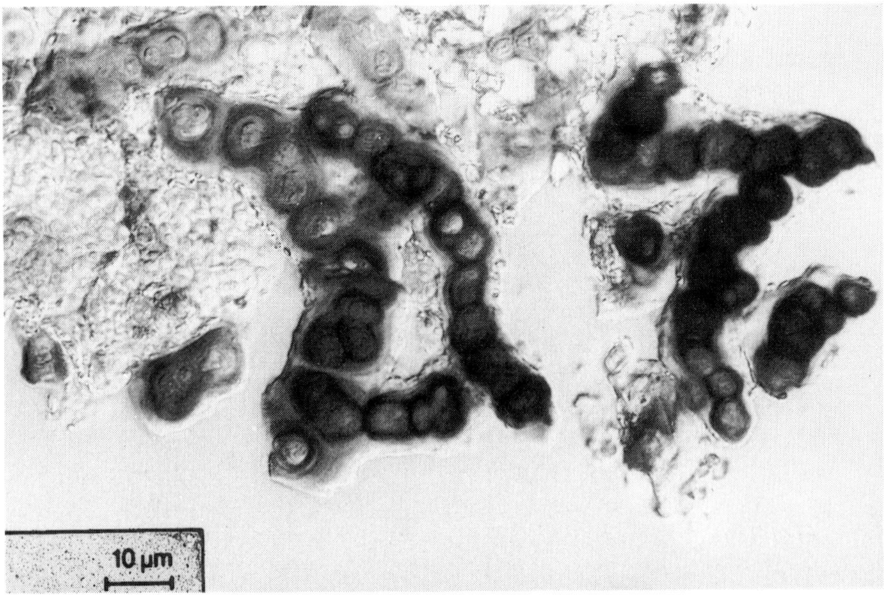

Figure 5.6. *Rhizoplaca peltata,* hyphae dividing into thallospores from the thallus margin.

families of crustose lichens (Poelt and Obermayer, 1990; Fig. 5.6). Here again it is necessary to ask "are those thallospores analogous structures with a similar type of dispersal, or expressions of a potential widely distributed in Lecanorales but normally suppressed?"

Secondary metabolites

Lichen compounds are regarded as essential elements of the lichen symbiosis, although their function is still in dispute. No part of lichenology has been treated with more modern and elaborate methodologies than the chemistry of secondary lichen products. But lengthy discussions, and even arguments over their taxonomic importance has raged in lichenology for over 100 years. Why? Because also in this case it is very difficult, and may even be impossible, to discriminate between homologies and analogies. The following examples illustrate this.

The well-known boreal-subarctic lichen *Parmelia centrifuga* (syn. *Arctoparmelia centrifuga*) is normally coloured yellowish green by usnic acid. Within large populations of the species a greyish type rarely occurs, which lacks usnic acid. Fries (1870, p. 129) knew it from only one locality and named it var. *dealbata*. Hasselrot (1953, p. 80) reported 70 collections from Sweden, and was also of the opinion that this taxon, independently described as the species *P. aleuritica* by Nylander (1875), was merely a depigmented form. He also mentioned that thalli can sometimes be found which are partly yellowish green and partly grey, but the reason for this loss of colour is unknown. Santesson (1970), using sensitive chemical methods, was able to show that var. *dealbata* differs from *P. centrifuga* only in the lack of usnic acid. Pigments have always fascinated taxonomists and *P. centrifuga* var. *dealbata*, which obviously developed independently in different localities from *P. centrifuga*, was treated by Hale (1986) as a distinct species under the name *Arctoparmelia aleuritica*.

A somewhat similar case occurs within the well-known species *Evernia prunastri*, which is also usually coloured by usnic acid. Material not containing this yellowish green compound has been described as *E. herinii*. This taxon is rare when it occurs in Central Europe, and is found only as small colonies. It seems to have originated many times in different localities by the loss of the ability to synthesize usnic acid. It has therefore been treated as var. or f. *herinii*. It is clearly not a good species. Similar mutants characterized by losses in chemical abilities are the yellow-green (versus yellow) taxa within the genus *Candelariella* (Gilbert *et al.*, 1981) which lack calycin within the stictaurin-complex. It is easy to understand that these very distinct looking taxa have been described as independent species. They occur very sporadically within colonies of the normal type, and seem to be derived independently in their different localities.

The genus *Evernia* contains another, but somewhat different, example. *E. divaricata*, a widely distributed and often common species in boreal and montane forests, has a parallel grey and usnic acid deficient taxon, *E. illyrica*, that has a quite different distribution within the montane forests of the mediterranean region. *E. illyrica* could have evolved by the loss of usnic acid and a higher production of atranorin, but nevertheless seems to be genetically stable. A similar case in the genus *Lecanora* is *L. subaurea* and *L. handelii*. These taxa are morphologically identical and share the same ecology; both are restricted to acid rocks containing heavy metals. *L. subaurea* contains atranorin as well as rhizocarpic acid, whereas *L. handelii* is characterized by usnic acid (Huneck, 1966*a*, *b*). In this case two biogenetically unrelated compounds replaced each other, and again the paler taxon is the rarer.

Within the small, well-known genus *Parmeliopsis*, two widely distributed boreal species have been recognized for a long time: *P. hyperopta* which is grey and contains atranorin in its cortex, and an otherwise identical species *P. ambigua* which is yellowish green due to usnic acid which replaces atranorin. Both species were once considered as chemotypes of a single species (Culberson, 1955, p. 336). Ecologically both species are quite similar, except that also in this case the usnic acid containing

P. ambigua has a much wider ecological spectrum. Both taxa always seem to be distinct and as no intermediates have been found they can be treated as separate species. In modern times, within *Parmelia* s.lat a large number of segregate genera has been described; several of these differ only by the substitution of usnic acid by atranorin which causes differences in the colour of the thallus or *vice versa*. The questions which arise from the above considerations should be more thoroughly discussed, especially with regard to the generic taxonomy of *Parmelia* s.lat.

Anthraquinones were generally considered as restricted to lichens belonging to the Teloschistaceae, in so far as they are deposited as surface pigments. Subsequently we have learned that principally the same chemistry occurs on single occasions in other families, as in *Edrudia* (Lecanoraceae) and *Protoblastenia aurata* (Psoraceae); in Psoraceae anthraquinones were already known as apothecial pigments in several species.

Much as the crystalline lichen compounds and their pathways have been discussed, another chemical phenomenon, which has possibly played a major role in the evolution of different lichen groups remains hardly known, the replacement of crystalline, superficially deposited compounds by amorphous ones included in and colouring the hyphal walls. This problem can be discussed using examples from the Teloschistaceae. Most of the species of *Caloplaca*, which includes by far most of the species of the family, are characterized by yellow or orange colours caused by anthraquinones. In a number of species, especially those exposed to extreme stress in often arctic-alpine habitats, changes in colour can be observed within the thalli, or even within a single apothecium, ranging from yellow-orange through a mixture of colours to blackish and black (Hansen *et al.*, 1987, p. 7). During this change the crystalline pigments decompose, as can be seen under the microscope, and in their place non-crystalline pigments of a greyish purple or even greenish or bluish colour develop within the hyphal walls; if greyish purple, these walls produce a violet reaction with the addition of potassium hydroxide, which seems to point towards a relationship of these colouring matters to anthraquinones. But in others, including some widely distributed *Caloplaca* species such as *C. variabilis*, this type of colouring without crystallized pigments is developed from the first and is clearly genetically fixed. The so-called "black" *Caloplaca* species are probably the result of parallel evolution from several groups of bright coloured ones.

In this connection, I wish to draw attention to a phenomenon which must have impressed anyone who has studied lichens on high alpine and high arctic siliceous rocks. This is the predominance of species with dark, sometimes even black, thalli which all are coloured by amorphous wall pigments. At least in some of the species, the black colour predominates because of the predominance of the black prothallus over the thallus itself. Those "black" species belong to quite different taxonomic groups. One is tempted to attribute this phenomenon to the ability to survive under such extreme conditions by the uptake of heat by developing dark thalli.

Lichen biology

Since Schwendener in the 1860s, lichens have been considered as fungi that have become autotrophic by enslaving algae ("rent symbiosis" of Stocker, 1975). In many discussions of lichen biology it has almost been forgotten that the lichen fungi have never lost their fungal abilities. Only in this way can you explain that a few decades ago, nobody could have imagined how great a number of lichen species were parasitic (in the broadest sense) on other lichens and, in some cases, also on bryophytes. Amongst these are predators of algae, which take up their photobionts from the host lichen photobionts, which are often rarely found free-living. They begin their own development as parasites, later on becoming independent lichens. Furthermore, there are many species which remain in connection with their hosts during their entire life-span; if the host dies or is completely taken over by the parasite, the parasite also soon expires. With a progressive adjustment to parasitism, evolution follows the general rules of evolutionary processes in parasitic groups; the thalli become continuously reduced, become more and more specialized to distinct hosts, and also produce smaller fruiting bodies. At the end of this sequence, in some cases an externally visible thallus disappears completely, and the reduction ends in a biological type, which has been called "parasymbiosis". However, sometimes an "endocapylic" internal thallus is developed.

Unfortunately the anatomical relations between parasitic lichens and their hosts have been scarcely studied. One may argue whether the evolution of such parasitic taxa is purely analogous, or in view of the lower or higher parasitic ability of the different lichen fungi, the result of convergence. In *Rhizocarpon* five parasitic species were known in the 1950s. In a recent survey (Poelt, 1990) 28 such species are recognized, which are parasitic on other lichens, including other *Rhizocarpon* species. They are partly interrelated, belong to different biological types of parasites and are mostly known only from Europe. Surely more parasitic members of *Rhizocarpon* will be found outside this continent. In *Rhizocarpon* we may assume that convergence has played a leading evolutionary role.

Growth form

Wherever we look at lichens, we can see analogies or convergences, where parallel processes arise from the same basic pattern. Finally I will cite an example which is usually explained by simple analogy: the evolution of foliose (and fruticose) lichens from crustose predecessors. Within the Lecanorales this evidently took place several times. *Parmelia* s.lat., *Physcia* s.lat., and *Xanthoria* are all constructed according to the same growth pattern. All have an "exoskeleton", formed by a true cortex, which is more or less identical in all three genera. From where does this cortex originate as viewed phylogenetically? It is connected in all three examples to the thalline exciple cortex of the apothecia. The foliose thallus is nothing but a precociously formed excipular structure where the development of the hymenium has been delayed. There is no time to discuss this idea completely here and I will return to it in a forthcoming paper.

References

Culberson, W.L. (1955) Note sur la nomenclature, repartition et phytosociologie du *Parmeliopsis placorodia* (Ach.) Nyl. *Revue bryologique et lichénologique* 24, 334–337.

Eriksson, O.E. and Hawksworth, D.L. (1990) Outline of the ascomycetes – 1989. *Systema Ascomycetum* 8, 119–318.

Fries, Th. (1870) *Lichenographia Scandinavica.* Pars prima. Berling, Uppsala.

Gilbert, O.L., Henderson, A. and James, P.W. (1981) Citrine-green taxa in the genus *Candelariella. Lichenologist* 13, 249–251.

Hale, M.E. (1986) *Arctoparmelia*, a new genus in the Parmeliaceae. *Mycotaxon* 25, 251–254.

Hansen, E.S., Poelt, J. and Søchting, U. (1987) Die Flechtengattung *Caloplaca* in Grönland. *Meddelelser om Grønland, Bioscience* 25, 1–25.

Hasenhüttl, G. and Poelt, J. (1978) Über die Brutkörner bei der Flechtengattung *Umbilicaria. Bericht der Deutschen botanischen Gesellschaft* 91, 275–296.

Hasselrot, T.E. (1953) Nordliga lavar i syd- och mellansverige. *Acta phytogeographica Suecica* 33, 1–200.

Hawksworth, D.L. and Poelt, J. (1986) Five additional genera of conidial lichen-forming fungi from Europe. *Plant Systematics and Evolution* 154, 195–211.

Hawksworth, D.L. and Poelt, J. (1990) A second lichen-forming species of *Cheiromycina* from Austria. *Lichenologist* 22, 219–224.

Hertel, H. (1967) Revision einiger calciphiler Formenkreis der Flechtengattung *Lecidea. Beiheft zur Nova Hedwigia* 24, 1–155.

Hertel, H. (1987) Progress and problems in taxonomy of Antarctic saxicolous lecideoid lichens. *Bibliotheca lichenologica* 25, 219–242.

Hertel, H. and Rambold, G. (1985) *Lecidea* sect. *Armeniaca*: lecideoide Arten der Flechtengattungen *Lecanora* und *Tephromela* (Lecanorales). *Botanischer Jahrbucher für Systematik und Pflanzengeographie*, 107, 469–501.

Hestmark, G. (1990) Thalloconidia in the genus *Umbilicaria. Nordic Journal of Botany* 9, 547–574.

Huneck, S. (1966a) Über die Inhaltsstoffe von *Lecanora hercynica* Poelt & Ullrich, *Lecidea silacea* (Ach.) Ach. und *Acarospora montana* H. Magn. (XXII. Mitteilungen über die Flechteninhaltsstoffe). *Zeitschrift für Naturforschung* 21b, 80–81.

Huneck, S. (1966b) Über Flechteninhaltsstoffe, XXVI. Die Inhaltsstoffe von *Lecanora handelii* Steiner, und *Stereocaulon nanodes* Tuck. *Zeitschrift für Naturforschung* 21b, 199–200.

Laundon, J.R. (1974) *Leproplaca* in the British Isles. *Lichenologist* 6, 102–105.

Nylander, W. (1875) Addenda nova ad lichenographiam europaeam. Continuatio vicesima. *Flora* 58, 102–106.

Poelt, J. (1977) Die Gattung *Umbilicaria* (Umbilicariaceae) (Flechten des Himalaya 14). *Ergebnisse des Forschungsunternehmens Nepal Himalaya, Khumbu Himal* 6, 397–435.

Poelt, J. (1986) Morphologie der Flechten. Fortschritte und Probleme. *Bericht der Deutschen botanischen Gesellschaft* 99, 3–29.

Poelt, J. (1990) Parasitische Arten der Flechtengattung *Rhizocarpon*: eine weitere Übersicht. *Mitteilungen der Botanischen Staatssammlung München* 29, 515–538.

Poelt, J. and Obermayer, W. (1990) Über Thallosporen bei einigen Krustenflechten. *Herzogia* 8, 273–288.

Poelt, J. and Romauch, R. (1977) Die Lagerstrukturen placodialer Küsten und Inlandsflechten. In: W. Frey, H. Hurka and F. Oberwinkler (eds), *Beiträge zur Biologie der niederen Pflanzen.* G. Fischer, Stuttgart, pp. 141–153.

Santesson, J. (1970) Neuere Probleme der Flechtenchemie, *Vorträge aus dem Gesamtgebiet der Botanik, Deutschen botanischen Gesellschaft, N.F.* 4, 5–21.

Scutari, N.S. (1990) Studies on foliose Pyxinaceae (Lecanorales, Ascomycotina) from Argentina, II: Anatomical-ontogenetic studies in *Physcia tribacia* (Ach.) Nyl. *Nova Hedwigia* 50, 451–461.

Stocker, O. (1975) Prinzipien der Flechtensymbiose. *Flora* 164, 359–376.

Tibell, L. (1984) A reappraisal of the taxonomy of Caliciales. *Beiheft zur Nova Hedwigia* 79, 597–713.

Vobis, G. (1980) Bau und Entwicklung der Flechten-Pycnidien und ihrer Conidien. *Bibliotheca lichenologica* 14, 1–141.

Webster, J. (1987) Covergent evolution and the functional significance of spore shape in aquatic and semi-aquatic fungi. In: A.D.M. Rayner, C.M. Brasier and D. Moore (eds), *Evolutionary Biology of the Fungi.* Cambridge University Press, Cambridge, pp. 191–201.

Importance in Ecosystems and to Man

6

Mycorrhizas in Ecosystems – Nature's Response to the "Law of the Minimum"

D.J. Read, *Department of Animal and Plant Sciences, University of Sheffield, Sheffield S10 2UQ, UK.*

ABSTRACT The almost universal association between fungi and plant roots to form "mycorrhizas" is seen as being nature's response to Liebig's "Law of the Minimum". Selective advantages accrue to those fungi able to form compatible associations with roots since, through the association, they gain access to that nutrient, carbon, which most limits the growth of heterotrophs in soil. Similarly autotrophic plants, the bulky roots of which are ill-equipped to exploit the nutrient resources of soil, gain advantages from associations with filamentous fungi, the hyphae of which provide access to nutrients otherwise unavailable to the plant.

The nature of the nutrient limitation varies with ecosystem. In heathland, as in most boreal and temperate forests, nitrogen is in "minimal" supply, but progressively, along an environmental gradient this is replaced by phosphorus as the growth limiting nutrient. In each ecosystem a predominant mycorrhizal type is recognized in which selection has favoured fungi with the ability to mobilize or capture the growth-limiting nutrient characteristic of that part of the gradient.

The most recalcitrant residues are produced by ericaceous plants, and are exploited by fungi of ericoid mycorrhizas here shown to have ligninolytic, polyphenolytic, chitinolytic, and proteolytic capabilities. Substrates of intermediate qualities, explored by ectomycorrhizal fungi, are produced by trees in the boreal and temperate forest zone. These fungi differ in their abilities to cleave polymeric resources, those in which such capabilities are best developed predominating in residues with the highest carbon nitrogen (C:N) ratios. At the warm end of the gradient ectomycorrhizal systems are replaced by grassland and tropical forest biomes in which plants with vesicular-arbuscular fungi predominate. These fungi lack the ability to cleave organic substrates but form an extensive

Frontiers in Mycology. Honorary and General Lectures from the Fourth International Mycological Congress, Regensburg 1990. Edited by D.L. Hawksworth.

mycelial network which effectively scavenges for phosphate ions.

It is concluded that the nutrient dynamics of terrestrial ecosystems can be understood only in terms of the activities of their mycorrhizal fungi, and that in any system these have been selected to provide their host plants with access to the key elements which, as predicted by Liebig, would otherwise limit plant development.

Introduction

Liebig (1843), in what became known as the "Law of the Minimum" observed that when all other nutrients are available, the growth of plants is regulated by supplies of that nutrient element which is present in the smallest amount. In communities of vascular plants either nitrogen or phosphorus are normally the primary growth-limiting elements, while in fungal communities, to which Liebig's Law can be applied as effectively as to those of autotrophs, carbon is almost universally the major limiting factor.

In view of these limitations, it is perhaps not surprising that natural selection has favoured the development of intimate associations between fungi and the roots of plants which provide the heterotroph with direct access to carbon and the plant with enhanced ability to capture key mineral elements. Mycorrhizal infection might indeed be seen as nature's response to the Law of the Minimum, and the fact that structures reminiscent of present day vesicular arbuscular mycorrhizas are seen in the poorly developed roots of the earliest land plants adds strength to the suggestion that mycorrhizal mutualisms have been involved in the development of terrestrial ecosystems from the outset. The task now is to determine the extent and nature of this involvement.

Until recently, despite the relatively broad approach to mycorrhizal research encouraged by pioneers such as Frank (1894), studies of this mutualism have dealt largely with individual plants or isolated roots and have concentrated upon the role of the fungus in the enhancement of phosphorus uptake. While providing a sound scientific basis for understanding the mechanisms of absorption, storage and transfer of phosphate ions these emphases may, inadvertently, have signalled to plant ecologists a belief that mycorrhizal fungi were of little significance in ecosystems other than those suffering phosphorus deficiency and thus led them to ignore the symbiosis in most studies of plant nutrition and ecosystem biology.

According to the Law of the Minimum, an ecologically sound approach to the question of mycorrhizal function would be to determine which nutrient or combination of nutrients most limits growth in a particular ecosystem and then to ask whether infection can enhance access to that resource. Using this approach it is recognized that, particularly in warmer regions of the world where high pH can lead to the precipitation of mineral salts as insoluble residues, phosphorus is indeed the major plant growth-limiting nutrient. However, much of the land surface of the Northern, as of cooler parts of the Southern Hemisphere, supports forest or heath ecosystems in which low temperature, evapotranspiration and pH combine to inhibit the mineralization and release of that nutrient, nitrogen, which is required by plants in quantities approximately ten times greater than those of any other element. The Law of the Minimum predicts that under these circumstances enhanced access to phosphorus will be of relatively little ecological significance and that selection pressures will favour adaptations which provide for mobilization or capture of nitrogen. However, after the initial suggestion of Frank (1894) that the ectomycorrhizas of forest trees might be capable of releasing the nitrogen of the largely organic residues in which they proliferate, the possibility that mycorrhizas were also involved in nitrogen mobilization has been largely overlooked. Only now is experimental evidence emerging which suggests that the ericoid mycorrhizal fungi which dominate the soils of heathland ecosystems, and many of the ectomycorrhizal fungi which infect trees of the boreal and temperate forests, function in accordance with Liebig's Law to provide their hosts with access to nutrient elements, notably nitrogen, the limited availability of which would otherwise impose severe limitations on plant growth and survival. This evidence is examined here and is used to emphasize the fundamentally important role of mycorrhizal fungi in the nutrient economies of individual plants and plant communities.

Ecosystems dominated by plants with ericoid mycorrhizas

Heathland communities, often almost entirely consisting of ericaceous species, become prominent in those environments, typically of high latitude and altitude, in which low temperature and evapotranspiration rates lead to inhibition of general microbial activity, and hence to sequestration of nutrients in an acidic organic matrix which Muller (1884) termed "mor" humus. Muller observed that this material consisted largely of the residues of the plants and fungi which grew upon "mor" humus and emphasized the important contribution made by fungi with dark-walled hyphae to the overall structure. Subsequent analyses have confirmed that the chemical characteristics of this soil type are inextricably linked to those of the vegetation which it supports and that mycorrhizal fungi with dark hyphal walls play a key role in the processes of nutrient cycling in these ecosystems.

The shoots of ericaceous plants characteristically contain very large amounts of phenolic and lipid materials. Levels of total phenols as high as 28% of dry weight (Jalal *et al.*, 1982) and of total lipids of 13% dry weight (Tschager *et al.*, 1982) have been recorded. Such tissues yield, on decomposition, a range of aromatic and aliphatic acids which contribute to the maintenance of low pH and act as phyto- and fungi toxins which are known to express their maximal activities when in the undissociated form under acidic conditions (Jalal and Read, 1983*a,b*). The plants are thus directly involved in the production of a "syndrome" (Read, 1984) in which low pH, the organic acid and metal toxicity associated with acidity, together with the co-precipitation of mineral elements in organic polymers, combine to impose severe stresses. Perhaps the major above-ground manifestation of this syndrome is poverty of plant species, members of ericaceous genera such as *Calluna, Empetrum, Erica, Gaultheria, Kalmia, Loiseleuria, Rhododendron* or *Vaccinium* often occurring as pure stands of low productivity over large areas.

While these dwarf shrubs show a considerable range of shoot morphologies the feature which they all have in common is a root structure of remarkable uniformity in which the hair-like distal portions are extensively occupied by the ascomycete *Hymenoscyphus ericae* to form "ericoid" mycorrhizas (Read, 1983). Collectively these mycorrhizal hair roots form dense mats under the litter of the parent plant which the hyphae of the mycorrhizal fungus extend to colonize the organic residues through which the roots are permeating. Knowledge of the function of the ericoid mycorrhizal association has been achieved by a combination of studies involving pure culture of the fungal endophyte, aseptic synthesis of mycorrhizas in artificial and soil media, and, latterly, the examination of interspecific interaction between host species and competitors in microcosms inoculated with the mycorrhizal fungus.

Early analysis of the effects of mycorrhizal infection upon the nutrition of ericaceous plants indicated that *Hymenoscyphus ericae* provided enhancement of nitrogen capture (Read and Stribley, 1973) and that access to ammonium ions was improved when ammonium nitrogen (NH_4N) was suppled at low concentrations (Stribley and Read, 1976). However, experiments in which ^{15}N labelled ammonium was added to heathland soils, showed that despite having significantly higher total nitrogen contents the proportion of the total attributable to the labelled ammonium ion was smaller than in uninfected plants (Stribley and Read, 1974) suggesting that the mycorrhizal infection was providing access to organic sources of nitrogen. Studies were therefore carried out of the relative abilities of mycorrhizal and non-mycorrhizal plants to utilize such sources of nitrogen. It was soon evident that infection provided access to amino acids (Stribley and Read, 1980), and that it also enabled the utilization of peptide (Bajwa and Read, 1985), and protein (Bajwa *et al.*, 1985), nitrogen. These substrates were completely inaccessible to uninfected plants. Thus, infection was not simply improving the efficiency of absorption of those mineral ions which are of low mobility in soil, but was providing the capability to use an otherwise largely unavailable resource. These observations necessitated an analysis of the mechanisms whereby nutrient mobilization was achieved.

Studies of *Hymenoscyphus ericae* in pure culture have confirmed its ability to utilize a wide range of amino acids as well as peptides and proteins. Of direct relevance to the question of nutrient mobilization in ericaceous litter and "mor" humus was the observation that lignin (Haselwandter *et al.*, 1990) and a range of phenolic compounds (Leake, 1987) were attacked and used as carbon sources by *H. ericae.* Lignin contents of roots and shoots of *Calluna* when deposited as litter can be as high as 49% and 39% respectively (Heal *et al.*, 1978). In combination, lignin and polyphenols are responsible for immobilizing much of the nitrogen of "mor" humus and, together with lipids, help to produce the extremely high C:N ratios of ericaceous litter (Fig. 6.1). The breakdown of these polymers will provide the fungus with access to nitrogen which is otherwise masked. Direct evidence that *H. ericae* has some access to protein nitrogen coprecipitated with tannin has how been obtained (Leake and Read, 1989*a*). These experiments thus add support to the hypothesis, first raised after the observation of dilution of $^{15}NH_4^-N$ in infected plants, that mycorrhizal infection enabled the mobilization of complex sources of organic nitrogen.

In addition to plant litter, the fungal component of heathland litter and humus represents a significant potential source of nitrogen. Bååth and Söderstrom (1979) estimated that up to 20% of total nitrogen in a Swedish pine heath was localized in fungal mycelium. Of this nitrogen the most complex form will be that in the chitin of hyphal walls and the early reports of large concentrations of brown mycelia in "mor" humus (Muller, 1884) is testament to the recalcitrance of such structures. Bajwa and Read (1986) showed that N-acetyl glucosamine, the primary building unit of the chitin polymer was readily assimilated by *Hymenoscyphus ericae*; more recently it has been demonstrated that pure chitin itself can be used as a sole source of nitrogen by *H. ericae* (Leake and Read, 1990*b*). Melanization of hyphal walls, such as is seen in *H. ericae* itself, is known to provide resistance to wall breakdown (Bloomfield and Alexander, 1967) and it is likely that this attribute both reduces the susceptibility of *H. ericae* to autolysis and provides the endophyte with an advantage when growing with ectomycorrhizal fungi, many of which have hyaline walls. While the fungus may be protected from autolysis the question arises as to the mechanism whereby its diverse enzyme activities are regulated so as not to threaten the host. Analysis of the control of proteinase production has thrown some light on this matter (Leake and Read, 1990*a*). The production and activity of the enzyme are modulated by pH, being both maximal in the range of pH 2–5 and almost entirely inhibited at pH 6 and above which is typical of that prevailing in the epidermal cells of the ericoid root. Host proteins are thus protected. Modulation of proteinase production by pH represents a sound strategy in ecological terms. In soil of pH greater than 5.0, mineralization will normally be sufficiently rapid to ensure that organic nitrogen does not accumulate. There is a strong possibility, yet to be investigated, that the lignase, polyphenol oxidase and chitinase activities of *H. ericae* are also regulated by this ecologically expedient mechanism.

We know less about the role of ericoid mycorrhizas in the mobilization of phosphorus which, like nitrogen, is present largely in organic form in heathland soils. The endophyte has been shown to produce an acid phosphatase (Pearson and Read, 1975; Straker and Mitchell, 1986) and it can utilize organic phosphates when added in the form of Ca, Fe and Al-phytates to culture media (Mitchell and Read, 1981). A recently revealed attribute of the enzyme, which is likely to be of ecological significance, is its resistance to inhibition by the metal ions Al^{3+} and Fe^{2+} (Shaw and Read, 1989) both of which occur in solution at the pH prevailing in "mor" humus. As with organic acids, the potential for metallic ions to exert toxic effects is greater at low soil pH, in their case because of increased solubility under acid conditions. It has been shown previously that ericoid infection can confer resistance to copper and zinc toxicity by reducing the inflow of the metals to the shoots (Bradley *et al.*, 1982).

Of greater importance as potential toxins in natural environments are the metals aluminium and iron. The inherent toxicity of the former element has long been recognized as a factor involved in the exclusion of some species from acid soils while the role of the latter is complicated due to it being an essential element which can be toxic at high concentrations. Recent studies have shown that the ericoid endophyte has remarkable resistance to aluminium, no inhibition of yield being found even in media containing 800 mg l^{-1} Al^{3+} (Burt *et al.*, 1986). In solutions containing in excess of 400 mg l^{-1} the metal strongly inhibits root development in non-mycorrhizal plants, characteristic "stilt" roots being formed. While these plants survive under protected growing conditions, it is unlikely that such root systems would be adequate to sustain them in the field. Mycorrhizal infection enables roots to develop normally and leads to a reduction of inflow to the shoots.

Studies of iron uptake by mycorrhizal and non-mycorrhizal ericaceous plants have shown that infection provides for significantly increased absorption of the element over that range of exogenous concentration (6–18 mg l^{-1} = 120–360 μM) typically found in soil solution in heathlands (Shaw *et al.*, 1990). Significantly greater root yields and root iron concentrations were found over this range in mycorrhizal than in non-mycorrhizal plants grown in sand culture. In short-term studies comparing the rates of uptake of Fe^{59} by infected and uninfected plants from solution culture it was shown that the mean uptake rates of 3.9 nmol Fe per mg root h^{-1} achieved by mycorrhizal roots of *Calluna* were more than double those of non-mycorrhizal plants. Hydroxamate siderophores, the production of which has now been confirmed in a number or ericoid mycorrhizal endophytes (Schuler and Haselwandter, 1988) are likely to be involved in the process of scavenging for iron when this element is present at very low concentrations.

While the dominance of plants with ericoid mycorrhizas in the most extreme edaphic circumstances appears to be at least in part a result of their possessing an inherently greater tolerance of the stresses encountered, there are many less extreme situations in which competition with other types of plant is a factor determining community structure. Amongst the commonest competitors in heathland environments of Europe are grasses such as *Festuca ovina* and trees such as *Betula* spp. The role of ericoid mycorrhizas in competitive interactions between *Calluna* and *F. ovina* has been examined (Leake *et al.*, 1989). Aseptically grown mycorrhizal (M) or non-mycorrhizal (NM) *Calluna* plants, and axenically germinated seedlings of *F. ovina* were grown together or separately in sterile sand moistened with dilute mineral solution. A mixture of phenolic acids comparable with that obtained by Jalal and Read (1983*a*) from heathland soil was added to half of the flasks.

The largest yield of *Festuca* was obtained when it was grown alone in the absence of phenolic acids. The presence of *Calluna* in this treatment reduced yield of the grass, a reduction that was particularly dramatic when the *Calluna* was mycorrhizal. When phenolic acids were added the inhibition of *Festuca* was even more pronounced, the grass producing virtually no roots when grown alone. Infection provided pronounced advantages to *Calluna* in the phenolic treatments, in particular in the form of stimulation of root growth.

The production of protein-binding aromatic compounds by ericaceous plants is likely to play a significant role in the maintenance of their dominance. Both directly through toxic effects and indirectly through nutrient conservation it will restrict the growth of competitors which lack the specialized ability, conferred by the ericoid endophyte, to resist toxicities and to mobilize organic nitrogen. It is notable that the most widespread of the small number of plants capable of co-existing with ericaceous species in upland heaths are legumes such as *Ulex*, *Genista* and *Sarothamnus* and insectivorous species such as *Drosera* which have their own specialized mechanisms for nitrogen capture. Conversely, as demonstrated graphically by the decline of the Dutch heaths, the addition of mineral nitrogen in that case from pollutant sources, can lead to replacement of heathland by more productive grassland ecosystems.

Ecosystems dominated by plants with ectomycorrhizas

As conditions ameliorate with decrease of altitude or latitude, ectomycorrhizal tree species form progressively more important components of the vegetation eventually producing forests which replace heathlands as the natural climax communities.

The leaf litter produced by most of the ectomycorrhizal tree species while having a carbon:nitrogen ratio lower than that produced by ericaceous plants is still relatively slow to decompose (Fig. 6.1) and so forms a distinctive layer of acidic organically enriched material at the soil surface. The quality of the humus layer arising from the decomposition processes varies from the most acidic raw humus or "mor" on the poorest substrates, upon which plants with ericoid infection also often occur as an understorey, through an intermediate "moder" formed even over mull soils, to a mull humus over more base-rich substrates. Of these humus types, "mor" and "moder" predominate over much of the forest area of the Northern Hemisphere. Since the pioneering studies of Hesselman (see Romell and Malmström, 1945) in the boreal forests of Scandinavia, and of Mitchell and Chandler (1939) in the temperate forests of the north-eastern USA, it has been recognized that in these, as in heathland ecosystems, it is nitrogen rather than phosphorus which is the major growth-limiting nutrient.

Even within these forest types the availability of nitrogen varies greatly, it being least in the most recalcitrant residues be they of leaf, root or fungal origin. While lignin content appears to be a primary determinant of recalcitrance in plant residues (Fogel and Cromack, 1977; Meentemeyer, 1978), the carbon–nitrogen (C:N) ratio (Fig. 6.1) provides a more broadly based index of nitrogen availability since it is applicable to fungal and fine root as well as to leaf residues.

The failure of some ectomycorrhizal species, in particular of spruce, to grow when planted into heathland environments is almost certainly attributable to the inability of its fungal associates to mobilize nitrogen from the recalcitrant ericaceous residues (Fig. 6.1) which its roots encounter after outplanting. Some workers have considered that antagonism between ericoid and ectomycorrhizal fungi may be involved in this growth inhibition (Handley, 1963; Robinson, 1971). However, the fact that spruce trees previously held in "check" by ericaceous plants such as *Calluna* can be rapidly induced to grow by application of nitrogen fertilizer to heathland (Zehetmayr 1970; Baule and Fricker, 1970) and that they can be "nursed" into growth by interplanting with legumes (Weatherell, 1953, 1957) strongly suggests that accessibility of nitrogen is the main factor determining the outcome of interactions.

As in the case of heathland environments analysis of the distribution of mycorrhizal roots provides vital clues as to their possible functional role.

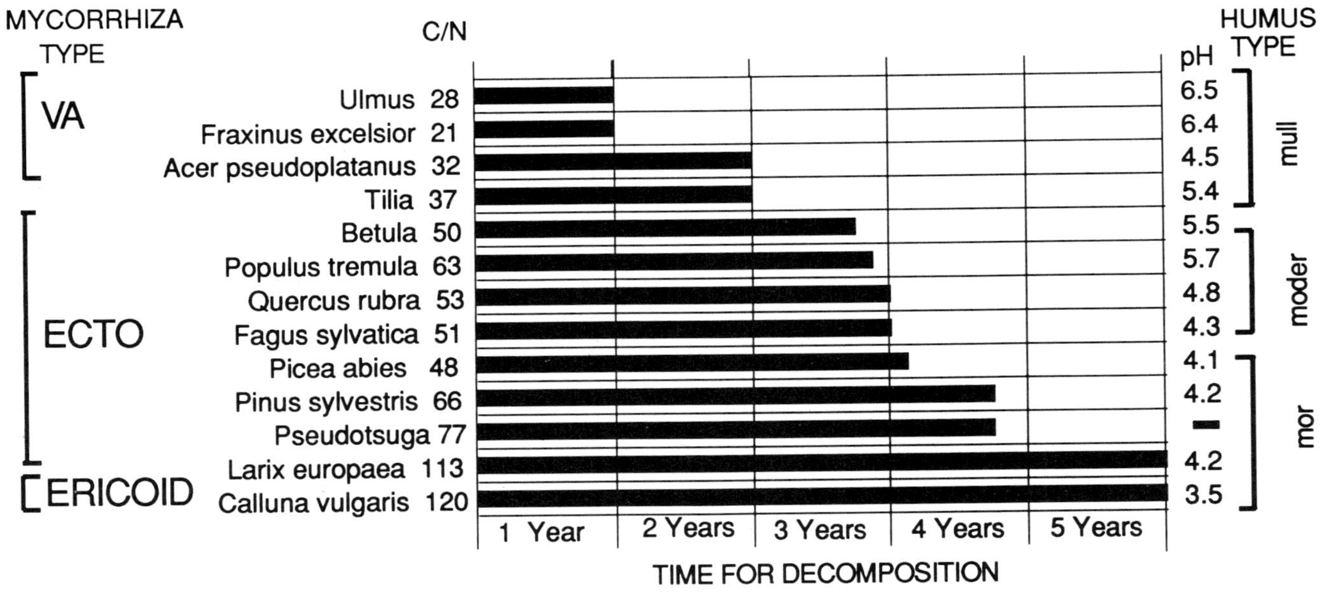

Figure 6.1. Relationship between the quality of leaf litter produced by a range of ecologically distinctive shrub or tree taxa, the humus types arising from it, and the predominant mycorrhizal type exploiting the resource. Resource quality is expressed in terms of carbon-nitrogen ratio (C:N), pH, and time taken for decomposition. A gradient is revealed extending from litter of the lowest resource quality which is exploited by ericoid fungi, through intermediate forms colonized by ectomycorrhizal fungi to material of relatively high quality, which is readily mineralised, yielding environments exploited by VA fungi. Modified from Scheffer and Ulrich (1960) and Ellenberg (1988).

Detailed studies in boreal (Ogawa, 1977; Persson, 1980; Harvey *et al.*, 1976), temperate (Harley, 1940; Meyer, 1973), and tropical (Newbery *et al.*, 1988; Alexander, 1989) forests, confirm that ectomycorrhizas are concentrated in the surface layers. Here their fungal mycelia are intimately associated with the substrates of the "F" or decomposition horizon. Since total amounts of the major nutrients in the litter layer can be high relative to those present in mineralized form, it is evident that in these ecosystems, as in those dominated by ericaceous plants, it is the quality of the resource rather than its quantity which is of greatest importance.

Forests of ectomycorrhizal trees in boreal and temperate regions are characteristically of low species diversity. As a consequence, the mycorrhizal fungi proliferate most extensively in litter produced by their own host species. The quality and chemical composition of this resource will determine the range of fungal activities required for nutrient mobilization and hence exert a primary selective influence upon partnerships formed in forests dominated by a given host species. Unfortunately, while differences between populations of mycorrhizal fungi occurring under for example pine, spruce, birch or beech have been recognized, they have hitherto been considered largely from the point of view of host rather than substrate compatibility and, as a result, an important ecological perspective has been missed.

The very strong association between the occurrence of ectomycorrhizal roots and soils with organic surface horizons has led, from the time of Frank (1894), to speculation that the ectomycorrhizal fungi were themselves directly involved in mobilization of the organic resources. However, until recently the experimental evidence has supported the view that these fungi have limited ability to mobilize nutrients from complex polymers. The studies of Lindeberg (1944), Norkrans (1950) and Lundeberg (1970) concluded that in relation to fungi of saprotrophic habit mycorrhizal species had very limited ability to degrade organic polymers. While recent studies have largely confirmed that most ectomycorrhizal fungi, unlike their ericoid counterparts, fail to achieve significant breakdown of the most complex polymers such as lignin (Trojanowski *et al.*, 1984; Haselwandter *et al.*, 1990) and chitin (Leake and Read, 1990*b*) it has become increasingly clear that there is, in this group, a wider capacity for mobilization of nutrients contained in organic polymers than previously realized.

Most importantly, the production in genera such as *Amanita*, *Boletus*, *Paxillus*, *Piloderma*, *Suillus* and *Thelephora* of an acid proteinase enzyme similar in properties to that found in the ericoid system provides these fungi with the potential to mobilize the growth-limiting element, nitrogen, which is the key to success in forest ecosystems. The characteristic spectrum of pH dependent activity shows an optimum between pH 3 and 4 which is broadly representative of the range over which organic matter accumulates. There is progressive inhibition of activity above pH 5.0 (Abuzinadah and Read,

1986*a*,*b*). The failure of Lundeberg (1970) to detect significant amounts of proteolytic activity in ectomycorrhizal fungi is attributable to his assays being carried out at pH 5.6, a value high enough to inhibit production and activity of these enzymes.

Studies of plants grown aseptically with and without their fungal associates have shown that only when in association with an appropriate fungal partner does the host have access to organic nitrogen (Abuzinadah and Read, 1989*a*). This observation makes possible a reinterpretation of the function of ectomycorrhizas which have hitherto been seen simply as structures which improve the efficiency of capture of mineral nutrient ions. While a large number of ectomycorrhizal fungi are now known to have the potential directly to mobilize organic nitrogen it is acknowledged that within this group of fungi there is a very wide range of proteolytic capability. In addition to the very active forms there are some, the so-called "non-protein" fungi (Abuzinadah and Read, 1986*a*) that have little proteolytic potential. Amongst these *Laccaria laccata* and *Pisolithus tinctorius* are important examples. Such fungi are dependent upon mineralization processes, initiated by other organisms, for the release of nitrogen which is largely assimilated in the form of ammonium. Practically all ectomycorrhizal fungi retain the ability to utilize ammonium as a source of nitrogen but whereas species such as *L. laccata* and *P. tinctorius* are largely restricted to mineralizing environments, proteolytic potential will provide the majority with the ability to assimilate organic nitrogen directly from polymeric sources.

Further differences occur between the fungi in the effectiveness with which they transfer nitrogen assimilated from protein to their host plants. Thus, it has been shown that whereas *Paxillus involutus* and *Hebeloma crustuliniforme* show comparable abilities to degrade protein, the latter transfers significantly more nitrogen to its host tree under standardized conditions of protein supply (Abuzinadah and Read, 1989*a*). Determination of the differences between fungi in the rates at which they release assimilated nutrients from their mycelia to the host plant may ultimately provide a better test of the effectiveness of a mycorrhizal association than will measurements of rates of substrate breakdown. The distinctive patterns of distribution of ectomycorrhizal fungal species which have been observed to occur at both temporal and spatial levels are likely to be a reflection of the physiological and biochemical diversity of their group of organisms.

Temporal successions of fungal species associated with ageing of forest stands have been reported by several workers (Chu-Chou, 1979; Lamb, 1979; Mason *et al.*, 1983) and distinctions have been drawn between "early" and "late-stage" fungi (Deacon *et al.*, 1983; Dighton and Mason, 1985). Such successions have been considered to be, at least in part, a consequence of changes in resource quality (Last *et al.*, 1987) and hence of

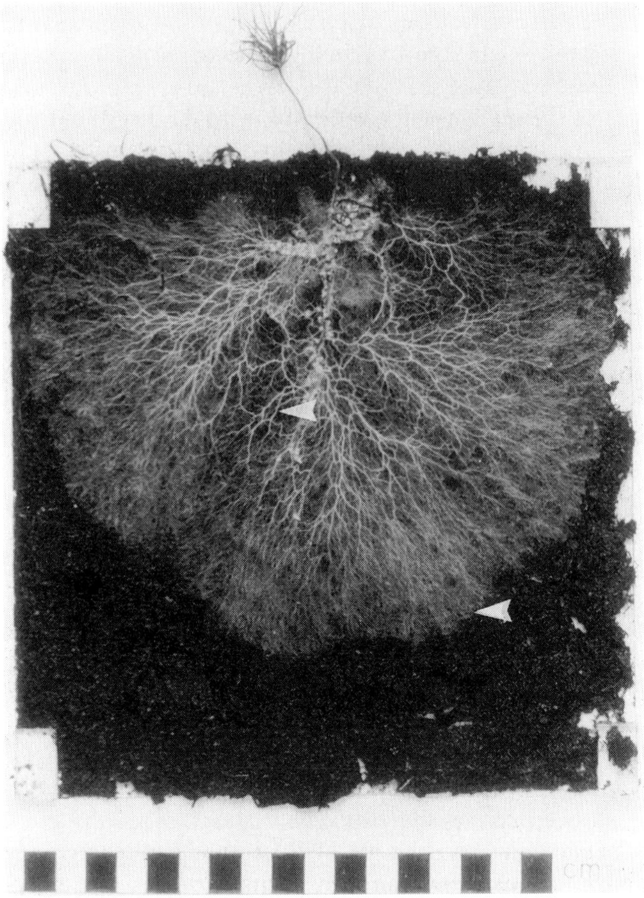

Figure 6.2. The ectomycorrhizal mycelial system of *Suillus bovinus* developing in non-sterile forest soil in association with *Pinus*. The effectiveness of the exploitation of the environment achieved by the fan-shaped growth of the mycelium is evident. Note differentiation between the diffuse hyphal front (lower arrow), which advances at a rate of 2–3 mm a day, and mycelial strands (upper arrow), which connect the front to the mycorrhizal root.

selection pressures which favour fungi with differing biochemical attributes as forests become older. Ageing leads to a progressive increase in the proportion of the nutrient fund that is in organic combination (Grier *et al.*, 1981; Heal and Dighton, 1985), a feature which would be expected to select for fungi with the abilities to exploit such resources. The studies of proteolytic capability (Abuzinadah and Read, 1986*a*, *b*) suggest that such selection occurs. All of the so called "late-stage" fungi examined so far have a well-developed proteolytic potential. They are also largely species with vigorous mycelial systems which are differentiated into strands.

The distinction between "early-" and "late-stage" fungi is blurred in natural circumstances where the "late-stage" fungi are the most vigorous colonists of seedlings. When Fleming *et al.* (1986) examined the patterns of colonization of young seedlings growing with mature trees in a birch wood, he found that they were largely colonized by "late-stage" fungi. A consequence of this mode of infection, which is likely to be of considerable ecological significance, is that, young plants are rapidly integrated into the absorptive network of mycorrhizal mycelia. Furthermore, since the seedlings are growing in the litter produced by adult trees it is appropriate that they should be infected by those fungi which are biochemically adapted to grow in these substrates.

The use of transparent observation chambers containing unsterile natural substrates distributed in a uniform or patchy manner has enabled the non-destructive analysis of the spatial distribution and patterns of resource exploration by ectomycorrhizal mycelia (Brownlee *et al.*, 1983; Finlay and Read, 1986*a,b*; Coutts and Nicol, 1990). The major strand-forming fungal species of ectomycorrhizal forests such as *Suillus bovinus*, *Amanita muscaria*, *Paxillus involutus* and *Thelephora terrestris* produce mycelia which grow in fan-shaped formation from infected roots. A diffuse mycelial front covers the substrate growing at rates between 2.4 and 5.6 mm per day. Behind the front, hyphal elements aggregate by collateral growth into linear organs, the "strands" or "cords" (Fig. 6.2) through which translocation occurs (Fig. 6.3A,B). Coutts and Nicoll (1990) have shown that extension growth of such mycelia in *Thelephora terrestris* continues, albeit at a slower rate, even in winter.

The structure and development of these mycelial systems is comparable with that seen in many wood-rotting fungi (Rayner *et al.*, 1985). The main difference between the two groups of fungi is that in the latter the food base is wood whereas the resource unit captured by the extending mycorrhizal mycelium is another mycorrhizal root. The absence of host specificity shown by most ectomycorrhizal fungi facilitates the formation of extensive mycelial interconnections between trees at both inter- and intra-specific levels (Fig. 6.3A) which provide the potential for inter-plant transfer of nutrients (Fig. 6.3B).

The conventional view of the role of the fungal mycelium was that by growing beyond depletion zones around the roots it enhanced capture of phosphate ions, and there is evidence that phosphate (Finlay and Read, 1986*b*) ammonium (Finlay *et al.*, 1988) and alanine (Read *et al.*, 1989) can be absorbed by the mycelium from forest soil and translocated through mycelial strands over considerable distances (Fig. 6.3B). These observations substantiate those made under sterile conditions in sand by Melin and Nilsson (1952, 1953).

Confirmation that the mycelium fulfills absorptive functions leaves unanswered the more ecologically important question as to whether the fungi are directly involved in the mobilization of nutrients from organic resources. Increasingly, the evidence suggests that the potentials for phosphatase and protease production, recognized in the laboratory, are realized in natural substrates. Earlier studies using observation chambers indicated that even when the forest soil or peat used as a substrate was mixed to produce a resource of seemingly uniform quality, certain areas were more intensively exploited by the hyphae, producing "patches" of relatively dense mycelial growth. It was shown (Finlay and Read, 1986*a,b*) that these areas acted as sinks in which carbon, originally fed as $^{14}CO_2$ to the shoots of the host plant, and phosphate, fed as $^{32}PO_4$ to the hyphal front, accumulated. While attempts to induce such mycelial proliferation by the localized application of mineral salts failed, addition of discrete blocks of partially decomposed leaf litter within the uniform matrix leads to "patch" development (Fig. 6.4).

That "patch" formation was associated with nutrient absorption and transfer was first suggested by the observation that seedlings of *Pinus* and *Larix* which had been chlorotic as a result of nutrient starvation prior to the stage at which the blocks of organic matter were colonized, rapidly became green as colonization proceeded. Subsequently it was shown that the re-greening of the host plant was associated with increases of the tissue nitrogen contents (Fig. 6.5). It remains to be ascertained whether the mycorrhizal fungi are themselves mobilizing the resources, but the extensive investment of carbon and phosphorus in the patches is indicative of a direct attack upon the substrates.

Proliferation of absorptive structures in localized areas of resource enrichment, which has been called "active foraging" (Grime, 1979), can be induced in the root systems of herbaceous plants (Wiersum, 1958; Drew, 1975; Crick and Grime, 1987) where it is associated with capture of mineral ions. Intensive development of ectomycorrhizal mycelium in localized areas is a feature of natural forest soils. Large-scale examples of such exploitation are to be seen in the fungal wefts associated with the

Figure 6.3. A, Mycelial system of *Suillus bovinus* which has grown from a mycorrhizal root of a 'donor' plant of *Pinus* to colonize a medium made up of unsterile sieved forest soil and to infect the radicles of two germinating seedlings (arrowed) and thus incorporate them into the hyphal network. Distal portions of some of the mycelial strands are being fed with a radioactively labelled organic N source – ^{14}C alanine. B, Autoradiograph of observation chamber shown in "A" taken 48 hours after commencement of feeding of ^{14}C alanine. Transport of the amino acid is seen to have occurred, largely through the strands, and radioactivity has accumulated in the mycorrhizal roots of both the 'donor' and 'receiver' plants (arrowed).

A

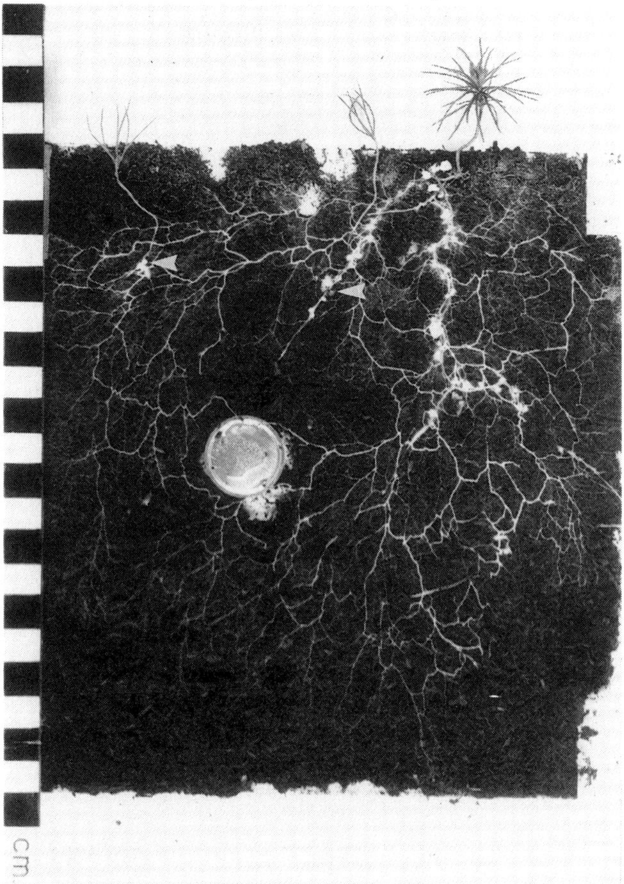

B

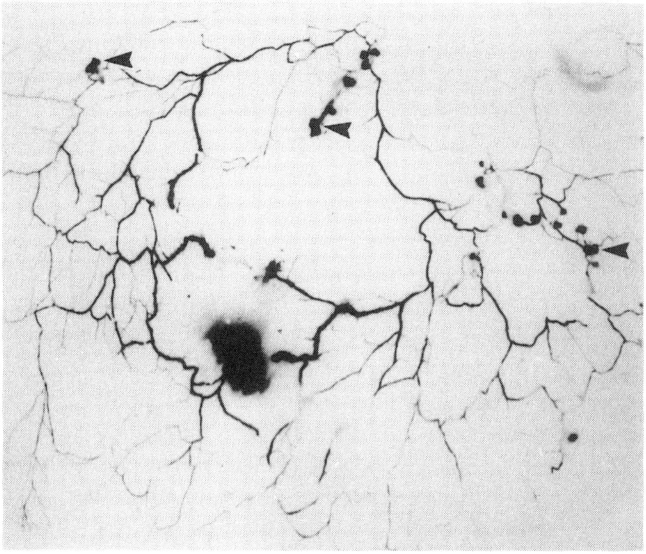

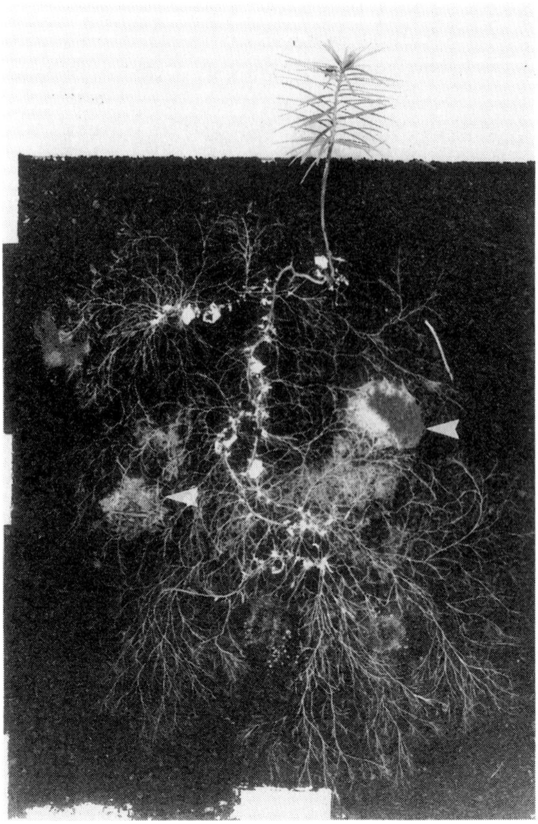

Figure 6.4. Formation of "patches" of mycelium in association with introduced leaf litter of larch (left arrow) and beech (right arrow). The mycelium is that of the mycorrhizal fungus *Boletinus cavipes* which is growing from its host plant *Larix*.

mycorrhizal roots proliferating in the decomposition horizons of coniferous forest soils (Ogawa, 1977) and in the mycelial mats formed by *Hysterangium* and related species underneath the organic horizons of Douglas fir forests (Griffiths *et al.*, 1987, 1990). It has been estimated that the mycelium of *H. crassa* can occupy 9.6% of the A horizon of a forest soil to a depth of 10 cm (Cromack, 1988). It is known that in addition to being sites of elevated enzyme and respiratory activity (Griffiths *et al.*, 1990), these mats show enhanced phosphatase, protease and peroxidase enzyme activities relative to those occurring in soil outside the mats.

In ectomycorrhizal forests the fungal mycelia themselves constitute a significant component of the total nitrogenous resource of the soil. Fogel and Hunt (1979) calculated that the mycorrhizal sheaths of the trees contributed 50% or more of the total nitrogen inputs to a Douglas fir ecosystem. As these fungi senesce they will constitute a major source of relatively labile organic nitrogen which, along with the other organic residues, should be available for attack by the proteolytic enzymes of the living mycorrhizal mycelium.

Recent advances in understanding of the capabilities of ectomycorrhizal fungi have led to a reevaluation of the nitrogen cycle in the ectomycorrhizal forest ecosystem (Read *et al.*, 1989). Infected trees are no longer envisaged as being entirely dependent upon the activities of a separate population of decomposers for the release of nitrogen in the form of ammonium ions. Through the provision of carbon to their own fungal partners they can themselves be involved in the primary mobilization events which, studies in pure culture suggest, will lead to the release and assimilation of amino compounds. The subsequent assimilation of these amino compounds in turn provides supplementary carbon as well as nitrogen to the host plant.

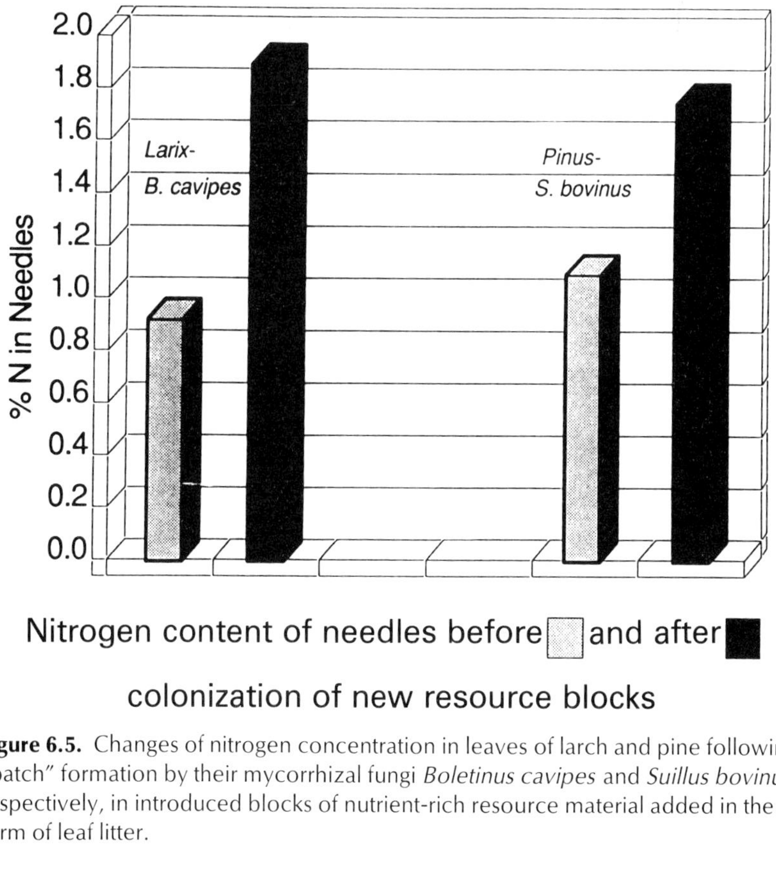

Figure 6.5. Changes of nitrogen concentration in leaves of larch and pine following "patch" formation by their mycorrhizal fungi *Boletinus cavipes* and *Suillus bovinus* respectively, in introduced blocks of nutrient-rich resource material added in the form of leaf litter.

In a study of assimilation of ^{14}C-labelled protein by mycorrhizal plants of *Betula pendula*, Abuzinadah and Read (1989*b*) showed that up to 9% of the carbon assimilated by the host plant over a period of 55 days was derived heterotrophically from the protein by way of its mycorrhizal fungi. It is likely that in nature the provision by the host plant of "starter" carbon in the form of assimilate will enable the ectomycorrhizal fungi to compete successfully with other sections of the soil microbial community, the activities of which are normally carbon-limited. Assimilation of the products of enzyme attack upon organic substrates will subsequently reduce the demands upon host photosynthate and thus provide a positive feedback of carbon and nitrogen.

Such feedback will be of importance to the carbon economy of trees, particularly when growing in shaded environments since it will provide some compensation for the considerable carbon drain now known to be imposed by the mycorrhizal mycelium. Using divided observation chambers in which the respiration rates of the intact mycelia could be measured independently of that of the host (Söderstrom and Read, 1987) it has been shown that mycelial respiration, maintenance of which is dependent upon inputs of host assimilate, can be equivalent to 29% of that occurring in roots at the same temperature. This adds a new dimension to the already large estimate of the total allocation of carbon to below-ground components of forest ecosystems, and emphasizes the importance of ectomycorrhizal mycelial systems as primary sources of carbon input to the whole soil microbial community.

Some studies (Fogel, 1980; Fogel and Hunt, 1979, 1983) have estimated that up to 50% of the total annual throughput of a Douglas fir stand can be accounted for by fungi, while others (Vogt *et al.*, 1983) suggest that around 15% of net primary production is consumed by the mycorrhizas. Harley (1971, 1978) calculated a likely carbon drain imposed by mycorrhizas upon trees of around 10%. None of these calculations took account of the external mycelium of the mycorrhizas, or of the possibility of "feedback" mechanisms involving assimilation of soil organic carbon.

Ecosystems dominated by plants with vesicular-arbuscular mycorrhizas

The enhancement of input of bases arising, for example, from the proximity to the surface of lime-rich substrates, or most notably at low latitudes, from increases of evapotranspiration, leads to increase of soil pH, and to the propensity for nitrification. As nitrogen, in the form of the relatively mobile nitrate ion, becomes progressively more available, phosphorus replaces nitrogen as the major growth-limiting nutrient. Liebig's Law predicts that under these circumstances selection will increasingly favour mutualisms which enhance the ability to capture phosphate ions at the expense of those which scavenge for nitrogen. Such predictions are fulfilled at both local and global levels in nature. Plant communities dominated by species infected by VA fungi become locally prominent wherever calcareous soils occur as "islands" within areas of acidic soil supporting ectomycorrhizal hosts. Thus, throughout Europe, tree species such as *Fraxinus* and *Taxus*, which are characteristically hosts only to VA fungi, can be seen to replace ectomycorrhizal species where calcareous soils outcrop even at high altitudes (Ellenberg, 1988). Interestingly in the warm moist tropics where local conditions favour accumulation of litter layers, "islands" of ectomycorrhizal species can occur in areas otherwise exclusively dominated by VA species (Newbery *et al.*, 1988).

At the global level the gradient of decreasing organic matter and increasing base status of soils towards the warmer regions is associated firstly with increases in the proportion of trees in families such as the

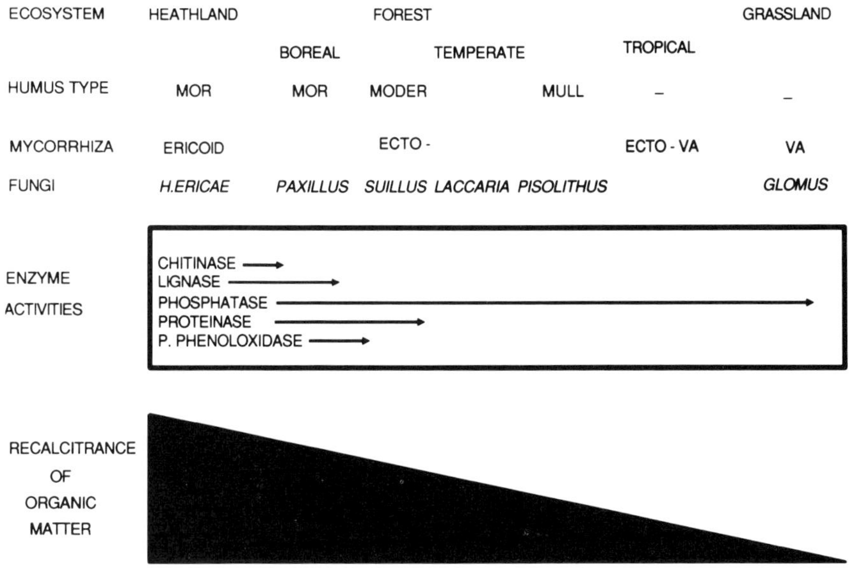

Figure 6.6. The biochemical attributes of mycorrhizal fungi, expressed as activities of exo-enzymes, in relation to their occurrence in ecosystems and the recalcitrance of the organic matter with which they are typically associated. Representative ectomycorrhizal genera are chosen to demonstrate the relationship between resource quality at any point on the environmental gradient and selection of fungi with the attributes necessary to enable nutrient acquisition from that resource. Phosphatase activity in VA systems may be associated with the infected root rather than the fungus.

Cupressaceae which are hosts to VA fungi at the expense of ectomycorrhizal families such as the Pinaceae, and eventually to the complete exclusion of the ectomycorrhizal habit.

Clearly, in addition to changes in the nature of the elemental limitation, such gradients involve changes in the quality of the nutrient resource (Fig. 6.6). Enhanced rates of nitrogen supply are well known to increase the nitrogen content of leaves and hence to decrease their C:N ratio and their recalcitrance (Fig. 6.1). More rapid rates of mineralization of organic matter will contribute to the relaxation of nitrogen limitation and to the weakening in the selective advantages previously obtained by ectomycorrhizal fungi.

In contrast to the majority of ectomycorrhizal fungi, the VA fungi which replace them appear to have little ability to mobilize nutrients from polymeric sources and are, as a group, much less physiologically diverse. However, laboratory studies confirm that they can provide infected plants with increased access to phosphate. This provision appears to arise largely as a result of the superior ability of the mycelium to exploit soluble phosphate ions in soil, though the possibility that VA fungi are directly involved in mobilization of recalcitrant sources of phosphorus cannot be excluded (Gianinazzi-Pearson and Gianinazzi, 1989).

Since, in nature, plants with VA mycorrhizas characteristically occur as heavily infected individuals growing in communities of mixed species (Read *et al.*, 1976), it is necessary to determine the role of infection in these circumstances. To date, most studies of intra- and inter-specific interactions have been carried out using simple combinations of plants grown together with or without infection. Fitter (1977) showed that the presence of infection changed the outcome of competitive interaction between two grass species. Similar results were obtained by Hall (1978) and Buwalda (1980) using mixtures of *Trifolium repens* and *Lolium perenne.* Infection improved the ability of *Trifolium* to compete with the grass especially when the soil phosphorus contents were low. Dramatic effects of the presence of VA inoculum upon the outcome of competition between two prairie grasses *Andropogon gerardii* and *Koeleria pyramidata* have been reported (Hetrick *et al.*, 1989). *A. gerardii*, which was highly responsive to infection dominated when grown in the presence of inoculum, while *K. pyramidata*, a less responsive grass, was dominant in the absence of infection.

In natural ecosystems, seedling establishment takes place in small gaps or "regeneration niches" (Grubb, 1977) within communities of more mature plants the roots of which already extensively and even exhaustively exploit the soil domain. It is at this stage in the development of the plant that the impact of VA infection is first seen. Harvests of seedlings from the field have revealed that infection takes place at an early stage in their development (Read *et al.*, 1976; Gay *et al.*, 1982; Fitter and Nichols, 1988).

Read *et al.* (1976) predicted that early integration of the seedling radicle into the mycelial network would, as in the case of ectomycorrhizal systems, provide the plant with an enormous increase in absorptive surface at relatively low energy cost, thus giving it access to resources otherwise tapped exclusively by the established plants. Significant inflows of phosphorus have been observed to occur soon after infection. By determining the ratios of phosphorus to nitrogen (P:N) in shoots of seedlings in the course of sequential harvesting from the field Read and Birch (1988) showed that after an initial fall, presumably as seed reserves of P were used, the P:N ratio stabilized as infection developed. In the cases of *Festuca ovina* and *Plantago lanceolata*, the P:N ratio stopped falling at the time of appearance of the first infections in the lateral roots. The infection process can thus be seen to be of potentially profound importance not only for the individual plant but, through effects upon recruitment of seedlings, upon the structure of the community as a whole. Subsequent studies have largely confirmed these predictions.

The role of VA mycorrhizas at the community level has been examined using microcosms in which ecologically realistic assemblages of herbaceous plants were grown with or without the presence of mycorrhizal fungi for one year (Grime *et al.*, 1987). The mixture contained the grass *Festuca ovina* as the dominant species and included plants such as *Arabis hirsuta* and *Rumex acetosa* which are not normally mycorrhizal. Prior to harvesting, $^{14}CO_2$ was fed to shoots of *Festuca* in selected microcosms and the pattern of distribution of label was determined. Significant amounts of transfer of radioactivity between plants occurred only in the microcosms containing mycorrhizal fungi and in these only the plant species which were actually infected contained high levels of activity. This is further evidence that plants in natural communities are functionally interconnected by their mycorrhizal fungi, and demonstrates that as in the case of the ectomycorrhizal mycelium, the network of VA hyphae is a major sink for current assimilates which are continuously flowing into and through the soil ecosystem. The significance of these inputs of carbon for the maintenance of the soil microbial community has been as much overlooked in grassland as in forest ecosystems.

The presence of mycorrhizal infection caused a shift in the distribution of biomass from the dominant *Festuca ovina* in favour of subordinate plants (Fig. 6.7). Those herbaceous species such as *Centaurium erythraea* with small seeds and which were receptive to infection benefited most from the presence of VA fungi in terms of survivorship, the result being to provide significant increases of species diversity in the inoculated microcosms. It appears that the characteristically high diversity of species found in so many communities dominated by plants with VA infection may be attributable to the presence of VA fungi which facilitate nutrient capture by the most vulnerable species at a critical phase in their life-cycle.

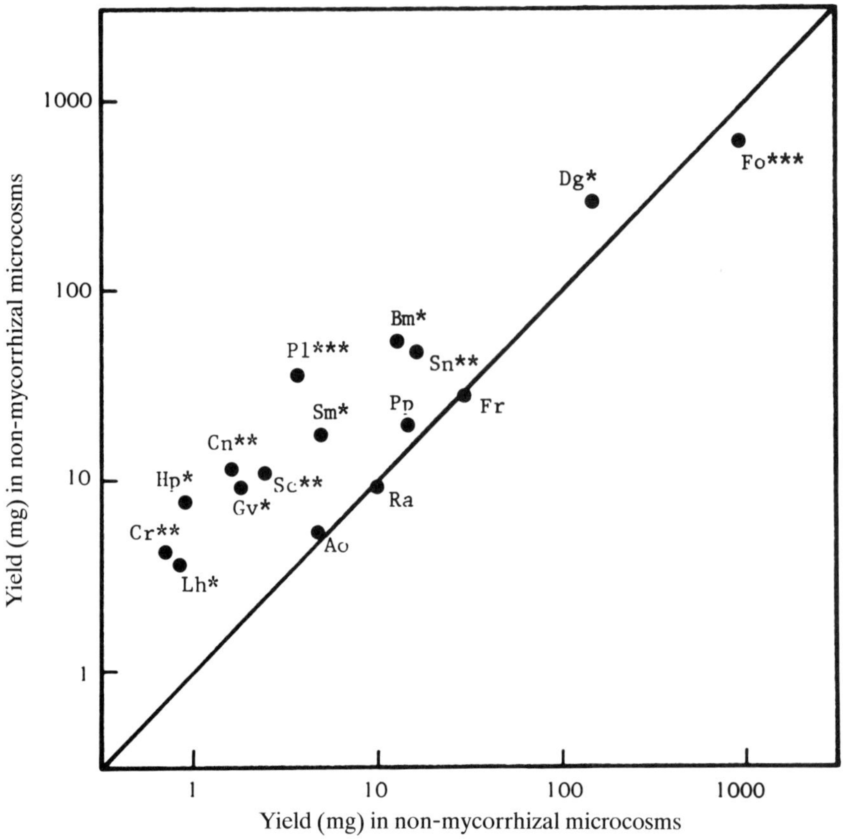

Figure 6.7. Effects of mycorrhizal infection on the average shoot weight of individuals of various grassland species grown together for one year in laboratory microcosms. The statistical significance of changes in weight associated with infection are indicated as follows: *, $p < 0.05$; **, $p < 0.01$; ***, $p < 0.001$. Key to species: Ao, *Anthoxanthum odoratum*; Bm, *Briza media*; Cr, *Campanula rotundifolia*; Cn, *Centaurea nigra*; Dg, *Dactylis glomerata*; Fo, *Festuca ovina*; Fr, *F. rubra*; Gv, *Galium verum*; Hp, *Hieracium pilosella*; Lh, *Leontodon hispidus*; Pp, *Poa pratensis*; Ra, *Rumex acetosa*; Sm, *Sanguisorba minor*; Sc, *Scabiosa columbaria*; Sn, *Silene nutans*.

Since plants with VA mycorrhizas dominate some of the most drought stressed habitats of the world it is not surprising that the role of this type of infection in plant–water relations has been the subject of some debate. The view has been expressed (Safir *et al.*, 1971, 1972; Nelsen and Safir, 1982) that the often observed increase of water throughput arising from VA infection is only a secondary product of enhanced access to phosphate.

However such an interpretation is challenged by studies which clearly demonstrate greater transpiration rates in infected plants which are of comparable size and P status to those that are uninfected (Augé *et al.*, 1986). In studies of the impact of mycorrhizal infection upon the water relations of two grasses of xeric habitats it has been shown (Allen and Allen, 1986) that one of the species *Agropyron smithii* had lower stomatal resistance and increased leaf water potential when infected, but only during the driest part of the growing season. Low stomatal resistance was maintained in infected plants of this species even when it was grown in competition with annuals. These workers postulate that under field conditions mycorrhizal infection is of importance under particular circumstances of stress which were described as "ecological crunches". Such results again emphasize the need to assess the role of mycorrhizas over the life of the plant, rather than over a short period of time at some stage after establishment or between episodes of stress.

Conclusions

It emerges from an analysis of the environments and activities of the three distinctive types of mycorrhiza described above that each is adapted to the particular suite of conditions in which it and its host plants predominate. The function of mycorrhizas in ecosystems can thus only realistically be understood in the context of the nutritional resources which they exploit. Since these substrates are, in natural communities largely derived from the host plants themselves, the specialized abilities of their mycorrhizal fungi to mobilize and transport key nutrients provides for conservation of nutrients within the ecosystems. Conservation is further enhanced by the formation of semi-permanent mycelial systems which, as a result of the low host specificities shown by most of the common mycorrhizal fungi in a given community, form extensive absorptive networks interlinking plants at both intra- and inter-specific levels (Read *et al.*, 1985; Read, 1990). The importance of such networks lies in the fact that the nutrient dynamics of those plant communities most susceptible to mycorrhizal infection, which grow on infertile soils, are characterized by seasonal pulses of resource mobilization rather than by continuous release (Gupta and Rorison, 1975; Abuarghub and Read, 1988). In these circumstances the mycorrhizal network is able rapidly to exploit resources as they become available and thus to maintain them in circulation.

There seems little doubt that mycorrhizal fungi play a central role in the nutrient dynamics of the heathland, forest, and grassland ecosystems of the world. In heath and forest environments the primary function of the mutualism appears to be the facilitation of nitrogen release, while in grassland and many tropical ecosystems phosphorus capture is probably its most important role. Clearly, emphasis on single nutrients represents an oversimplification but consideration of Liebig's Law has enabled us to identify key functional attributes of the mycorrhizal symbiosis and so to broaden our appreciation of its role at the ecosystem level. The challenge for the future is to evaluate the full range of attributes shown by the different types of mycorrhiza and to integrate the results of such studies into a more broadly based and realistic understanding of ecosystem dynamics.

The difficulties involved will be great but the evidence obtained so far suggests that the rewards obtained in terms of understanding of ecosystem function will justify the effort.

References

Abuarghub, S.M. and Read, D.J. (1988) The biology of mycorrhiza in the Ericaceae XII. Quantitative analysis of individual "free" amino acids in relation to time and depth in the soil profile. *New Phytologist* 108, 433–441.

Abuzinadah, R.A. and Read, D.J. (1986*a*) The role of proteins in the nitrogen nutrition of ectomycorrhizal plants. I. Utilization of peptides and proteins by ectomycorrhizal fungi. *New Phytologist* 103, 481–493.

Abuzinadah, R.A. and Read, D.J. (1986*b*) The role of proteins in the nitrogen nutrition of ectomycorrhizal plants. III. Protein utilization by *Betula*, *Picea* and *Pinus* in mycorrhizal association with *Hebeloma crustuliniforme*. *New Phytologist* 103, 507–514.

Abuzinadah, R.A. and Read, D.J. (1989*a*) The role of proteins in the nitrogen nutrition of ectomycorrhizal plants. V. Nitrogen transfer in birch (*Betula pendula*) grown in association with mycorrhizal and non-mycorrhizal fungi. *New Phytologist* 112, 61–68.

Abuzinadah, R.A. and Read, D.J. (1989*b*) Carbon transfer associated with assimilation of organic nitrogen sources by silver birch (*Betula pendula* Roth.). *Trees* 3, 17–23.

Alexander, I. (1989) Mycorrhizas in tropical forests. In: Procter, J. (ed.), *Mineral Nutrients in Tropical Forest and Savannah Ecosystems.* Blackwell, Oxford, pp. 169–188.

Allen, E.B. and Allen, M.F. (1986) Water relations of xeric grasses in the field: interactions of mycorrhizas and competition. *New Phytologist* 104, 559–571.

Augé, R.M., Schekel, K.A. and Wample, R.L. (1986) Greater leaf conductance of well-watered VA mycorrhizal rose plants is not related to phosphorus nutrition. *New Phytologist* 103, 107–116.

Baath, E. and Söderström, B. (1979) Fungal biomass and fungal immobilization of plant nutrients in Swedish coniferous forest soils. *Review of the Ecology and Biology of Soil* 16, 477–489.

Bajwa, R. and Read, D.J. (1985) The biology of mycorrhiza in the Ericaceae. IX. Peptides as nitrogen sources for the ericoid endophyte and for mycorrhizal and non-mycorrhizal plants. *New Phytologist* 101, 459–467.

Bajwa, R. and Read, D.J. (1986) Utilization of mineral and amino N sources by the ericoid mycorrhizal endophyte *Hymenoscyphus ericae* and by mycorrhizal and non mycorrhizal seedlings of *Vaccinium*. *Transactions of the British Mycological Society* 87, 269–277.

Bajwa, R., Abuarghub, S. and Read, D.J. (1985) The biology of mycorrhiza in the Ericaceae. X. The utilization of proteins and the production of proteolytic enzymes by the mycorrhizal endophyte and by mycorrhizal plants. *New Phytologist* 101, 469–486.

Baule, H. and Fricker, C. (1970) *The Fertiliser Treatment of Forest Trees.* Bayerischer Landwirtschafts Verlag, Munich.

Bloomfield, B.J. and Alexander, M. (1967) Melanin and resistance of fungi to lysis. *Journal of Bacteriology* 93, 1276–1280.

Bradley, R., Burt, A.J. and Read, D.J. (1982). The biology of mycorrhiza in the Ericaceae. VIII. The role of mycorrhizal infection in heavy metal resistance. *New Phytologist* 91, 197–209.

Brownlee, C., Duddridge, J.A., Malibari, A. and Read, D.J. (1983) The structure and function of mycelial systems of ecto-mycorrhizal roots with special reference to their role in forming inter-plant connections and providing pathways for assimilate and water transport. *Plant and Soil* 71, 433–443.

Burt, A.J., Hashem, A.R., Shaw, G. and Read, D.J. (1986) Comparative analysis of metal tolerance in ericoid and ectomycorrhizal fungi. In: Gianinazzi-Pearson, V. and Gianinazzi, S. (eds), *Physiological and Genetical Aspects of Mycorrhizae.* INRS, Paris, pp. 683–687.

Buwalda, J.G. (1980) Growth of clover-rye grass association with vesicular-arbuscular mycorrhizas. *New Zealand Journal of Agricultural Research* 23, 379–383.

Chu-Chou, M. (1979) Mycorrhizal fungi of *Pinus radiata* in New Zealand. *Soil Biology and Biochemistry* 11, 557–562.

Coutts, M.P. and Nicoll, B.C. (1990) Growth & survival of shoots, roots & mycorrhizal mycelium in clonal Sitka spruce during the first growing season after planting. *Canadian Journal of Forestry Research* 20, 861–868.

Crick, J.C. and Grime, J.P. (1987) Morphological plasticity and mineral nutrient capture in two herbaceous species of contrasted ecology. *New Phytologist* 107, 403–414.

Cromack, K. (1988) Below ground processes in forest succession. In: West, D.C., Shugart, H.H. and Botkin, D.B. (eds), *Forest Succession – Concepts and Applications.* Springer, New York, pp. 140–146.

Deacon, J.W., Donaldson, S.J. and Last, F.T. (1983) Sequences and interactions of mycorrhizal fungi on birch. *Plant and Soil* 71, 257–62.

Dighton, J. and Mason, P.A. (1985) Mycorrhizal dynamics during forest tree development. In: Moore, D., Casselton, L., Wood, D.A. and Frankland, J.C. (eds), *Developmental Biology of Higher Fungi.* Cambridge University Press, Cambridge, pp. 117–139.

Drew, M.C. (1975) Comparison of the effects of a localised supply of phosphate, nitrate, ammonium and potassium on the growth of the seminal root system, and the shoot in barley. *New Phytologist* 75, 479–490.

Ellenberg, H. (1988) *Vegetation Ecology of Central Europe.* Cambridge University Press, London.

Finlay, R.D. and Read, D.J. (1986*a*) The structure and function of the vegetative mycelium of ectomycorrhizal plants. I. Translocation of ^{14}C-labelled carbon between plants interconnected by a common mycelium. *New Phytologist* 103, 143–156.

Finlay, R.D. and Read, D.J. (1986*b*) The structure and function of the vegetative mycelium of ectomycorrhizal plants. II. The uptake and distribution of phosphorus by mycelial strands interconnecting host plants. *New Phytologist* 103, 157–165.

Finlay, R.D., Ek, H., Odham, G. and Söderström, B. (1988) Mycelial uptake, translocation and assimilation of nitrogen from ^{15}N-labelled ammonium by *Pinus sylvestris* plants infected with four different ectomycorrhizal fungi. *New Phytologist* 110, 59–66.

Fitter, A.H. (1977) Influence of mycorrhizal infection on competition for phosphorus and potassium by two grasses. *New Phytologist* 79, 19–125.

Fitter, A.H. and Nichols, R. (1988) The use of benomyl to control infection by vesicular-arbuscular mycorrhizal fungi. *New Phytologist* 110, 201–206.

Fleming, L.V., Deacon, J.W. and Last, F.T. (1986) Ectomycorrhizal succession in a Scottish birch wood. In: Gianinazzi-Pearson, V., and Gianinazzi, S. (eds), *Physiological and Genetical Aspects of Mycorrhizae.* INRA, Paris, pp. 259–264.

Fogel, R. (1980) Mycorrhizae and nutrient cycling in natural forest ecosystems. *New Phytologist* 86, 199–212.

Fogel, R. and Cromack, K. (1977) Effect of habitat and substrate quality on Douglas fir litter decomposition in Western Oregon. *Canadian Journal of Botany* 55, 1632–1640.

Fogel, R. and Hunt, G. (1979) Fungal and arboreal biomass in a western Oregon Douglas fir ecosystem: distribution patterns and turnover. *Canadian Journal of Forestry Research* 9, 265–256.

Fogel, R. and Hunt, G. (1983) Contribution of mycorrhizae and soil fungi to nutrient cycling in a Douglas-fir ecosystem. *Canadian Journal of Forestry Research* 13, 219–232.

Frank, A.B. (1894) Die Bedeutung der Mykorrhizapilze für die gemeine Kiefer. *Forstwissenschaftliches Centralblatt* 16, 1852–1890.

Gay, P.E., Grubb, P.J. and Hudson, H.J. (1982) Seasonal changes in the concentrations of nitrogen, phosphorus and potassium and in the density of mycorrhiza, in biennial and matrix-forming perennial species of closed chalkland turf. *Journal of Ecology* 70, 571–594.

Gianinazzi-Pearson, V. and Gianinazzi, S. (1989) Phosphorus metabolism in mycorrhizas. In: Boddy, L., Marchant, R. and Read, D.J. (eds), *Nitrogen, Phosphorus and Sulphur Utilisation by Fungi.* Cambridge University Press, Cambridge, pp. 227–243.

Grier, C.C., Vogt, K.A., Keyes, M.R. and Edmonds, R.L. (1981) Biomass distribution and above and below ground production in young and mature *Abies amabilis* zone ecosystems of the Washington Cascades. *Canadian Journal of Forestry Research* 11, 155–167.

Griffiths, R.P., Caldwell, B.A., Cromack, K. and Morita, R.Y. (1987) A study of chemical and microbial variables in forest soils colonised with *Hysterangium setchellii* rhizomorphs. In: Sylvia, D.M., Hung, L.L. and Graham, J.H. (eds), *Proceedings of the Seventh North American Conference on Mycorrhizas.*

University of Florida, Gainesville, p. 96.
Griffiths, R.P., Caldwell, B.A., Cromack, K. and Morita, R.Y. (1990) Microbial dynamics and chemistry in Douglas fir forest soils colonised by ectomycorrhizal mats: I. Seasonal variation in nitrogen chemistry and nitrogen cycle transformation rates. *Canadian Journal of Forestry Research* 20, 211–218.
Grime, J.P. (1979) *Plant Strategies and Vegetation Processes.* John Wiley & Sons, New York.
Grime, J.P., Mackey, J.M.L., Hillier, S.H. and Read, D.J. (1987) Floristic diversity in a model system using experimental microcosms. *Nature, London* 328, 420–422.
Grubb, P.J. (1977) The maintenance of species richness in plant communities: the importance of the regeneration niche. *Biological Reviews* 52, 107–145.
Gupta, P.L. and Rorison, I.H. (1975) Seasonal differences in the availability of nutrients down a podsolic profile. *Journal of Ecology* 63, 521–534.
Hall, I.R. (1978) Effects of endomycorrhizas on the competitive abilities of white clover. *New Zealand Journal of Agricultural Research* 21, 509–515.
Handley, W.R.C. (1963) Mycorrhizal associations and *Calluna* heathland afforestation. *Forestry Commission Bulletin* 36, 1–70.
Harley, J.L. (1940) A study of the root system of the beech in woodland soils with especial reference to mycorrhizal infection. *Journal of Ecology* 28, 107–117.
Harley, J.L. (1971) Fungi in ecosystems. *Journal of Ecology* 59, 653–668.
Harley, J.L. (1978) Ectomycorrhizas as nutrient absorbing organs. *Proceedings of the Royal Society of London, Series B,* 203, 1–21.
Harvey, A.E., Larsen, M.J. and Jurgenson, M.F. (1976) Distribution of ectomycorrhiza in a mature Douglas fir/larch forest system in Western Montana. *Forest Science* 22, 393–398.
Haselwandter, J., Bobleter, O. and Read, D.J. (1990) Utilisation of lignin by ericoid and ectomycorrhizal fungi. *Archives of Microbiology* 153, 352–354.
Heal, O.W. and Dighton, J. (1985) Nutrient cycling and decomposition in natural terrestrial ecosystems. In: Mitchell, M.J. and Nakas, J.P. (eds), *Soil Microflora-Macrofauna Interactions.* Martinus Nijhoff, The Hague, pp. 14–73.
Heal, O.W., Latter, P.M. and Howson, G. (1978) A study of the rates of decomposition of organic matter. In: Heal, O.W. and Perkins, D.F. (eds), *Production Ecology of British Moors and Montane Grasslands.* John Wiley & Sons, New York, pp. 136–160.
Hetrick, B.A.D., Wilson, G.W.T. and Hartnett, D.C. (1989) Relationship between mycorrhizal dependence and competitive ability of two tallgrass prairie grasses. *Canadian Journal of Botany* 67, 2608–2615.
Jalal, M.A.F., Read, D.J. and Haslam, E. (1982) Phenolic composition and its seasonal variation in *Calluna vulgaris. Phytochemistry* 21, 1397–1401.
Jalal, M.A.F. and Read, D.J. (1983*a*) The organic acid composition of *Calluna* heathland soil with special reference to phyto- and fungi-toxicity. I. Isolation and identification of organic acids. *Plant and Soil* 70, 257–272.
Jalal, M.A.F. and Read, D.J. (1983*b*) The organic acid composition of *Calluna* heathland soil with special reference to phyto- and fungi-toxicity. II. Monthly quantitative determination of the organic acid content of *Calluna* and spruce dominated soils. *Plant and Soil* 70, 273–286.
Lamb, R.J. (1979) Factors responsible for the distribution of mycorrhizal fungi of *Pinus* in eastern Australia. *Australian Journal of Forest Research* 9, 25–34.

Last, F.T., Dighton, J. and Mason, P.A. (1987) Successions of sheathing mycorrhizal fungi. *Trends in Ecology and Evolution* 2, 157–161.

Leake, J.R. (1987) Metabolism of phyto- and fungitoxic phenolic acids by the ericoid mycorrhizal fungus. In: Sylvia, D.M., Hung, L.L. and Graham, J.H. (eds), *Proceedings of the Seventh North American Conference on Mycorrhizas*, pp. 332–333.

Leake, J.R. and Read, D.J. (1989*a*) The effects of phenolic compounds on nitrogen mobilisation by ericoid mycorrhizal systems. *Agricultural Ecosystems and Environment* 29, 225–236.

Leake, J.R. and Read, D.J. (1989*b*) The biology of mycorrhiza in the Ericaceae. XIII. Some characteristics of the extracellular proteinase activity of the ericoid endophyte *Hymenoscyphus ericeae. New Phytologist* 112, 69–76.

Leake, J.R., Shaw, G. and Read, D.J. (1989) The role of ericoid mycorrhiza in the ecology of ericaceous plants. *Agricultural Ecosystems and Environment* 29, 237–250.

Leake, J.R. and Read, D.J. (1990*a*) Proteinase activity in mycorrhizal fungi. I. The effect of extracellular pH on the production and activity of proteinase by ericoid endophytes from soils of contrasted pH. *New Phytologist* 115, 243–250.

Leake, J.R. and Read, D.J. (1990*b*) Chitin as a nitrogen source for mycorrhizal fungi. *Mycological Research* 94, 993–995.

Liebig, J. (1843) *Chemistry in its Application to Agriculture and Physiology.* Third edn. Taylor and Walton, London.

Lindeberg, G. (1944) Über di Physiologie Ligninabbauender Boden/hymenomyzeten. *Symbolae botanicae Upsaliensis* 8(2), 1–183.

Lundeberg, G. (1970) Utilisation of various nitrogen sources, in particular bound nitrogen, by mycorrhizal fungi. *Studia Forestalia Suecica* 79, 1–95.

Mason, P.A., Wilson, J., Last, F.T. and Walker, C. (1983) The concept of succession in relation to the spread of sheathing mycorrhizal fungi on inoculated tree seedlings growing in unsterile soils. *Plant and Soil* 71, 247–56.

Meentemeyer, V. (1978) Macroclimate and lignin control of litter decomposition rates. *Ecology* 59, 465–472.

Melin, E. and Nilsson, H. (1952) Transport of labelled nitrogen from an ammonium source to pine seedlings through mycorrhizal mycelium. *Svensk Botanisk Tidskrift* 46, 281–285.

Melin, E. and Nilsson, H. (1953) Transfer of labelled nitrogen from glutamic acid to pine seedlings through the mycelium of *Boletus variegatus* (S.W.) Fr. *Nature, London* 171, 434.

Meyer, F.H. (1973) Distribution of ectomycorrhizae in native and man-made forests. In: Marks, G.C. and Kozlowski, T.T. (eds), *Ectomycorrhizae: their ecology and physiology.* Academic Press, New York, pp. 79–105.

Mitchell, H.L. and Chandler, R.F. (1939) The nitrogen nutrition and growth of certain deciduous trees of Northeastern United States. *Black Rock Forest Bulletin* 11, 1–94.

Mitchell, D.T. and Read, D.J. (1981) Utilization of inorganic and organic phosphates by the mycorrhizal endophytes of *Vaccinium macrocarpon* and *Rhododendron ponticum. Transactions of the British Mycological Society* 76, 255–260.

Muller, P.E. (1884) Studier over Skovjord, som bidrag til Skovdyrkningens Theori II Om Muld og Mor i Egeskove og paa Heder. *Tidskrift for Skog* 7, 1–12.

Nelsen, C.E. and Safir, G.R. (1982) Increased drought tolerance of mycorrhizal onion plants caused by improved phosphorus nutrition. *Planta* 154, 407–13.

Newbery, D.M., Alexander, I.J., Thomas, D.W. and Gartlan, J.S. (1988) Ectomycorrhizal rain-forest legumes and soil phosphorus in Korup National Park, Cameroon. *New Phytologist* 109, 433–450.

Norkrans, B. (1950) Studies on growth and cellulolytic enzymes of *Tricholoma. Symbolae botanicae Upsaliensis* 11(1), 1–126.

Ogawa, M. (1977) Ecology of higher fungi in *Tsuga diversifolia* and *Betula ermani – Abies mariesii* forests of subalpine zone. *Transactions of the Mycological Society of Japan* 18, 1–19.

Pearson, V. and Read, D.J. (1975) The physiology of the mycorrhizal endophyte of *Calluna vulgaris. Transactions of the British Mycological Society* 64, 1–7.

Persson, J. (1980) Spatial distribution of fine-root growth, mortality, and decomposition in a young Scots Pine stand. *Oikos* 34, 77–87.

Rayner, A.D.M., Powell, K.A., Thompson, W. and Jennings, D.H. (1985) Morphogenesis of vegetative organs. In: Moore, D., Casselton, L.A., Wood, D.A. and Frankland, J.C. (eds), *Developmental Biology of Higher Fungi.* Cambridge University Press, Cambridge, pp. 249–279.

Read, D.J. (1983) The biology of mycorrhiza in the Ericales. *Canadian Journal of Botany* 61, 985–1004.

Read, D.J. (1984) Interactions between ericaceous plants and their competitors with special reference to soil toxicity. *Symposia of the Association of Applied Biologists* 5, 195–209.

Read, D.J. (1990) Ecological integration by mycorrhizal fungi. In: Nardon, P. (ed.), *Endocytobiology* IV. INRA, Paris, pp. 99–106.

Read, D.J. and Stribley, D.P. (1973) Effect of mycorrhizal infection on nitrogen and phosphorus nutrition of ericaceous plants. *Nature, London* 244, 81.

Read, D.J. and Birch, C.P.D. (1988) The effects and implications of disturbance of mycorrhizal mycelial systems. *Proceedings of the Royal Society of Edinburgh* 94B, 13–24.

Read, D.J., Koucheki, H.K. and Hodgson, J. (1976) Vesicular-arbuscular mycorrhiza in natural vegetation systems. I. The occurrence of infection. *New Phytologist* 77, 641–655.

Read, D.J., Francis, R. and Finlay, R.D. (1985) Mycorrhizal mycelia and nutrient cycling in plant communities. In: Fitter, A.H., Atkinson, D., Read, D.J. and Usher, M.B. (eds), *Ecological Interactions in Soil: plants, microbes and animals.* Blackwell Scientific Publications, Oxford, pp. 193–217.

Read, D.J., Leake, J.R. and Langdale, A.R. (1989) The nitrogen nutrition of mycorrhizal fungi and their host plants. In: Boddy, L.L., Marchant, R. and Read, D.J. (eds), *Nitrogen, Phosphorus and Sulphur Utilization by Fungi.* Cambridge University Press, Cambridge, pp. 181–204.

Robinson, R.K. (1971) Importance of soil toxicity in relation to the stability of plant communities. In: Duffey, E. and Watt, A.S. (eds), *The Scientific Management of Animal and Plant Communities for Conservation.* Nature Conservancy Council, Huntingdon, pp. 105–113.

Romell, L.G. and Malmstrom, C. (1945) Henrik Hesselmans tallhedsförsök aren 1922–1942. *Meddelanden Statens Skogsförsöksanst* 34, 543–625.

Safir, G.R., Boyer, J.S. and Gerdemann, J.W. (1971) Mycorrhizal enhancement of water transport in soybean. *Science, New York* 172, 581–583.

Safir, G.R., Boyer, J.S. and Gerdemann, J.W. (1972) Nutrient status and mycorrhizal enhancement of water transport in soybean. *Plant Physiology* 49, 700–703.

Scheffer, F. and Ulrich, B. (1960) *Lehrbuch der Agriculturechemie und Boden Kunde. III Humus und Humusdungung.* Vol. 1. Gustav Fischer, Stuttgart.

Schuler, R. and Haselwandter, K. (1988) Hydroxamate siderophore production by ericoid mycorrhizal fungi. *Journal of Plant Nutrition* 11, 907–913.

Shaw, G. and Read, D.J. (1989) The biology of mycorrhiza in the Ericaceae. XIV Effects of iron and aluminium on the activity of acid phosphatase in the ericoid embophyte *Hymenoscyphus ericae* (Read) Korf & Kernan. *New Phytologist* 113, 529–533.

Shaw, G., Leake, J.R., Baker, A.J.M. and Read, D.J. (1990) The biology of mycorrhiza in the Ericaceae XVII. The role of mycorrhiza in the regulation of iron uptake by ericaceous plants. *New Phytologist* 115, 251–258.

Straker, C.J. and Mitchell, D.T. (1986) The activity and characterization of acid phosphates in endomycorrhizal fungi of the Ericaceae. *New Phytologist* 104, 243–256.

Stribley, D.P. and Read, D.J. (1974) The biology of mycorrhiza in the Ericaceae. IV. The effects of mycorrhizal infection on the uptake of ^{15}N from labelled soil by *Vaccinium macrocarpon* Ait. *New Phytologist* 73, 1149–1155.

Stribley, D.P. and Read, D.J. (1976) The biology of mycorrhiza in the Ericaceae. VI. The effects of mycorrhizal infection and concentration of ammonium nitrogen on growth of cranberry (*Vaccinium macrocarpon* Ait.) in sand culture. *New Phytologist* 77, 63–72.

Stribley, D.P. and Read, D.J. (1980) The biology of mycorrhiza in the Ericaceae. VII. The relationship between mycorrhizal infection and the capacity to utilize simple and complex organic nitrogen sources. *New Phytologist* 86, 365–371.

Söderstöm, B. and Read, D.J. (1987) Respiratory activity of intact and excised ectomycorrhizal mycelial systems growing in unsterilized soil. *Soil Biology and Biochemistry* 19, 231–236.

Tschager, A., Hilscher, H., Franz, S., Kull, V, and Larcher, W. (1982). Jahreszeitliche Dynamik der Fettspeicherung von *Loiseleuria procumbens* und anderen Ericaceen der alpinen zuergstrauch heide. *Acta Oecologia* 3, 119–130.

Trojanowksi, J., Haider, K. and Hutterman, A. (1984) Decomposition of ^{14}C-labelled lignin, holocellulose and lignocellulose by mycorrhizal fungi. *Archives of Microbiology* 139, 202–206.

Vogt, K.A., Moore, E.E., Vogt, D.J., Redlin, M.J. and Edmonds, R.L. (1983) Conifer fine root and mycorrhizal root biomass within the forest floors of Douglas fir stands of different ages and site productivities. *Canadian Journal of Forest Research* 13, 429–37.

Weatherell, J. (1953) The checking of forest trees by heather. *Forestry* 26, 37–41.

Weatherell, J. (1957) The use of nurse species in the afforestation of upland heaths. *Quarterly Journal of Forestry* 51, 298–304.

Wiersum, L.K. (1958) Density of root branching as affected by substrate and separate ions. *Acta botanica Neerlandica* 7, 174–190.

Zehetmayr, J.W.L. (1960) Afforestation of upland heaths. *Forestry Commission Bulletin* 32, 1–145.

7

The Significance of Mycology in Medicine

O. Male, *Department of Dermatology I, University of Vienna, Medical School, Alserstrasse 4, A-1090 Vienna, Austria.*

ABSTRACT The importance of fungi in medicine is reviewed, with particular emphasis on mycotic infections. Their frequency, etiology, natural habitats, pathogenesis (formal and causal), nomenclature and classification, and clinical aspects (mycoses caused by dermatophytes, yeasts, moulds and biphasic fungi; therapy) are outlined. The differences between mycotisations, mycoallergoses, mycotoxicoses, and mycetism are clarified.

General aspects

Outside the relatively small circle of specialists, the significance of mycology in medicine is mostly equated with that of the more or less superficial fungal infections of the skin and their adjacent mucous membranes. Besides these, endemic mycoses and opportunistic fungal infections in immuno-compromised patients are occasionally also taken into consideration. Such mycoses have a very high sociomedical significance, among other reasons because the first group of diseases is extremely frequent and the other two forms of mycoses mainly follow a severe and often lethal course. The medical significance of mycology, in reality, neither derives only from these nor a series of other fungal infections, but also – sometimes even more than these – from allergoses and intoxications caused by the allergens and toxins, respectively, of numerous fungi.

The most important fungal diseases are therefore discussed here. As the Congress and readers of its proceedings are mainly non-medical, it is appropriate to survey the entire field rather than provide in-depth accounts of single aspects. This survey can only be superficial, nevertheless, the

Frontiers in Mycology. Honorary and General Lectures from the Fourth International Mycological Congress, Regensburg 1990. Edited by D.L. Hawksworth.

principal aspects are discussed as thoroughly as circumstances permit. The main subject of the survey is mycotic infections; the mycotisations, mycoallergoses, mycotoxicoses, and mycetism will only be touched upon since they are dealt with at some length by another contributor (Chapter 8).

Definition of concepts and nomenclature

Collective nouns for all forms of impaired health which may be caused by fungi themselves, their toxins and/or their allergens are *mycopathies*, *fungal diseases*, *mycotic diseases* and – less precisely – just *mycoses* (in the broad sense of the term). These mycopathies are subdivided into the following five forms:

1. *Mycoses* (in the strict sense). Definition: The attack and destruction of living cells/tissue/organs by pathogenic fungi.
2. *Mycotisation*, nosoparasitism by fungi. Definition: Saprobic/nosoparasitic colonization of dead or damaged tissues, such as necroses of burns or ulcers, cavities or diverticula, by fungi which may be nonpathogenic or potentially pathogenic (opportunistic), such as *Aspergillus* or *Mucor* species.

Table 7.1. Seasonal frequency of the most important allergenic fungi in the atmosphere in Central Europe.

Fungi	Spores (m^{-3} air)	Season (Months) J F M A M J J A S O N D
Alternaria	1000 – 1500	- - - - - - - - - - - - ———— - - - -
Aspergillus	500 – 1000	- - - - - - ———— - - - - - - - -
Aureobasidium	50 – 100	- - -
Botrytis	500 – 800	- - - - - - - - ————
Cladosporium	4000 – 6000	- - - - - - - - - - - ———— - -
Entomophthora	30 – 80	- ———— -
Epicoccum	50 – 300	- ———— - - -
Erysiphe	50 – 150	- ———— - -
Fusarium	50 – 100	- - ———— - - -
Neurospora	1500 – 2500	————
Penicillium	100 – 300	- - - - - - - ———— - - - -
Puccinia	10 – 100	————
Ustilago	8000 – 12000	————
ascospores	4000 – 6000	- - - - - - - - - - ———— - - - -
basidiospores	2000 – 3000	- - - - - - - ———— - - -

Table 7.2. Synopsis of the most important mycotoxicoses and the appropriate microbiological data.

Fungus	Biotope/Habitat	Toxins/Foods	Effects	Diseases
Aspergillus flavus	peanuts, maize, legumes	aflatoxin B1, B2, G1, G2, rice, manioc, animal feed, oils, margarine, milk	antimitotic, teratogenic, carcinogenic, hepato-, nephrotoxic	aflatoxicoses turkey-X-disease
A. ochraceus	maize	maltoryzin, ochratoxin A, B, C, maize flour	hepatotoxic	
A. versicolor	grain	sterigmatocystin (mouldy) flour	hepato-, nephrotoxic	
Penicillium patulum, etc.	grain	patulin	carcinogenic, antibiotic, antidiuretic, central respiratory paralysis	
P. citrinum	rice	citrinin	nephrotoxic, parasym-pathomimetic	
P. chrysogenum	grain	notatin	methaemoglobin formation	
P. islandicum	rice, legumes	luteosykrin	hepatotoxic, carcinogenic	
Fusarium graminearum	maize, legumes	zearalenone	oestrogenic, habitual abortion	
F. sporotrichoides	grain	fusaridine flour	disturbance of growth, arthritism	Kashin-Beck disease alimentary toxic anaemia
Sporidesmium bakeri	ryegrass	sporodesmin, phylloerythrin	photosensitizing, cytotoxic, (mitochondrial membranes), hepatotoxic	"facial eczema" of sheep
Paecilomyces variotii	fruits, grain	byssochlaminic acid fruit juices, flour	haemorrhaghias	haemorrhagic fowl syndrome

3. *Mycoallergoses.* Definition: Allergic reactions caused by the allergens of nonpathogenic or pathogenic fungi which come into contact with a – usually predisposed – person (or animal) by inhalation, ingestion, or immediate contact; the latter may take place directly on the skin or the adjacent mucous membranes (mainly the genitals) or within a mycotic infection or a mycotisation. The most frequent allergenic fungi are listed in Table 7.1.
4. *Mycotoxicoses.* Definition: Alimentary poisonings by the toxins of certain fungi, mainly moulds, which have contaminated foods or feeds; this usually occurs when foodstuffs are not sufficiently dried or stored in a humid environment. The poisoning of man occurs most frequently by consuming the products (meat, milk, eggs, etc.) of animals which are affected by contaminated feeds (mainly hay, silo-stored food and food used for fattening), less often by eating products of contaminated grains and only exceptionally by eating the contaminated food itself (e.g. vegetables and nuts; predominantly those transported overseas by ship). The mycotoxins are highly thermostable; they can tolerate temperatures up to 300°C, and are therefore not inactivated by preservation, cooking, or baking. A survey of the most important pertinent data is given in Table 7.2.
5. *Mycetism.* Definition: Alimentary poisoning by the toxins (alkaloids) of macromycetes resulting from a confusion of edible with poisonous mushrooms (mostly *Agaricus* species with *Amanita* species).

Mycoses (sensu strictoire)

The mycoses are a group of diseases which is extraordinarily extensive and heterogeneous, etiologically as well as nosologically. The number of organisms involved exceeds 200 species from all parts of the fungal kingdom, and therefore variations occur in their metabolic and pathogenic properties and in their sensitivity/resistance with respect to environmental factors and therapeutic measures.

The spectrum of clinical manifestations of mycoses is extraordinarily wide, ranging from largely harmless discolorations and scalings of the most superficial layers of the skin, through hyperphlegmasic purulent lesions of mucous membranes, infiltrations and abscesses in the subcutaneous tissue to extensive destruction in the bones, the internal organs, the brain, and/or to mycotic sepsis. With the exception of the adamantin of teeth, every individual tissue or organ of the human body may be infected.

Frequency and spectrum

The frequency and the spectrum of the relevant mycoses have changed significantly in recent decades. Before the end of the last World War, mycoses had only a very subordinate significance in medicine: apart from some tropical and subtropical endemic areas – mainly those of histoplasmosis, coccidioidomycosis, and blastomycoses – mycoses were distinctly rare; in more than 99% of cases they affected the skin and the adjacent mucous membranes and usually took a more or less harmless course. Mycoses of viscera did of course occur, but extremely rarely. From the 1960s, as a result of certain (side-)effects of civilization and medical therapy (e.g. antibiotics, hormonal contraceptives, steroids, immunosuppressants, allografts), which led to profound alterations in microecology as well as in the immunological and hormonal status, the situation has changed. Superficial mycoses burgeoned and now, with an incidence of 20 to 25% of the world population, represent the most frequent of all infectious diseases. (These forms of mycoses are relatively banal in the majority of cases but, nonetheless, mostly take a recalcitrant course and tend to become recurrent. Apart from the virtual nosologic degree of the diseases, some of them, such as onychomycoses of fingers or a pityriasis of the trunk, are extremely detrimental aesthetically.)

At the same time the percentage of mycoses caused by yeasts, mainly in

Table 7.3. Synopsis of the most important pathogenic fungi.

Dermatophytes	**Moulds**
Epidermophyton floccosum	*Aspergillus* species
Trichophyton species	*A. fumigatus, A. niger, A. flavus*, etc.
anthropophilic species:	*Alternaria alternata*, etc.
T. rubrum, T. interdigitale,	*Acremonium* species
T. violaceum	*Cladosporium werneckii*, etc.
zoophilic/geophilic species:	*Fusarium oxysporum*, etc.
T. mentagrophytes, T. verrucosum,	*Fonsecaea* species
T. schoenleinii, T. gallinae,	*Drechslera* species
T. equinum, T. tonsurans,	*Hormodendrum pedrosoi*, etc.
T. soudanense, etc.	*Penicillium marneffei*
Microsporum species	*Phialophora* species
M. canis, M. gypseum, M. rivalieri, etc.	*Rhinocladium* species
	Sepedonium species
Yeasts	
	Biphasic fungi
Candida species	
C. albicans, C. tropicalis,	*Blastomyces dermatitidis*
C. parapsilosis, C. guilliermondii, etc.	*Paracoccidioides brasiliensis*
Torulopsis species	*Coccidioides immitis*
T. glabrata, T. famata, etc.	*Histoplasma capsulatum*
Cryptococcus neoformans	*H. duboisii*
Pityrosporon orbiculare	*Sporothrix schenckii*
Trichosporon cutaneum	

mycoses of mucous membranes, has risen from less than 1% to 12–15%. Moreover, the number of systemic and visceral mycoses caused by opportunistic yeasts and moulds (above all *Candida* and *Aspergillus* species and *Cryptococcus neoformans*), the so-called *survival mycoses*, i.e. mycoses resulting from the progress of medicine and the mycoses associated with AIDS, have risen, and continue to rise, dramatically.

Etiology

For everyday medical purposes, the causative organisms of mycoses are not critically identified, but grouped in a simplified scheme based mainly on therapeutic aspects. According to this scheme they are divided into the three groups of *dermatophytes*, *yeasts*, and *moulds*. Some authors place a fourth group of organisms between the last two groups, the so-called *dimorphic* or *biphasic fungi*. The most important pathogenic fungi are listed in Table 7.3.

Dermatophytes and moulds produce filaments; yeasts produce mainly budding cells, whereas dimorphic fungi may develop both.

Dermatophytes attack only the skin and its appendages (hair, nails); yeasts, moulds and dimorphic fungi are able to attack every tissue of the body. All dermatophytes and moulds are world-wide, while some of the dimorphic fungi are endemic in certain areas.

Natural habitats

By far the most important natural habitats of all fungi, especially dermatophytes and moulds, are soil and decaying organic material, especially plants. The main habitat of yeasts is the mucous membranes, or, more strictly the secretions, excretions, and excrements of man and animals.

A further important habitat of numerous yeasts and several moulds is water. Not, of course, natural pure water, but brackish and waste water of sewers, baths, and industrial plants polluted not only by faeces, which have a particularly high yeast content, but also by the side-effects of civilization, such as organic remnants, detergents, and other chemicals. The latter favour the growth of numerous fungi to a high degree.

The biological function of the dermatophytes in the soil is the destruction of keratinized materials such as hides, furs, claws, nails, and horns of dead animals. In the soil the dermatophytes live in their teleomorphic (=sexual) stage in the form of cleistothecia or perithecia, whereas in keratinized material they live in an anamorphic (=asexual) stage in which they develop only a very simple morphology. The keratinophilic nature of dermatophytes makes it possible to isolate them from soil by implanted hair, the "hair-bating" method.

The biological function of many moulds, especially *Alternaria*, *Asper-*

gillus, *Penicillium*, *Cladosporium* and *Sepedonium* species is the organic destruction of foliage and needles and also other dead material of plants, such as branches, stems, and roots. (Under certain conditions, some of these moulds may also cause numerous mycoses in living plants, fruits or trees.)

Biphasic fungi also exist in the soil and on plants. The arthrospores of *Coccidioides immitis* occur in the sand of the endemic area of the coccidiomycosis, mainly California (Death Valley), the southwestern USA, northern Mexico and foci in Venezuela, Colombia, Paraguay, and Argentina; *Ceratocystis stenoceras*, the sexual morph of *Sporothrix schenckii*, has been found on moss and in moulding wood.

Pathogenesis

The knowledge of pathogenesis is indispensable to the understanding of the course of a mycosis, its treatment, and prognosis as well as the prevention of reinfection. It is therefore discussed in greater detail here.

We have to distinguish between formal and causal pathogenesis. Formal pathogenesis represents the way and mode of infection, whereas causal pathogenesis refers to the mechanism of the disease.

Formal pathogenesis

The following forms are of practical interest:

1. Immediate contact between the fungus and host on the skin and/or the adjacent mucous membranes. In such cases the fungus occurs only rarely in a free form; it lives mostly in infected or contaminated material of man or animals (scales, crusts, hairs, secretions, excretions, or water, towels, toothbrushes, dental prostheses, hay, straw, and soil, respectively). In this way all dermatophytoses of epidermis, hair and nail, also all candidoses of skin and genitals, as well as *pityriasis versicolor*, *tinea alba* and *tinea nigra* originate.

2. Infection by trauma. Superficial wounds and burns are preferentially infected by opportunistic moulds, such as *Aspergillus*, *Rhizopus*, and *Mucor* species. Injuries reaching into deeper regions, such as the subcuticular layer, muscles, and bones – especially if they have been caused by wooden objects, such as stakes, splinters, thorns, pricks of which particles remain in the tissue – lead rather to infections with another group of moulds, the *dematiaceous* (dark-spored) hyphomycetes (*Alternaria*, *Fusarium*, or *Drechslera* species or, mainly in temperate climates, to sporotrichosis or to *mycetoma*. A special way infection comes about is by the perforation of the walls of the sinuses in the course of dental treatment causing the unnoticed implantation of prosthetic material into the mucous membrane of the sinuses, which favours

the settlement of microbes, mainly zygomycetes and *Aspergillus* and the development of mycotic sinusitis.
3. Infection by open surgery, particularly of the heart. The most frequent pathogens of such infections, which are very seldom, are *Candida* and *Torulopsis* species.
4. Transmission of the pathogenic agents by catheters. This occasionally follows long-term urethral catheterization and in individual cases can be caused by vein catheters.
5. Inhalation of organisms and infection of the lungs, which may lead to haematogenic propagation of the pathogens. This happens mainly with the conidia of moulds and biphasic fungi (*Aspergillus, Cladosporium, Mucor, Rhizopus, Histoplasma* species, and *Coccidioides immitis*).
6. Ingestion of organisms and propagation via persorption. In this way, largely unrecognized up to now, the majority of all candidoses and torulopsidoses (and probably also of cryptococcosis) develop. Yeasts which originate from the mucous membrane of the mouth, or have been ingested in small amounts with contaminated foods, may multiply rapidly and to a high degree in the intestines where they are favoured by pathological influences, above all by a treatment with certain antibacterial antibiotics, such as aminoglycosides and some newer cephalosporins. If the concentration of the yeasts reaches values higher than about 10^4, a considerable part of the yeast cells is persorbed through the healthy wall of the intestine, then reaches the bloodstream via the lymphatic vessels. In immuno-compromised patients, this almost inevitably leads to a fungal sepsis.

Causal pathogenesis

The precondition for the establishment of a mycosis is an intimate contact of a pathogen able to infect with the tissue of a host. “Able to infect” means that the pathogen can tolerate the physical, chemical, and immunological conditions of the host environment on the one hand, and that it can break down enzymatically and utilize the tissue of the host on the other. The fatty acids of a seborrhoeic area or the low pH value of the stomach juice, for example, are not tolerated by most fungi, but quite well by *Aspergillus* or *Candida* species respectively. Keratinized tissue (epidermis, hair, nail), may be metabolized only by dermatophytes but not by yeasts, moulds, and biphasic fungi. Conversely, dermatophytes are unable to attack non-keratinized tissue, unlike the pathogenic representatives of the other three groups of fungi.

Contact between a pathogen which is able to infect and an “appropriate” tissue of a host, however, leads to an infection only in cases in which the pathogenic potential of the pathogen overcomes the resistance of the host. Such fungi (or other microbes) are called *primarily* or *obligatorily pathogenic* and the pertinent infections *primary mycoses* (or other microbial diseases). Among the pathogenic fungi which occur in Europe, only a small number of zoophilic and geophilic dermatophytes are primarily pathogenic.

The resistance of a host, however, is not a fixed value, but may be reduced by numerous influences. Thus, it becomes relatively smaller than the pathogenic potential of such pathogens whose pathogenic potentials are smaller than the resistance of a host with a normal resistance. Such pathogens are called *secondarily* or *potentially pathogenic* or *opportunistic* and the respective infections *secondary mycoses.*

With regard to pathogenic potential and behaviour respectively, of a fungus (or microorganism) three variants exist in principle: *nonpathogenic*, *pathogenic* (in a more or less obligatory degree), and *potentially pathogenic*; the pertinent data and relationships to resistance on the part of the host are depicted schematically in Figure 7.1.

The degree of pathogenic potency of the fungi is largely constant among representatives of the same species, but differs to a considerable degree

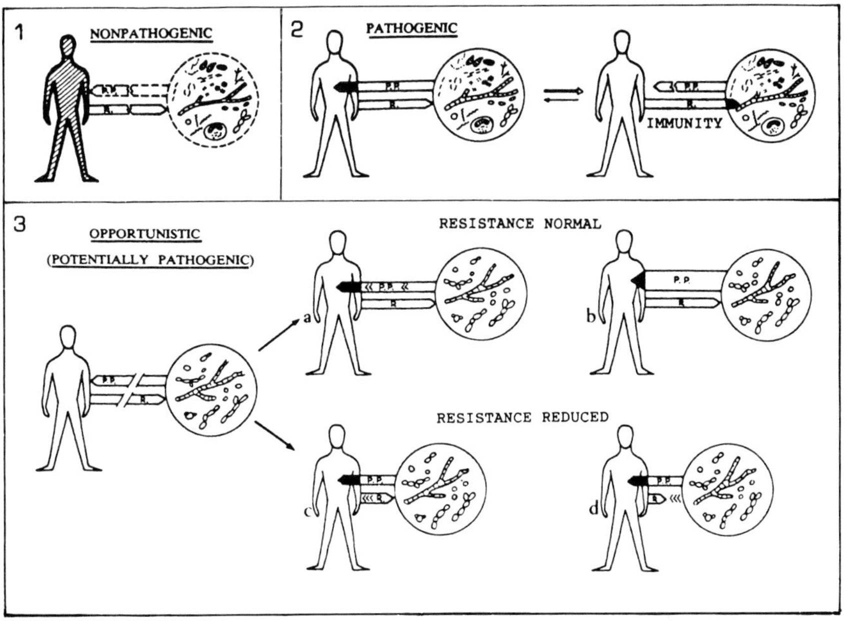

Figure 7.1. Pathogenic, host-parasite relationships. (1) Environment of the host incompatible to the pathogen; (2) pathogenic potential of the pathogen (p.p.) are intensified by stimulating factors on the part of the host; (3b) p.p. prevail over resistance of the host (r.) by an extraordinarily high quantity of pathogens; (3c) r. is reduced by endogenous host factors; (3d) r. is reduced by exogenous factors on the part of the host.

between the members of different species; similarly, the impairment of the resistance of the host may vary greatly. The pathogenic potency of some potentially pathogenic fungi is only minimally smaller than the resistance of the healthy host and may, therefore, already gain the upper hand if the latter is only minimally reduced. Conversely, the resistance of a patient may be reduced to such an extreme degree that even fungi whose pathogenic potential are only minimal, and which are not pathogenic to a healthy person, achieve the capacity to infect. Such conditions/fungi can be found above all in aleucemic patients or in cases of advanced AIDS.

In rare cases potentially pathogenic fungi may gain the capacity to also infect patients with a normal resistance. This happens when the fungi occur in an extraordinarily high quantity (e.g. in candidoses of the genitals), or when the pathogenic potencies of the fungus are stimulated (e.g. by hormones such as gestagenes, and steroids).

The factors which lead to an impairment of the resistance of a person are both very numerous and heterogeneous. They may be caused exogenously, endogenously, or indirectly endogenously. The former exert their effect only locally, the two latter universally, and occasionally also locally (above all on the mucous membranes of the mouth). A synopsis of the most important of these factors and influences is given in Table 7.4.

Nomenclature and classification

Before discussing the clinical aspects of mycoses, for a more precise understanding it is necessary to outline the main principles of mycological systematics and nomenclature.

As a consequence of the manifold heterogeneity of the mycoses in etiological, nosological, epidemiological, pathogenetic and therapeutic respects, their nomenclature and classification can, and indeed must, be approached from numerous points of view, the most important of which are:

1. *Etiologically.* The criteria for nomenclature and classification are the causative organisms or their group: *aspergillosis*, *candidosis*, *cryptococcosis*, *histoplasmosis*, *dermatophytosis*, *levuroses*, etc. The terms are constructed by attaching the suffix *-osis* to the stem of the term of the causative organism in question.
2. *Nosologically.* The criterion is the affected tissue or organ: *cerebromycosis*, *dermatomycosis*, *hepatomycosis*, *oculomycosis*, *onychomycosis*, *pneumomycosis*, etc. The terms are constructed by attaching the suffix *-mycosis* to the name of the tissue/organ in question; a simplified variant is to speak of *mycosis of*... (epidermis, hair, nail, brain, liver, etc.).

   Dermatophytic infections of skin, hair and nail are also termed *tinea.* This term, although imprecise, is well established.

   Subcriteria for a more detailed nosological nomenclature are other clinical points of view such as localized–generalized, solitary–multiple, unilateral–bilateral, homolateral–heterolateral, acute–chronic, superficial–deepseated,

Table 7.4. Factors reducing resistance and predisposing for mycoses.

I. Exogenous

Environmental; retention of heat and moisture ("moist chamber" situations)
- caused by occupation or/and milieu (miners, stockers, workers in laundries, cellars, and lacqueries, bar keepers etc.)
- caused by clothing; occurs mainly on the feet (shoes, boots of rubber or plastic, stockings, socks of synthetic materials)

Interactions with chemicals (solvents, detergents, lubricants, bleaching agents, softening and retexturing agents etc.)
→ alteration of the local milieu (pH ↑, alkaline neutralization ↓)
→ inactivation of antifungals

Mechanic alteration, continuous compression; on the feet mainly caused by footwear, in cases of spondylopathies alteration of segmental nerves
→ ischaemia
→ hyper- and parakeratoses, onycholysis, onychodystrophy
→ disorders of trophic of skin and nail

Other local infections; preceeding or additional to those caused by yeasts and/or (mostly gram-negative) bacteria

II. Endogenous

Endocrinopathies
mostly diabetes, occasionally morbus Cushing, hypoparathyroidism, dysregulation of sexual hormones

Severe general diseases
lymphomas, leukaemias, hemoblastoses, immunopathies, viral infections, AIDS, debilitating diseases, malignomas, shock

Disorders of circulation
universal:
(mostly congenital) cardiac failures → acidosis
localized:
arterio(lo)pathies → ischaemia
phlebopathies → chronic venous stasis
lymphopathies → chronic lymphoedema

Disorders of thermoregulation
mostly acral hypothermia

III. Indirectly endogenous

Therapeutic:
corticosteroids and other hormones, particularly hormonal contraceptives, antibiotics, immunosuppressives, cytostatics, intravenous hyperalimentation, actinic effects

Accidental/habitual:
polytraumata, burns, intoxications, narcotics, above all heroin

---

infiltrative–ulcerative–suppurative–fistulating, etc.

3. *Geographic/climatic/epidemiologically.* The criterion is the origin of the organism or its endemic area or its degree of contageousness: *African histoplasmosis, North American blastomycosis, European mycoses, Extra–European mycoses*; *mycoses of tropical, subtropical* or *temperate zones*; *highly* or *marginally contagious mycoses* (*coccidioidomycosis, dermatophytoses* caused by *zoophilic* species, and *microsporia*, or *alternarioses, mucoralesmycoses*, and *dermatophytoses* caused by *anthropophilic* species, respectively).
4. *Pathogenically.* The criterion is the pathomechanism: primary or secondary, and in secondary mycoses the kind of predisposing factors.

It must be noted that the etiological terms give no information about the nosological aspects (site of infection, affected tissue, type of lesions, pathomechanism, etc.), and that the nosological terms imply nothing about the causative fungi (not even if it belongs to one of the four large groups of dermatophytes, yeasts, moulds, and biphasic fungi) because, as a consequence of the *hetero-* and *polypathogenicity* of numerous fungi, pathogens of the same species may cause different clinical lesions, and pathogens of different species may cause identical clinical lesions (*rule of nonspecificity of lesions*). The etiological terms do not therefore yield information about the nosological circumstances, and the nosological terms do not yield (particular) information about the etiology of the mycosis in question. A *candidosis*, for example, can be an infection of the skin, of mucous membranes, of viscera, of the eye, of the brain etc., and a *mycosis of the face* can be caused by *Trichophyton* or *Microsporum* species as well as by *Candida* species or by moulds; likewise, the lesions of a *candidosis* and a *dermatophytosis of the epidermis* may be clinically indistinguishable.

Clinical aspects

Of the numerous points of view from which mycoses may be classified, that of the appropriation of their pathogens to the four groups dermatophytes, yeasts, moulds, and biphasic fungi is taken as a basis here; besides this, their causal pathogenesis will be taken into consideration as far as possible since it is almost always of basic importance for their treatment.

Within the scope of this article the description of the mycoses cannot be exhaustive, but must be restricted to the most significant aspects.

Mycoses caused by dermatophytes (dermatophytoses, tinea)

EPIDERMAL DERMATOPHYTOSES, TINEA MANUUM ET PEDUM

By far the most numerous and widespread group of all mycoses are the superficial mycoses of the feet, above all the toe-webs, and the hands, mainly the palms (Fig. 7.2A). The clinical symptoms may include

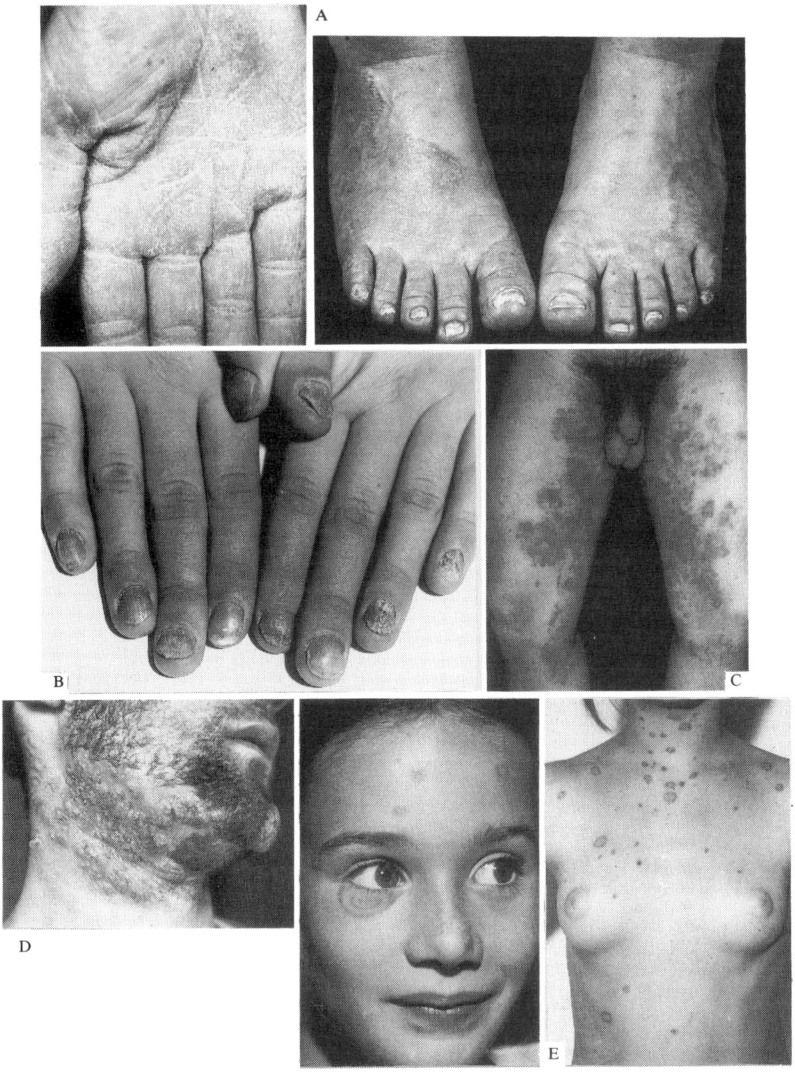

Figure 7.2. A, Dermatophytosis (*tinea*), (left) *palmaris*, (right) *pedis* (additionally onychomycosis); pathogen: *Trichophyton rubrum*. B, Onychomycosis (*tinea unguis*) manuum; pathogen: *Trichophyton rubrum*. C, Superficial dermatophytosis of hairy skin (*tinea inguino-cruralis*); pathogen: *Trichophyton rubrum* (the extraordinary extent of the mycosis is caused by a retention of heat and moisture by riding-breeches with inserts of plastic on the sites corresponding to the extent of the mycosis; the patient is a professional rider). D, Profound dermatophytosis (*tinea barbae profunda*); pathogen: zoophilic strain of *Trichophyton mentagrophytes*. E, Microsporia of face and trunk; pathogen: *Microsporum canis*; the source of infection was a cat.

reddening, scaling, hyperkeratoses, fissures, inflammation, and erosions (especially on the toe-webs), and itch. Pathogens are anthropophilic *Trichophyton* species, mainly *Trichophyton rubrum, T. interdigitale* and *Epidermophyton floccosum.* The site of the infection is the horny layer (the stratum corneum) of the skin; these fungi never attack deeper layers. Pathogenically this type of mycosis is a secondary infection; its predisposition results from a retention of heat and moisture. What essential role the latter plays can be recognized from the fact that native tribes who do not wear (occluding) footwear, do not have such mycoses.

The socio-medical significance of these mycoses results from their enormous frequency (20–25% of the world population are currently affected) on the one hand and from their refractoriness to therapy and their tendency to relapse on the other. The two latter problems result first of all from two facts: firstly, that the main factor which predisposes to the mycoses on the feet, the moist-chamber situation caused by footwear, is difficult to eliminate under conditions of civilization, and secondly, that the respective causative agents are so frequent in the human environment that there is more or less permanent exposure. This exposure leads almost inevitably to a reinfection, provided that the predisposition persists.

ONYCHOMYCOSIS, TINEA UNGUIS Onychomycoses occur in about 85% of cases on the toes, in about 12% on the fingers, and in about 3% on toes and fingers. The clinical symptoms are deformation and destruction of the nailplate which is thickened and has usually lost contact with the nailbed in its distal position.

The causative organisms are mainly *Trichophyton rubrum* and *T. interdigitale*; only occasionally, other anthropophilic *Trichophyton* species and *Epidermophyton floccosum.* Pathogenically, onychomycoses are also a standard type of secondary infection (it is not possible to produce an onychomycosis experimentally in healthy nails). The necessary predisposition on the toes results first of all from subungual scars, callosities, and hyper-parakeratoses which are the consequence of trauma or continuous pressure by the footwear; and on the hands it results predominantly from trophic disorders, which are a consequence of a mechanical alteration of the spinal nerves in cases of spondylopathies.

The socio-medical significance of onychomycoses is, despite their comparatively low frequency (only about 1% of all mycoses), relatively high, because cure rates of the disease are very low (not more than 20–25% of cases) and the appearance of the lesions is rather unaesthetic. Especially if localized on the fingers, onychomycoses are felt in general to be repulsive, often producing considerable social or occupational problems for the patient.

SUPERFICIAL MYCOSES OF HAIRY SKIN This type of dermatophytosis occurs mainly in intertriginous areas of the body: the inguinocrural, the perigenital and perianal region. The clinical manifestations consist of itchy, sometimes burning, round, oval, or polycyclic cutaneous lesions with reddening and superficial infiltration of the skin which is covered by numerous scales, papules, and pustules (Fig. 7.2D). The symptoms are pronounced at the margin of the lesions; in its central parts erosions may develop.

The causative agents are the same anthropophilic dermatophytes as on the feet. Mycoses of this type are predisposed by hyperhidrosis which may be caused by obesity, hormonal disorders (hyperthyroidism, diabetes), or conditions of the environment (occupation, geographic zone; in the Persian Gulf area, for example, between 40 and 55% of all diseases of the skin are mycoses of this type).

DEEP MYCOSES OF HAIRY SKIN This type of mycosis occurs mainly on the scalp, neck, face and, more rarely, on the extremities. The clinical lesions consist of painful, highly inflammatory, deep-reaching nodular or humpy infiltrations of skin and the subcutis which show on their surface scales, peripilar pustules, and crusts; occasionally there are abscesses (Fig. 7.2D). The causative organisms are zoo- or geophilic trichophytes: *T. mentagrophytes*, *T. tonsurans*, *T. verrucosum*, and *T. schoenleinii* and less often *Microsporum gypseum*. All these organisms are primarily pathogenic; therefore no predisposing factors are necessary for an infection. Most affected by such infections are farmers, zoo-keepers, veterinarians, and children who have contact with infected pet animals, such as guinea pigs, hamsters, and rabbits.

MICROSPORIA Microsporia are a special form of dermatophytosis. The clinical lesions are single or multiple, round, oval, or polycyclic reddened scaly lesions, occasionally with small peripilar pustules, on only lightly infiltrated skin (Fig. 7.2E). Sites of the lesions are above all the face, neck, and arms; in more than 95% of cases children are attacked, owing to the fact that children are particularly susceptible to the causative pathogen of this mycosis, *Microsporum canis*, and that they have frequent contact with the main carriers of this fungus, cats. Most (80–85%) of the infected cats are strays, but the remainder are well-kept pets. *Microsporum canis* is highly contagious for felids and certain other animals, especially when these are domesticated or kept in a zoo.

Mycoses caused by yeasts

CANDIDOSES The pathogen is *Candida albicans* in 85–95% of the cases; the remainder are due to several other *Candida* species, above all *C. tropicales*, *C. pseudotropicalis*, *C. guilliermondii*, *C. parapsilosis*, and

C. intermedia. The spectrum of clinical forms is the largest of all mycoses and can be recapitulated here, therefore, only in rough outline.

CANDIDOSES OF SKIN Localization: intertriginous areas, particularly inguinocrural, perigenital, submammary region, and big folds on the abdomen (Fig. 7.3A), and in babies the skin occluded by plastic pants and nappies (Fig. 7.3B).

Symptoms: intensive reddening of the skin concerned, densely arranged pustules, scales, debris, serocrusts, often extended erosions, itch, and burning.

The source of infection is above all the digestive tract (saliva, faeces).

Predisposing factors: hyperhidrosis, obesity, diabetes; in babies maceration of the skin by urine and faeces (often containing pathogens), together with a moist-chamber situation under air-tight nappy pants.

CANDIDA PARONYCHIA Affected almost always are the nails of the fingers; the sites of the infection are the deep-lying parts of the nail fold.

Symptoms: erythema, heavy swelling and infiltration of the nail wall (Fig. 7.3C), which is very painful; whitish or purulent secretion out of the nail fold. Alteration of the neighbouring nail matrix leads to disorders in the growth of the nail plate, mainly transversal defects.

Predisposing factors: maceration of skin, angiospasms through cold water, and local contact with carbohydrates (sugar, fruit juices, residuals of beer in rinse water, etc.). These influences are mostly occupationally caused (confectioners, workers in fruit canneries, bartenders, etc.).

CANDIDOSES OF MOUTH Affected sites: lips, mainly in the corners of the mouth, mucous membranes of mouth, tongue, and throat.

Symptoms: erythema and swelling of the mucous membranes which are covered with rather firmly adhesive blotchy or membraneous whitish furs, the so-called thrush or muguet (Fig. 7.3D).

Predisposing factors extend from a harmless transitory weakness of the immune system (common in the newborn, or in adults in cases of

Figure 7.3. A, Intertriginous submammary candidosis of skin; pathogen: *Candida albicans*; predisposing factors: obesity, diabetes. B, Cutaneous perianal candidosis in a baby; pathogen: *Candida albicans*; source of infection: faeces; predisposition: maceration of skin, moist-chamber situation in the nappy pants. C, *Candida paronychia* in a barmaid; pathogen: *Candida albicans*; predisposition: maceration of skin. D, Candidosis of the mouth, (i) in a child with a hereditary immune defect, (ii) in a patient suffering from AIDS; pathogen: *Candida albicans*. E, Chronic mucocutaneous candidosis, (i) in a child with an hereditary immune defect, (ii) in an adult suffering from an acrodermatitis enteropathica; pathogen: *Candida albicans*.

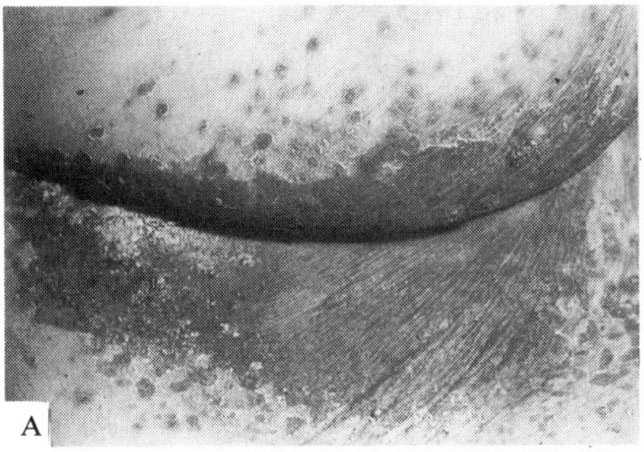

A

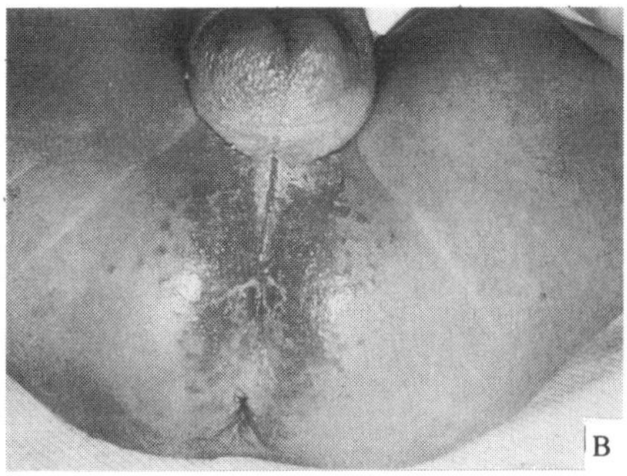

B

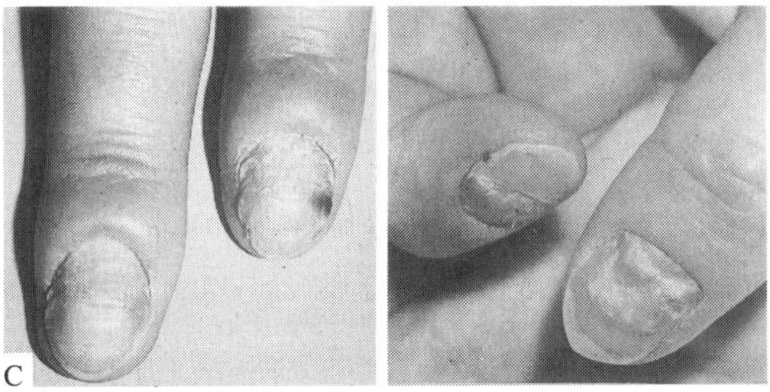

C

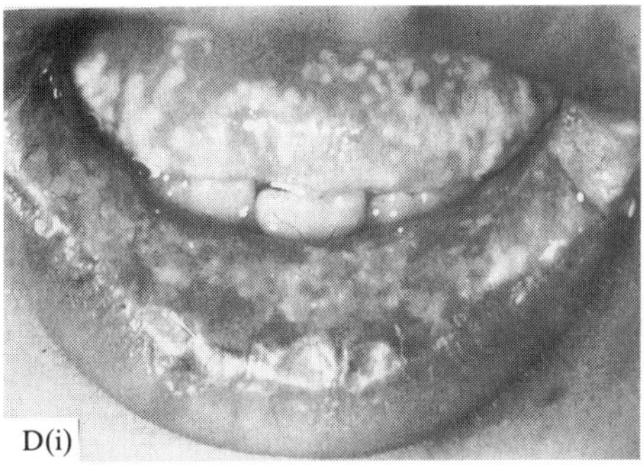

D(i)

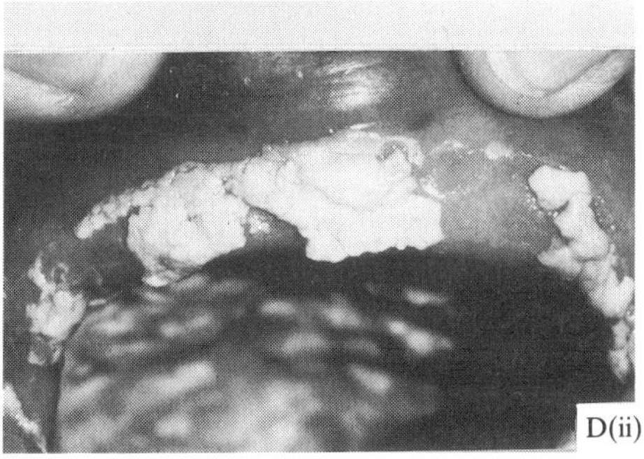

D(ii)

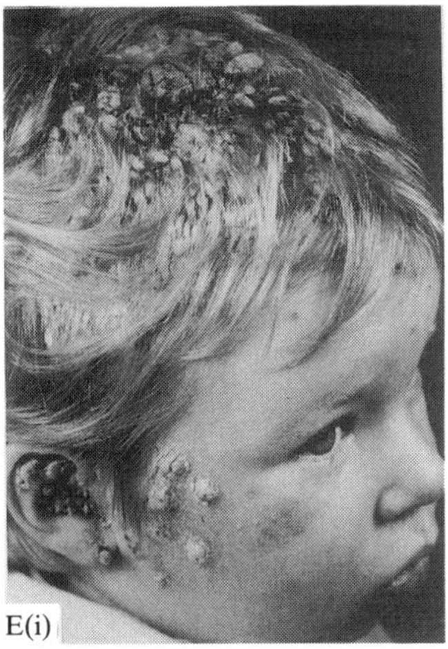

E(i)

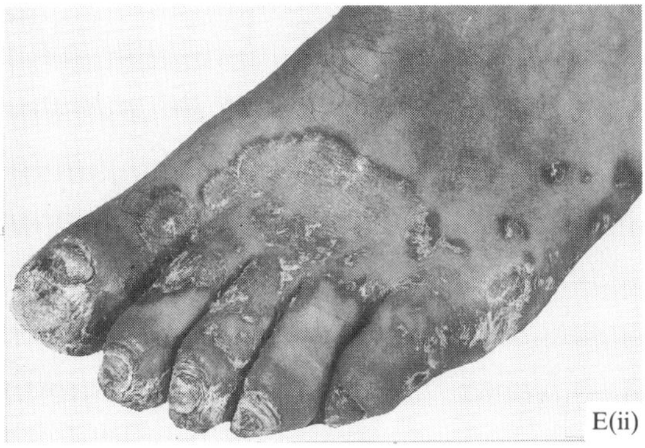

E(ii)

influenza, or as a side effect of antibiotic therapy) to severe immunopathies, such as cases of B- and T-lymphocyte defects, lymphoma, leukaemia, malignomas, etc. (Fig. 7.3E). As a rule of thumb, the degree of severity, the extent of the candidosis and its persistence are proportional to the severity of the underlying disease. A severe candidosis, therefore, is considered a *signum mali ominis*; often it is also the first symptom (a *precursor symptom* or an *indicator disease*) of a severe occult disease unnoticed until then. The best characterization of the pathogenetic relationships was given already in 1848 by the Swedish pediatrician F.Th.Berg: “Candida is a ‘little mushroom’ which attacks only a sick tree but makes this tree even sicker”.

In particularly severe cases the oral candidosis usually proceeds to the pharynx, larynx, and oesophagus, which in most cases represents the transition to an infection of internal organs.

CANDIDOSIS OF GENITALS Clinical forms: *Vulvo-vaginitis,* and *balano-posthitis candidamycetica.*

Symptoms: analogous to those of the candidosis of the mouth; additionally there is vehement pruritus and in severe cases of vulvo-vaginitis strong whitish fluor and burning.

Predisposition: the main factors have been characterized by the World Health Organization as the so-called “three p’s”: pill-porno-promiscuity. (Indeed the frequency of the candidoses in question increased exponentially after the introduction of hormonal contraceptives.)

Other predisposing factors may be pregnancy and diabetes.

CANDIDOSIS OF VISCERA Preferentially concerned organs; kidneys, liver, spleen, bone, brain, lungs.

Predisposing factors: all severe underlying diseases listed in Table 7.2.

PITYRIASIS VERSICOLOR Pathogen: *Pityrosporon ovale.*

Symptoms: numerous patches of 0.5 to 1 cm diam covered with light-brown (cafe au lait-coloured) fine scales on trunk and neck (Fig. 7.4A), only exceptionally on the arms. The skin under the scales is depigmented (the pathogen produces a metabolite (acelaic acid) which impairs melanogenesis).

Predisposing factors: above all a hereditary metabolic defect, additionally hyperhidrosis.

Pityriasis versicolor is very frequent in tropical zones (up to 40% of population); in Europe the percentage is between 3 and 5%. This mycosis has a primarily recurrent course which generally extends over years and decades. The discoloration of skin which usually persists for several months, and not seldom for years, often represents an essential problem for the persons concerned.

PITYROSPORON SEPSIS In immature newborn babies nourished intraveneously by infusions too rich in lipids, *Pityrosporon orbiculare* may cause lethal mycoses of the lungs and even generalized septic infections.

CRYPTOCOCCOSIS Pathogen: *Cryptococcus neoformans.*

Infection sources: the main habitat of *C. neoformans* is the digestive tract of birds, especially pigeons, parrots, and canaries living in the human environment. From dried bird droppings fungal cells get into the air; if they are inhaled by man, it depends on his immune system whether an infection occurs, and, if so, which type and to which degree. In healthy persons the fungal cells seem to be inactivated; in persons with only a slight deficiency of the immune system, and also in cases of local abnormalities of the lungs, (e.g. bronchiectases and cavities) a solitary localized infection develops; and only if there are severe immunopathies, above all in AIDS, extensive infections originate. The latter occur less often in the lungs (or viscera, bones, and skin) than in the central nervous system to which the pathogen has a particular affinity ("neurotropy"). The greatest part of these cryptococcoses is lethal.

Mycoses caused by moulds

The spectrum of the pathogens potentially able to cause this type of mycosis comprises several dozen species; the most important are listed in Table 7.5. The most frequent by far are *Aspergillus* species; their percentage is between 80 and 90% of all pathogenic moulds.

The common characteristic of the fungi in question, which are "opportunists" throughout, is that their pathogenic potential is particularly weak; therefore they may cause a mycosis only if the predisposing factors are particularly severe. (Such mycoses are *secondary mycoses*, as mentioned above.) These types of predisposing factors or *underlying diseases* always have an endogenous or indirectly endogenous character; their detailed list is given in Table 7.4.

The practical significance of the individual factors varies considerably; the most important are leukaemias, especially aleukaemias (25–30%), organ transplants (20–25%), lymphomas (12–15%), and solid tumors (5–7%).

With respect to the tissue/organ which can be attacked by opportunistic moulds and also to the type of lesions/mycoses which they can cause, no principal differences exist; provided that there is a relevant disposition, every individual fungus can, in principle, behave pathogenically in more or less the same fashion. In practice, however, certain organs are attacked preferentially: above all the lungs (50–80% depending on the type of underlying disease), the brain (8–15%), the kidneys (5–15%), and the liver (5–10%).

Table 7.5. Spectrum of mycoses and mycetes related to AIDS.

Mycoses	Causative organisms/saprophytes	Main target tissues	Incidence %
Dermatophytoses	anthropophilic dermatophytes: *Trichophyton rubrum, Epidermophyton floccosum* and others	skin and appendages	80–90
Candidoses	*Candida albicans, C. tropicalis*	oral cavity	70–90
	C. parapsilosis, C. guilliermondii	skin	25–30
	C. krusei and other species	vagina	20–25
		oesophagus	10–15
Torulopsidoses	*Torulopsis glabrata, T. candida*	intestinal tract	
		parasitic	1–2
		saprobic	70–90
Trichosporosis	*Trichosporon cutaneum*	systemic; mainly brain	< 1
Cryptococcosis	*Cryptococcus neoformans*	brain (lungs, skin)	5–7
Histoplasmosis			
'American'	*Histoplasma capsulatum*	lungs, lymphatic system	1 (–2)
'African'	*Histoplasma duboisii*	skin, lungs, lymph. syst.	1 (–5)
Coccidioidomycosis	*Coccidioides immitis*	lungs, brain	sporadic
Aspergillosis	*Aspergillus fumigatus, A. flavus, A. nidulans, A. glaucus, A. terreus* and other species	respiratory tract sinuses, intestinal tract, brain, liver, kidney	sporadic

Table 7.5. *Continued.*

Mycoses	Causative organisms/saprophytes	Main target tissues	Incidence %
Blastomycosis ('North American Blastomycosis')	*Blastomyces dermatitidis*	lungs, skin, bone	sporadic
Paracoccidioidomycosis ('South American Blastomycosis')	*Paracoccidioides brasiliensis*	(lungs) oral/nasal cavity gastrointestinal mucosa lymph vessels/nodes, skin	sporadic
Sporotrichosis	*Sporothrix schenckii*	skin, lymph vessels, brain	sporadic
Mycoses caused by opportunistic moulds	various species of *Fusarium*, *Paecilomyces*, *Alternaria*, *Drechslera*, *Mucor*, *Rhizopus*, *Absidia*, *Pseudallescheria*, *Penicillium* species and other moulds	various: lungs, brain bone, sinuses, skin and other tissues/organs	sporadic

In this connection it warrants separate mention that the spectrum of mycotic infections in patients suffering from AIDS is essentially different. As can be seen from Table 7.5, a great number of opportunistic moulds (and also yeasts) may cause mycoses in such cases; all these, however, occur only sporadically, whereas the majority of all severe systemic mycoses are caused by *Cryptococcus neoformans.*

Among the mycoses caused by opportunistic moulds three groups, the mycetomas, the zygomycoses and the alternarioses have a certain special status. The mycetomas are mostly caused by streptomycetes, actinomycetes, and *Nocardiae* and only exceptionally by moulds (*Acremonium*, *Phialophora* species, etc.) and, moreover, occur almost exclusively in tropical zones; they can, therefore, be omitted here, whereas the two other groups of mycoses will be discussed in more detail.

ZYGOMYCOSES Pathogens are mostly species of *Rhizopus*, *Rhizomucor*, *Mucor*, and *Absidia.* These fungi can, it is true, attack every tissue/organ in the same manner as the other members of this group of fungi; in about 70% of cases, however, they cause a nearly characteristic type of mycosis: an infection of the sinuses of the mouth and nose (Fig. 7.4B). They usually enter the *sinus maxillaris* as a result of dental treatment; the pathogenetic basis of the infection is the metabolic state of a derailed diabetes. The infection usually takes an extremely rapid course; it tends to proceed to the orbita, the base of the brain, then into the brain within only a few days, and is lethal in about two thirds of the cases.

ALTERNARIOSIS Pathogens: *Alternaria alternata* in about 95% of cases, in the remainder mainly *A. tenuissima*, *A. chartarum*, or *A. dianthicola.*

Clinical lesions: mostly solitary, occasionally two or three, and only exceptionally multiple (up to a dozen) pseudoepitheliomatous verrucous infiltrations of skin and subcutis of a size between some millimetres and about 5 centimetres (Fig. 7.4C). The lesions originate mostly from injuries of the deeper parts of the skin by wooden particles which are carriers of the pathogen on many plants. The predisposition of the patient, indispensable to an infection with this practically nonpathogenic fungus, results from a hypercorticism.

Mycoses caused by biphasic fungi

The causative pathogens are listed in Table 7.3. The blastomycoses, the coccidioidomycosis, and the histoplasmosis will be omitted here since they are not significant outside their endemic areas. Sporotrichosis occurs occasionally in Europe, and is therefore briefly outlined.

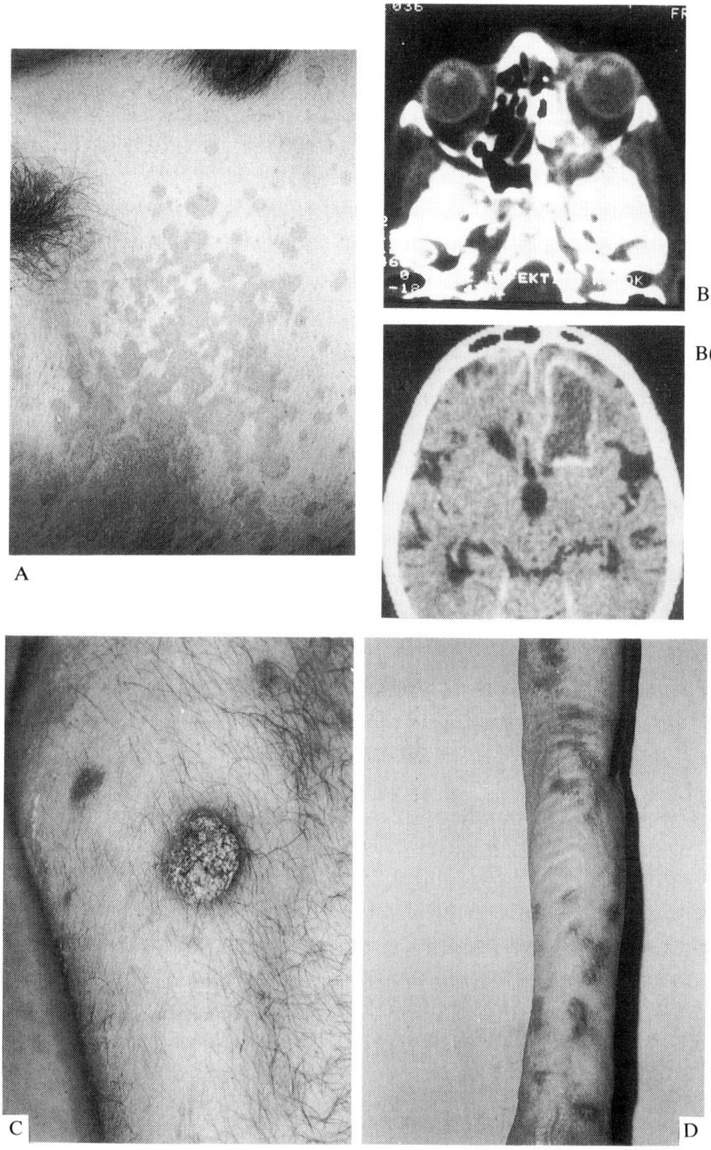

Figure 7.4. A, *Pityriasis versicolor* of trunk; pathogen: *Pityrosporon orbiculare*. B, Mycotic sinusitis, CT-pictures of, (i) sinuses, and (ii) the fronto-basis of brain; pathogen: *Rhizopus nigricans*; predisposing factor: ketoacidosis in a derailed diabetes. C, Cutaneous alternariosis, pathogen: *Alternaria alternata*; predisposing factor: hypercorticism (adenoma of hypophysis). D, Subcutaneous lymphangiopathic form of a sporotrichosis; pathogen: *Sporothrix schenckii*; mode of infection: insult with a wooden splinter.

SPOROTRICHOSIS Pathogen: *Sporothrix schenckii.* This lives in nature on plants, especially wood. The infection mostly occurs on extremities, in rare cases also in the mouth and intestines, and ensues in most cases from injury to the skin or mucous membranes by wooden particles (splinters, thorns, etc.). This leads locally to an inflammation and painful infiltration of the skin or mucous membranes and the adjacent deeper layers, to ulceration, fistulation and lymphangitis, as well as to regional lymphadenitis. The infection proceeds within the lymphatic vessels (Fig. 7.4D); when the main lymphatic glands (usually in the axilla or in the inguinal region) are reached, the progression ceases. Its further course depends on the patient's condition; if this is good, the infection may be cured or take an attenuated localized chronic course. In cases of immunodeficiency, it may transgress to (other) internal organs. In cases of AIDS, a special affinity to the brain (*neurotropy*) has been observed.

Therapy

For the topical therapy of mycoses there is a sufficiently large, or more exactly unnecessarily large, number of antifungal agents available, with adequate activity against the pertinent pathogens. That therapeutic results are altogether unsatisfactory, as pointed out above, has three principal reasons:

1. In many of the forms of mycosis under discussion, particularly onychomycoses, but also in hyperkeratotic mycoses of the plantae and toe-webs, it is extremely difficult, if not impossible, to achieve sufficient levels of the antifungal agent where this matters, i.e. at the subungual location of the pathogen or at the bottom of a homogeneous layer.
2. More than 90% of these mycoses are secondary, of which the underlying primary disorders often cannot be eliminated, since they represent consequences of civilization, are irreparable lesions caused by wear and tear, or are simply not taken into account in treatment.
3. Owing to the comparatively minor subjective symptoms usually provoked by such superficial mycoses, their treatment is often not carried out with the necessary rigour and persistence.

For the systemic treatment of superficial mycoses, mainly nail and hair infections, only four preparations are currently available, namely griseofulvin, ketoconazole, fluconazole, and itraconazole. Despite this, therapeutic results are better than those obtained under topical therapy. The reason for this is above all that: some of the mycoses, the hair mycoses caused by zoophilic or geophilic dermatophytes, are primary processes; and that a greater bioavailability of the antifungals can be achieved. Systemic therapy of secondary mycoses, above all onychomycoses, produces no markedly better results, in the long term at least, than topical treatment. This, too, is a consequence of predisposing factors, and serves to confirm their nosological significance.

The therapy of systemic mycoses is characterized by a major discrepancy between the variety and severity of the mycoses and the available therapeutic possibilities. To date there are only five more or less effective preparations available, of which one (amphotericin B) is highly nephrotoxic, while another (5-fluorcytosin) is associated with severe problems of resistance. No less than three (amphotericin B, miconazole, ketoconazole, and itraconazole) are unsatisfactory from the point of view of tissue penetration. (Better pharmacokinetic characteristics are found in the new triazole-antifungal fluconazole, but this has only recently been introduced into therapy.) These drawbacks are compounded as these few antifungal agents clearly cannot be active in a sufficient measure against all the heterogeneous fungal species responsible for systemic mycoses. A further unfavourable circumstance is that where systemic mycoses are concerned, the (principally) primary forms (especially histoplasmosis and coccidioidomycosis) pose therapeutic problems scarcely more tractable than those caused by the secondary forms (mainly aspergillosis, candidosis, and cryptococcosis).

The overall situation is consequently far from favourable, and there is an urgent need for further systematically applicable preparations. It seems too obvious to state that these should have the most favourable pharmacokinetic properties possible, be as active as possible, and well-tolerated. It should by now be equally self-evident that efficacy and tolerance should correlate inversely with the breadth of the spectrum of activity. These relationships are ignored by the demand, increasingly fashionable in recent years, for as broad a spectrum of activity as possible (the so-called "broad spectrum-antifungals"), without any regard for the underlying pharmaco-physiological and ecological relationships. Such a demand also ignores practical medical necessities, as mixed infections do not occur in systemic mycoses, furthermore the rapid diagnosis of the mycoses in question generally poses no difficulties. The trend of future development in the field of antifungal agents can thus only lead in the direction of as narrow a spectrum of activity as possible.

Summary and final remarks

The medical significance of mycoses results above all from the following facts: 85–90% of all mycoses, infections of skin and its appendages, and of mucous membranes, occur world-wide, are enormously frequent, take an extraordinarily chronic course, are highly refractory to therapy, and tend to relapse; moreover reinfections are very frequent.

The remaining mycoses are severe, primarily life-threatening infections of internal organs. The pathogenic basis of this type of mycosis is mainly in disorders of the microbial ecology caused by the use of antibacterial antibiotics or/and an impairment of immune defence mechanisms, which are predominantly therapeutically induced.

The main influences which favour the development of mycoses originate above all from the "side-effects" of civilization and modern medicine. Further progress to be expected in the latter two fields necessarily also implies a further increase in the frequency and the severity of mycoses.

8

Aerobiology and Health: the Role of Airborne Fungal Spores in Respiratory Disease

J. Lacey *AFRC Institute of Arable Crops Research, Rothamsted Experimental Station, Harpenden, Herts AL5 2JQ, UK.*

ABSTRACT Aerobiology is the study of airborne biological material, chiefly pollens and spores, and their impact on other organisms. Airborne dusts containing fungus spores are associated with a range of respiratory diseases. The nature of the disease depends on the type of spore, whether mycotoxins are produced, the nature of the exposure and the immunological reactivity of the subject. The study of airborne fungal spores enables their occurrence to be related to the incidence of respiratory symptoms in exposed individuals while the isolation of different species and collection of airborne antigens enables specific antibodies to be identified in their sera. Challenge with antigenic extracts can then confirm their involvement in disease. No role has yet been established for fungi in bronchitis and organic dust toxic syndrome although susceptibility to allergic sensitization (atopy) appears to contribute to incidence of bronchitis. However, rhinitis, asthma and extrinsic allergic alveolitis are clearly caused by fungi. Rhinitis and asthma are caused in atopic subjects by normal, everyday exposure and, sometimes, by occupational exposure to fungal spores which results in the formation of specific IgE antibodies. By contrast, allergic alveolitis is characteristically occupational and caused by intense exposure to spores from stored products, with 10^8 spores m^{-3} perhaps necessary for sensitization. IgG antibodies are often produced against organisms in the environment but their relevance to disease has to be tested. Rhinitis and asthma may be caused by species of *Cladosporium*, *Alternaria* and *Didymella*. The first two genera are common in the air by day in dry weather and the last at night and following rain, when particularly large numbers of airborne spores may be produced. Some forms of allergic alveolitis are characteristically caused by actinomycetes but examples caused by fungi include some cases of

Frontiers in Mycology. Honorary and General Lectures from the Fourth International Mycological Congress, Regensburg 1990. Edited by D.L. Hawksworth. © C · A · B International. 1991.

farmer's lung, malt worker's lung, suberosis, sequoiosis and cheese washer's lung. Fungal metabolites, such as enzymes and antibiotics produced by biotechnological processes, may also cause allergy in work environments. *Aspergillus fumigatus*, often common in moulded stored products, can also sometimes cause infection but the involvement of toxic secondary metabolities of fungi (mycotoxins) in lung disease has still to be confirmed. The relationship between aerobiology and lung disease in different environments is discussed.

Introduction

Fungal spores are almost always present in the air. Mostly, they cause no trouble but some spores can, on occasion, cause respiratory disease in some people, sometimes during outdoor exposure to the normal air spora, and occasionally related to intense occupational exposure. Aerobiology is the study of airborne microorganisms and pollens: their sources, release, dispersal and deposition, their impact on other living systems and the effect of environmental conditions on all these stages (Edmonds, 1979). This contribution reviews the use of aerobiology in the study of fungi that cause human respiratory disease.

The role of airborne fungal spores in respiratory disease

Inhalation of fungal spores, and perhaps fungal metabolites, can cause a range of respiratory symptoms depending on the species, the circumstances of exposure, and the immunological reactivity of the subject. These may range from infection to different types of non-infectious disease.

Allergic rhinitis and asthma

Allergic rhinitis and asthma occur in subjects who are constitutionally predisposed to allergy, are readily sensitized by normal, everyday exposure to airborne allergens, and are described as atopic. About 20–30% of the population are atopic and react to one or more environmental allergens. Occupational exposure to allergens often exceeds that usual out of doors, but the incidence of occupational asthma is variable, accounting for 5–15% of all asthma (Butcher and Salvaggio, 1986). Symptoms occur rapidly on exposure to the allergen with rhinitis, wheeze, and asthma caused by ventilatory obstruction but there are normally no systemic symptoms. Specific IgE antibodies are formed and skin tests with relevant allergens produce rapid wheal and flare reactions. A late asthmatic response after several hours may also sometimes occur.

Chronic bronchitis

The role of airborne spores in causing bronchitis and emphysema is uncertain and the reaction is perhaps non-specific or associated with endotoxins (lipopolysaccharides) from Gram-negative bacteria. Bronchitis is typically associated with smoking but incidence may also increase among atopic subjects, suggesting a possible allergic component.

Extrinsic allergic alveolitis (hypersensitivity pneumonitis)

Extrinsic allergic alveolitis is a T-lymphocyte-dependent granulomatous inflammatory reaction, predominantly of the peripheral gaseous exchange tissue of the lung that does not depend to any great extent on constitutional predisposition (Newman Taylor, 1987). It is caused by a range of dusts containing both fungal and actinomycete spores as well as other materials, for example, avian and animal serum proteins. It is characteristically an occupational disease and the names given to the different forms often reflect the environment in which it occurs or the source of the antigen. The classic example is farmer's lung. Allergic alveolitis is usually associated with repeated exposure to large concentrations of spores, mostly 1–5 μm in diameter, and typically exceeding 10^6 spores m^{-3} air. A minimum concentration of 10^8 may be necessary for sensitization (Rylander, 1986). Symptoms may also develop insidiously to give chronic allergic alveolitis, without previous acute symptoms, perhaps as a consequence of prolonged exposure to relatively small spore concentrations.

Acute symptoms of allergic alveolitis develop several hours after exposure to the offending dust and include chills, fever, a dry cough, malaise and, with repeated exposure, increasing breathlessness, weight loss, and, eventually, permanent lung damage. Diagnosis is based on a history of exposure to the relevant antigen, clinical, radiographic, and functional changes typical of the disease and the presence of precipitating IgG antibodies (precipitins) to the causal antigen. Other evidence may be provided by the presence of basal crepitant rales, impairment of pulmonary diffusing capacity, decreased arterial oxygen tension, a restrictive ventilation defect, increased lymphocytes and granulomatous infiltrations, and by reproducing the disease by inhalation challenge with a relevant antigen (Do Pico, 1986). Precipitins are indicative of exposure to an antigen, but their relevance to the disease has to be proved.

Organic dust toxic syndrome

Organic dust toxic syndrome, silo unloader's disease, or pulmonary mycotoxicosis shares some of the features of allergic alveolitis. Intense exposure to airborne dust is again necessary and this results in influenza-like symptoms with leukocytosis and fever, but without prior sensitization or the formation of antibodies. Respiratory symptoms may or may not occur and there are usually no radiographic changes. Also, many workers exposed at one time may be affected. Fungi, bacteria, mycotoxins, and endotoxins have all been suggested as possible causes (Do Pico, 1986). There is no evidence of the involvement of mycotoxins, but endotoxins from Gram-negative bacteria are known to produce febrile reactions on inhalation. However, neither the aerobiology of environments in which this syndrome has occurred nor the roles of the different microorganisms have been defined.

Infection

Respiratory infection by fungi is most commonly associated with *Aspergillus fumigatus*, but the nature of the disease that it produces depends, like allergy, on the immunological status of the individual. In atopic subjects, mucus plugs in the airways are commonly colonized saprophytically, causing allergic bronchopulmonary aspergillosis. In others, it may grow in cavities, such as those resulting from tuberculosis, to produce a fungus ball or aspergilloma. Invasive disease occurs when the immunological defences break down as a result of underlying disease, immunosuppressant drugs or radiation therapy. The fungus is widespread in nature so that inoculum is always readily available.

Mycotoxins

Mycotoxins are the poisonous secondary metabolites of fungi, chiefly species of *Fusarium*, *Penicillium*, and *Aspergillus*. There is evidence that they may become airborne, probably associated with airborne spores, but there is little evidence of their effects when inhaled. Ingested aflatoxins may cause liver cancer and other mycotoxins can affect other organs, or interfere with defences against infections through immune suppression or cytotoxic effects on alveolar macrophages (Gerberick *et al.*, 1984; Pier and McLoughlin, 1985).

Airborne spores in the outdoor environment

Occurrence of airborne spores

Numbers and types of fungal spores in the air change with the time of day, weather, season, geographical location, and the presence of local spore sources. These result from differences in environmental requirements for fungal growth, sporulation and spore release, on hosts and cropping practices and, together, they affect the occurrence and time of presentation of allergic symptoms.

Weather and spore release mechanisms

Fungal spores are released from their fruiting structures in many different ways. Some have active spore release mechanisms initiated either by drying or wetting. Sporophores of *Phytophthora infestans* and some hyphomycetes twist hygroscopically and shake off sporangia or conidia during rapid changes of the relative humidity in the early morning as the sun dries the dew from crop canopies. Drying causes the contraction of unevenly thickened cells in the spore or sporophore of other fungi, as in *Deightoniella torulosa*, *Zygosporium oscheoides*, and *Drechslera turcica*, increasing tension on the water until adhesion between the water molecules or between them and the cell wall breaks down and a rapidly expanding gas bubble is formed. This allows the sudden return of the cell to its original shape, jerking or catapulting off the spores. Active release mechanisms activated by wetting depend on the sudden rounding (e.g. *Conidiobolus coronatus*, *Epicoccum nigrum*, *Arthrinium cuspidatum*) or bursting (e.g. *Nigrospora sphaerica*, many ascomycetes) of turgid cells. Ballistospore and basidiospore release also require water but the exact mechanism is still not clear. The spores are shot 0.01 to 0.02 cm, immediately after the formation of a drop of liquid at the hilum end of the spore, and, in agarics and polypores, they then fall vertically from the fruiting bodies to be carried away by the wind.

Many spores are released by mechanical disturbance, especially by wind movement of leaves and litter. Wind speeds of 0.4–2.0 m s^{-1} are necessary to release spores from the conidiophores of different hyphomycetes but often they may be released in crops by winds slower than 0.5 m s^{-1} because the shaking of leaves generates sufficient acceleration to remove them and adjacent stems and leaves knock together (Bainbridge and Legg, 1976). Numbers released increase with increasing wind speed and turbulence, and with decreasing humidity. Falling raindrops also vibrate vegetation when they impact and as they spread out, initially at 70 m s^{-1}, they push a fast moving cushion of air before them that disturbs the laminar boundary layer and disperses spores into the air.

Rain has several other effects on the occurrence of airborne spores. Droplets falling on fruiting bodies or on water films on the leaf surface disperse spores in splashes. The largest droplets follow ballistic trajectories and hardly become airborne while the smallest can evaporate and allow any spores that they carry to be dispersed by wind. Besides dispersing spores, rain removes spores from the atmosphere by impaction. Washout proceeds exponentially so that only 1% of 30 μm particles but 72% of 4 μm particles remain airborne after 120 min of rain falling at 2 mm h^{-1}. Finally, rain provides water for ascospore release resulting in greatly increased numbers in the air after the rain has ceased (Harries *et al.*, 1985). This may be associated with epidemic occurrence of acute asthma following thunderstorms (Packe and Ayres, 1985).

Circadian periodicity

Different release mechanisms are associated with different times of spore dispersal. Species with mechanisms activated by drying give maximum numbers of airborne spores from 07.00 to 10.00 h (post-dawn pattern); those with spores released by mechanical disturbance give maxima from 10.00 to 16.00 h (midday pattern) or double maxima from 08.00 to 10.00 and 14.00 to 18.00 h (double peak pattern), probably when wind speeds, turbulence and convection at midday are high; species requiring water for spore release give maxima from 20.00 to 22.00 h (post-dusk pattern) or from 02.00 to 04.00 h (night pattern), the former perhaps requiring less water and activated by increasing humidity. These circadian periodicities are modified if rain occurs.

Season

Seasonal trends are often related to crop growth cycles or climatic factors. In temperate regions, airborne spores are usually fewest during winter and spring and most abundant in summer when *Cladosporium* usually predominates by day and *Sporobolomyces* by night. In tropical areas, basidiospores are most abundant during the wet season and *Cladosporium* during the following cool, dry season. Few fungi are abundant during the hot, dry season (Sreeramulu and Ramalingam, 1966).

Plant pathogenic fungi, in particular, have seasonal trends linked to crop growth cycles. For instance, in England, *Erysiphe* is most abundant in June–July when the disease is most abundant on cereals, *Ustilago* during the flowering periods of their grass hosts, and *Phytophthora infestans* in August–September when weather conditions and growth stage of the potato crop favour late blight epidemics. *Alternaria* and *Didymella* species are both most numerous close to harvest and a second peak of *Cladosporium* may result from abundant growth on cereal straws. Basidiospores become abundant during the autumn when heavy dews favour the fruiting of many agarics. Conversely, *Penicillium* and *Aspergillus* are often most abundant in cities in winter (Hamilton, 1959; Mullins *et al.*, 1976). In India, the incidence of *Cladosporium* and *Nigrospora* is correlated with similar growth stages of each of the two rice crops, but *Deightoniella torulosa* occurs on rice only in the rainy season. *Aspergillus* is present throughout the year, and *Alternaria* chiefly at the end of the hot season as the rains start. *Sphacelotheca sorghi* is associated chiefly with the earlier of the two sorghum crops, *Cercospora* with the later and *Drechslera turcica* with both. However, the occurrence of *Fusarium* is correlated only with rainfall (Sreeramulu and Ramalingam, 1966; Shenoi and Ramalingam, 1976).

Geographical location

The predominant spore types are remarkably widespread in their occurrence (Lacey, 1981). Regional differences are mainly between minor components of the air spora which tend to increase in number and variety from cooler to warmer climatic zones. *Cladosporium* is the most abundant spore type over the whole year in temperate and most tropical regions, even though it is exceeded by other spore types in some regions and seasons of the year. Concentrations of *Cladosporium* may reach 240 000 spores m^{-3} air and form 93% of the total air spora but mean daily concentrations are usually of the order of 5 000 spores m^{-3}. *Alternaria* spores are the second most abundant overall, and can exceed the concentrations of *Cladosporium* in warm dry regions. The mean daily concentration is usually only about 150 spores m^{-3}, although maxima up to 6 000 spores m^{-3} have been found over short periods. *Curvularia* and *Nigrospora* sometimes make large contributions to the air spora of tropical regions, with maxima of 4000–9 000 spores m^{-3} air, although mean daily concentrations over the year are only 50 m^{-3} or fewer. *Aspergillus* species are particularly characteristic of humid tropical regions. In Mysore, 37 *Aspergillus* species accounted for 5.6% of the colonies isolated and *Penicillium* for only 1.7% (Rati and Ramalingam, 1965), while in the United Kingdom, *Penicillium* accounts for 2.5–13% of catches and *Aspergillus* for only 0.9–3% (Hyde and Williams, 1953; Richards, 1954; Hudson, 1969). In subarctic and arctic zones, where spore production seasons are short, maximum concentrations are less than 10 000 spores m^{-3} air, and are dominated by ascospores and basidiospores (Rantio-Lehtimäki, 1977).

Local spore sources

Despite the similarity between different regions, the air spora locally can be greatly affected, even over quite short distances, by nearby sources of spores, by differences in the microenvironment, and in the levels of human activity. Many of the fungal allergens common out of doors originate from agricultural crops, especially close to harvest (e.g. *Cladosporium, Alternaria, Didymella*). Other crops and natural vegetation may be sources of other spores. M.E. Lacey (1962) found 2.6 times more spores in a valley, close to a stream, than on an exposed hill nearby; these included five times more ascospores and three times more basidiospores but only 1.4 times more conidia. Mowing, haymaking and harvesting can put vast numbers of spores into the air, giving concentrations locally up to 10^9 m^{-3} air. Such spore clouds will be diluted as they disperse downwind. Differences over a wide area may be similar to those found around Derby (Brown and Jackson, 1978). Concentrations of *Cladosporium* at eight sites within a 60 km radius ranged from 83 to 125% of the catch in the city and of *Alternaria* from 54–102%, but catches of some other spore types were only 25% of those found around the city of Derby. By contrast, catches in London were less than 50% of those 40 km north at Rothamsted (Hamilton, 1959). Similarly, catches in Cardiff were smaller than at a nearby rural site, and were especially small when winds blew from the sea (Harvey, 1967).

Allergenicity of airborne fungal spores

Hyde (1972) included representatives of every class of fungi as having been implicated in allergic reactions, but he prefaced his list with five postulates that should be satisfied before recognizing a fungus as an allergen:

1. The spore must be produced in large quantities in most years.
2. The spore must be sufficiently buoyant to become airborne.
3. The species must be widely and abundantly distributed.
4. The spore should contain an excitant of hay fever or asthma.
5. Symptoms must occur when the spores are numerous in the air.

Species listed by Hyde included many that are rarely, if at all, abundant in outdoor air and some more characteristic of indoor environments. However, there is still little information on the doses necessary for sensitization.

Indeed, the best known allergens are indeed those found most abundantly in the air, for example *Cladosporium, Alternaria, Didymella*, and *Sporobolomyces*, although it is uncertain whether numbers or volumes are more important. One grass pollen grain has the same volume as about 200

Cladosporium conidia and about 3000 *Aspergillus fumigatus* spores. In Cardiff, the average numerical concentrations of spores in the air through 1958–61, were *Cladosporium* 1204 spores m^{-3}, ascospores 544 spores m^{-3}, and basidiospores 764 spores m^{-3} air. Volume concentrations were thus *Cladosporium* 192×10^3 μ^3 m^{-3}, ascospores 61.1×10^3 μ^3 m^{-3}, and basidiospores 93×10^3 μ^3 m^{-3} air. Mean grass pollen concentrations were 16.9 grains or 289×10^3 μ^3 m^{-3} air. Thus, despite the vastly greater numbers of fungal spores, their volume is no more than that of grass pollens.

Most airborne fungal spores are 2–100 μm diam and have specific gravities of about 1. To be dispersed they must escape from the laminar boundary layer surrounding the surface on which they grow into the more turbulent air above. This is achieved through the dispersal mechanisms described earlier, and may be aided by growth high on plants. Many spores are trapped within the vegetation in which they are formed but a proportion, perhaps 5–10%, escape into the atmosphere to be dispersed for long distances. Once airborne, they obey Stokes' Law and fall with terminal velocities proportional to the squares of their radii. For many species this is less than 0.1 cm s^{-1}, but for *Alternaria* is about 0.3–0.55 cm s^{-1} and for *Erysiphe* 1.2 cm s^{-1} (Gregory, 1973). This downward movement is offset by upward air currents in turbulence and convection that can carry spores to heights of more than 2 km. Thus clouds of spores released during the previous two days and nights can be detected over the North Sea up to 600 km from the British coast. Over the sea, with increasing distance from the coast and no replenishment of spores from the ground, the clouds become more dilute and their bases eroded by the deposition of spores. Over land, deposition is offset by the addition of more spores, replacing those that were deposited. Several fungi reported as allergens are not readily dispersed through the air because their spores are produced in slimy masses better suited to rain splash dispersal. These are unlikely to become airborne in large numbers and are therefore unlikely to be important as allergens. However, some may represent the anamorphs of ascomycetes that readily become airborne. Thus, allergy attributed to *Phoma* probably results from inhalation of *Leptosphaeria* ascospores (Ganderton, 1968), to *Ascochyta* from *Didymella* ascospores, and to *Fusarium* from *Nectria* ascospores, although airborne mesospores can be produced by some *Fusarium* species (Pascoe, 1990).

The third postulate, almost a restatement of the first, was applied chiefly to vascular plants but it should not rule out the possibility of spores from local sources causing allergy in exposed individuals. Basidiospores, as those in *Ganoderma*, are likely to be more numerous close to forests and are more likely to be important allergens there (Hasnain *et al.*, 1984).

There has been a rapid development of methods that allow the detection of specific antibodies in patients' bloodstreams and the fractionation of antigenic material to permit the identification of specific allergenic components. Radio-allergosorbent tests (RAST) have been used to detect specific IgE antibodies against allergens and enzyme-linked immunosorbent assay (ELISA) has been used to detect IgG antibodies as well as to detect IgE antibody against mainly polysaccharide antigens that do not bind well to cellulose discs for RAST. Separation of antigenic components in extracts by immunoelectrophoresis has long been practised but this has been developed to allow identification of allergenic components through crossed immunoelectrophoresis (CIE) and crossed radioimmunoelectrophoresis (CRIE). Crude fungal antigen is initially electrophoresed horizontally through agarose gel and then vertically through gel containing the appropriate antifungal serum. This gives a number of precipitin arcs or cones that can either be stained or incubated with ^{125}I-anti IgE. Autoradiography on X-ray film then allows identification of specific allergenic components among the precipitins. Biochemical methods, such as sequential ammonium sulphate precipitation, DEAE ion-exchange chromatography, preparative flat-bed electrofocusing, Sephadex G-100 gel filtration, and high performance liquid chromatography (Yunginger *et al.*, 1980; Vijay *et al.*, 1985) may also be used to separate fungal antigens, and Western blotting can be used to identify and characterize allergens (Kroutil and Bush, 1987). Culture filtrates and mycelial extracts are often used to provide allergenic material for immunological tests. Their reactivity may be very similar to spore extracts but can differ with time between isolates (Burge, 1985). However, spore-specific antigens have been identified in *Alternaria* spores (Hoffman *et al.*, 1981).

Diaries recording when symptoms occur can be instructive in indicating the cause of allergy, although late asthmatic responses, several hours after exposure, may sometimes occur. Both time of day and weather should be noted and compared with spore-trap results. Such correlation of patients' records and the incidence of airborne spores helped to implicate *Didymella exitialis* in late summer asthma in Britain (Harries *et al.*, 1985). Symptoms were worst after rain when *D. exitialis* spores were numerous in the air. To obtain such correlations, a spore-trapping method must be used that will catch the spores of interest and allow time discrimination. Automatic volumetric spore traps have revolutionized our concept of the air spora, indicating both the existence and importance of the night-time damp air spora of ascospores, basidiospores, and ballistospores of Sporobolomycetaceae. These spores were not collected efficiently by gravity slide traps while settle plates were rarely exposed at an appropriate time or only grew colonies that were sterile or with an alternative spore type. Because of inefficient trapping methods, the roles of ascospores and basidiospores in asthma have not been fully evaluated, although recent publications show an increasing awareness of their importance (Hasnain *et al.*, 1984, 1985; Harries *et al.*, 1985; Santilli *et al.*, 1985).

Occupational exposure to spores out of doors

Occupational allergy out of doors is difficult to define because the boundary between everyday exposure and occupational exposure is indistinct. Recorded examples refer only to agriculture (Table 8.1). Exposure to fungal spores occurs continuously. Many of the species present in the normal air spora grow abundantly on agricultural crops, but their numbers locally can be greatly enhanced by agricultural operations, for example by hay making and harvesting. Harvesting can lead to concentrations up to 2×10^8 spores m^{-3} air near combine harvesters. Concentrations near the driver are smaller, reaching about 2×10^7 spores m^{-3}. *Cladosporium* accounted for 46–75% of the spores, *Alternaria* for 9–28%, and *Verticillium* for 1–9%. Rust diseases caused by *Puccinia* species within the crop contributed 3–4%, and smuts (*Ustilago* species) up to 10% of spores. Bacteria contributed less than 10% to the total. About 20% of a small group of British farmworkers complained they were affected by harvester dust. Symptoms ranged from irritation, through rhinitis and asthma, to allergic alveolitis; positive skin tests and precipitins were found to some fungi. Exposure of the workers can be decreased by using helmets or cabs with filtered air supplies that remove up to 98% of spores from the breathing zone (Darke *et al.*, 1976). Dust produced during haymaking is similar to that produced during cereal harvest but yeasts, particularly *Sporobolomyces*, contributed 43%, bacteria 29%, and *Cladosporium* 24% to the total spore content (J. Lacey, unpubl.).

Airborne spores indoors and occupational lung disease

Species in the air of occupational environments often differ markedly from those out of doors. In the absence of an indoor source, species present in the air will be similar to those outdoors, but there will be fewer, perhaps

Table 8.1. Fungi implicated in occupational asthma in outdoor environments.

Source of allergen	Fungus implicated	Reference
Cereal grains and straw	*Puccinia* spp.	Cadham (1924)
	Ustilago spp.	Harris (1939)
	Tilletia caries	Jiminez-Diaz *et al.* (1947)
	Verticillium lecanii	Darke *et al.* (1976)
	Aphanocladium album	
	Paecilomyces farinosus	
Reeds	*Apiospora montagnei*	Duché (1944)

Table 8.2. Fungi implicated in occupational asthma in indoor environments.

Source of allergen	Species implicated	Reference
Tomato growing	*Verticillium albo-atrum*	Davies *et al.* (1988)
Mushroom culture	*Aspergillus fumigatus*	Sakula (1967)
	Lentinus edodes	Kondo (1969)
	Pleurotus ostreatus	Zadrazil (1974), Schulz *et al.* (1974)
	Oidiodendron sp.	Olivier *et al.* (1975)
Soup manufacture	*Agaricus bisporus*	Symington *et al.* (1981)
	Boletus edulis	
Flour	*Alternaria* spp.	Klaustermeyer *et al.* (1977)
	Aspergillus spp.	
Cheese dairy	*Penicillium camembertii*	Gari *et al.* (1983)
Tobacco	*Scopulariopsis brevicaulis*	Lander *et al.* (1985)
Enzyme production (surface culture)	*Aspergillus flavus*	Izrailet and Feoktistova (1970)
	A. awamori	
Food protein culture	*Candida tropicalis*	Cornillon *et al.* (1975)
Chiropody	*Trichophyton rubrum*	Pepys (1986)

only two thirds (Burge *et al.*, 1987). Stored products form vast sources of fungal and actinomycete spores which overwhelm the small numbers from outdoor sources, but their numbers and types change with storage conditions, degree of disturbance, and ventilation. Aerobiological studies have been used to determine the causes of occupational lung diseases, the sources of antigens, and possibilities for their control. Many of the spores of storage fungi and actinomycetes from these sources have been implicated in occupational asthma or allergic alveolitis (Tables 8.2, 8.3). The fungi implicated are chiefly dry-spored, including *Aspergillus* and *Penicillium* species, producing abundant spores smaller than 5 μm. Exceptionally, allergic alveolitis may be caused by basidiospores of mushrooms which spore abundantly (e.g. *Pleurotus ostreatus*), or by slimy spored fungi (e.g. *Graphium* sp., *Aureobasidium pullulans*, *Phoma violacea*, *Acremonium* spp.), in aerosols formed by showers or humidifiers or by sawmills.

Some fungi may carry mycotoxins in their spores, and consequently possible hazards to health on inhalation. Aflatoxins, secalonic acid D, zearalenone, and trichothecenes have all been reported in airborne dusts.

The exposure of workers to fungal and actinomycete spores in occupational environments is seldom constant but changes both quantitatively and qualitatively with time and place. The way in which aerobiological studies have contributed to understanding occupational lung disease will be discussed in relation to specific environments.

The farm environment

When a crop is stored, the microflora changes from that found in the field to one typical of storage. Which species predominate is determined by storage conditions, especially the water content of the substrate. Preharvest fungi only survive in materials stored dry (water activity [a_w] about 0.65; about 12–13% water content in starchy cereal grains). As water content increases, other species grow and increasing metabolic activity causes spontaneous heating. Above 35% water content, temperatures of 65°C may occur, allowing growth of thermophilic and thermotolerant fungi and actinomycetes. These include species that cause infection (e.g. *Aspergillus fumigatus*, *Absidia corymbifera*, *Rhizomucor pusillus*), asthma (*Aspergillus fumigatus*), and allergic alveolitis (thermophilic actinomycetes, and various fungi). Most species produce abundant spores that easily become airborne when the substrate is disturbed to give up to 10^{10} spores m^{-3} air, well in excess of the concentration (10^8 m^{-3}) thought to be needed for sensitization in allergic alveolitis. The numbers and types of fungi in the air change as the microflora of the substrate changes.

Actinomycetes predominate in heated hays and grains in Britain, giving up to 10^9 spores g^{-1} dry weight and forming up to 98% of the airborne spores produced when the material is disturbed (Lacey and Lacey, 1964; Lacey, 1980; Zeitler, 1986). The remaining fungal spores usually still greatly exceed the concentrations usual in outdoor air. Actinomycete spores are the chief source of farmer's lung antigens, but fungi have sometimes been implicated in allergic alveolitis associated with hay and grain (Table 8.3). Such large concentrations occur not only when hay bales are fed to cattle, but also when grain is moved. Traditional methods of hay making in Finland, with long periods of drying in the fields on racks, decrease heating and the incidence of some actinomycetes, while *Eurotium* species (with anamorphs of the *Aspergillus glaucus* group) become abundant. Antibodies to *E. rubrum* have been found in Finnish farmer's lung patients (Terho and Lacey, 1979), and *E. rubrum* spores were more common in the air of farms with farmer's lung patients than of those without (Kotimaa *et al.*, 1987). In Canadian grain elevators, *Aspergillus* and *Penicillium* species predominate in the settled dust present (Lacey, 1980) and may contribute to respiratory disorders (Dennis, 1973). Mites may be attracted by the moulding of stored products and themselves can be potent allergens (Revsbech and Andersen, 1987; Leskinen and Klen, 1987). It is not known whether the fungi on which they feed contribute to their allergenicity.

The microflora of moist barley and maize silage stored in silos is controlled by changes in intergranular gas composition. Carbon dioxide from the respiring grain and microflora can reach concentrations sufficient to inhibit fungal growth in the bulk of the grain for most of the storage

Table 8.3. Fungi implicated in allergic alveolitis.

Source of antigen	Fungus implicated	Reference
Mouldy, heated grain	Unspecified	Ramizzini (1913) Dunner *et al.* (1946)
Oat grain	*Aspergillus fumigatus*	Yocum *et al.* (1976)
Maize grain	*A. flavus*	Patterson *et al.* (1974)
Hay	*Eurotium rubrum*	Terho & Lacey (1979)
Straw	*Aspergillus versicolor*	Rhudy *et al.* (1971)
Rabbit farm	*A. clavatus*	De Closets *et al.* (1985)
Compost	*A. fumigatus*	Vincken and Roels (1984)
Mushroom culture	*Cephalotrichum stemonitis* *Pleurotus ostreatus* *Pholiota nameko*	Gandy (1955) Schulz *et al.* (1974) Tochigi *et al.* (1982)
Malting barley	*Aspergillus fumigatus* *A. clavatus*	Vallery-Radot and Giroud (1928) Riddle *et al.* (1968)
Cheese	*Penicillium casei*	De Weck *et al.* (1969) Molina *et al.* (1971)
Sausage	*P. jensenii*	Morin *et al.* (1984) Gerault *et al.* (1984)
Apple store	*Penicillium* spp. *Alternaria alternata*	Kroidl *et al.* (1984)
Maple bark	*Cryptostroma corticale*	Emanuel *et al.* (1962)
Redwood sawdust	*Graphium* spp. *Aureobasidium pullulans*	Cohen *et al.* (1967)
Wood chips	*Rhizopus microsporus* var. *rhizopodiformis* *Penicillium* spp. *Aspergillus fumigatus* *Trichoderma viride*(?)	Belin (1980) van Assendelft *et al.* (1985) Belin (1987) Schmitz-Schumann *et al.* (1986)
Decaying timber	*Serpula pinastri*	Garfield *et al.* (1984)
Cork	*Penicillium glabrum*	Avila and Lacey (1974)
Mouldy walls	*P. casei* *Aureobasidium pullulans* *Cladosporium* sp.	Torok *et al.* (1981) Jacobs *et al.* (1986)
Shower curtain	*Phoma violacea*	Green *et al.* (1972)
Tobacco	*Aspergillus fumigatus*	Huuskonen *et al.* (1984)

Table 8.3. *Continued.*

Source of antigen	Fungus implicated	Reference
Humidifiers	*Penicillium* spp.	Solley and Hyatt (1980)
	Acremonium spp.	Patterson *et al.* (1981)
	Aspergillus fumigatus	van Assendelft (1985)
Citric acid fermentation	*Penicillium* spp.	Horejsi *et al.* (1960)
	Aspergillus fumigatus	
	A. niger	Topping *et al.* (1985)

period, although a mouldy layer usually develops at the grain surface in unsealed silos. Removal of the mouldy material can generate large concentrations of actinomycete and fungal spores, up to 10^{10} m^{-3}, creating the conditions in which organic dust toxic syndrome occurs. Once the mould is removed, subsequent heating and moulding is controlled by the rate at which grain is removed from the exposed upper surface of top-unloaded silos. At least 8 cm has to be removed daily to prevent heating and moulding. Slower removal allows colonization first by yeasts, and then in sequence *Penicillium roquefortii*, *P. rugulosum* and, with heating, *Absidia corymbifera*, *Rhizomucor pusillus* and finally *A. fumigatus* and actinomycetes (Lacey, 1980). Such moulding is reflected in the air spora and many species have the potential to cause asthma and allergic alveolitis.

Farmers are exposed to many of the same allergens as the general population; the incidence of atopy is likely to be similar and that of allergy can therefore be expected to be comparable in the two groups (Heinonen *et al.*, 1987). Bronchitis is well-known among farmers, especially among pig and poultry farmers and grain handlers, whether they smoke or not. Allergy also appears to have a role in bronchitis, as yet undefined, since atopy has an additive effect with smoking on its incidence (Terho *et al.*, 1987).

Aflatoxin concentrations up to 13000 ng m^{-3} have been found in air during the harvesting and pre-storage handling of maize infected with *Aspergillus flavus* in the southern United States (Burg and Shotwell, 1984). In such dust, aflatoxin was mostly in particles 7–11 μm in aerodynamic size (Sorensen *et al.*, 1981). Secalonic acid D, from *Penicillium oxalicum*, and zearalenone, from *Fusarium graminearum*, have also been found in settled grain dust (Ehrlich *et al.*, 1982; Palmgren, 1985).

Mushroom farms

The chief hazard on mushroom farms during the preparation and spawning of composts for the common mushroom (*Agaricus bisporus*), is allergic alveolitis caused by actinomycete spores. However, asthma from exposure to fungi is more likely in the growing houses. There are few *Agaricus* spores in the air when mushrooms are picked as 'buttons', but up to 10^6 spores m^{-3} can become airborne if the caps open. Moulds growing on the compost and on the moist timber of the boxes are also present in the air and are probably dispersed during picking. Some could contribute to asthma in atopic pickers. These include *Cephalotrichum stemonitis* (to 2.1×10^5 spores m^{-3} air), a suggested cause for mushroom worker's lung (Gandy, 1955), *A. fumigatus* (to 9×10^4 colony forming units (cfu) m^{-3} air), and *Penicillium* species (to 4×10^4 cfu m^{-3} air). The oyster mushroom (*Pleurotus ostreatus*), is grown widely in Europe and produces spores prolifically from an early stage of development, giving concentrations up to 4×10^7 spores m^{-3} air. These can cause allergic alveolitis (Cox *et al.*, 1988). Oriental mushrooms, such as *Lentinus edodes* (Shiitake), can also produce many spores and are reported to cause asthma and allergic alveolitis (Kondo, 1969).

Feed mills and maltings

Feed mills are contaminated with the same fungi as grain stores on farms, with *Aspergillus* species often predominant (Stallybrass, 1961). Mills are also subject to the same risks from mycotoxins in airborne and settled dust. Dust from groundnuts (*Arachis hypogea*) may contain up to 250–400 ng aflatoxin from *Aspergillus flavus* and *A. parasiticus* g^{-1}, with more than 8% of particles smaller than 5 μm. Baggers could inhale 0.04–2.5 μg aflatoxins in a 45 h week which could perhaps cause lung, liver, and colon cancer in exposed workers. However, this interpretation is complicated because the workers also ate groundnuts during their work (Dvořáčková, 1976; Deger, 1976; Alavanja *et al.*, 1987). Airborne dust from maize can contain up to 206 ng aflatoxins g^{-1} and give concentrations up to 107 ng m^{-3} air with 5–17% of particles smaller than 7 μm (Burg and Shotwell, 1984). Exposure to aflatoxins in the northern USA is less than in the south-east, but it has been suggested that chronic exposure over 20–30 years employment could still present a significant hazard (Zennie, 1984). Similarly, the risk of primary liver cancer in Swedish grain millers was significantly increased, perhaps through exposure to aflatoxins or pesticides, even though the overall incidence of cancer was unchanged (Alavanja *et al.*, 1987).

Poor germination of grain for malting can lead to heavy colonization by *Aspergillus clavatus*, the cause of malt-worker's lung. Turning the sprouting grain releases abundant spores into the air, making it impossible to see across the floor (Riddle *et al.*, 1968).

Food processing

Allergy to spores of *Agaricus bisporus* has been found in a soup processing factory (Symington *et al.*, 1981), but airborne fungal spores during food processing usually come from stored products. For instance, coffee beans prior to roasting usually carry fungal spores which can give up to 2.4×10^4 cfu m^{-3} air in containers when unloaded at the warehouse. They include *Eurotium* species, *Aspergillus fumigatus*, *A. flavus*, *A niger*, and *Wallemia sebi.* Exhaust ventilation decreases concentrations within the warehouse but workers emptying sacks are still heavily exposed. However, the chief allergen is castor bean dust contaminating the sacks (Thomas *et al.*, 1991).

Fungi, especially *Aspergillus*, *Penicillium* and *Mucor* species, are frequent in the air of bakeries, and flour and dried fruit are important sources. In a British bakery, up to 1.2×10^3 cfu m^{-3} air, mostly *Penicillium* species, were found where flour was weighed and up to 3.7×10^3 cfu m^{-3} air, mostly *Wallemia sebi* and yeasts, where dry fruit was handled (Crook *et al.*, 1988*c*). *Alternaria* and *Aspergillus* species, fungal amylase, and mites are possible causes of baker's asthma (Klaustermeyer *et al.*, 1977; Baur *et al.*, 1986; Revsbech and Dueholm, 1990), but the true aetiology of this condition is unknown.

Penicillium species are important in the maturation of some cheeses and sausages and grow abundantly on their surfaces. When these products are cleaned before sale, spores are dispersed to cause allergic alveolitis in exposed workers. However, exposure has rarely been quantified aerobiologically.

Forestry and forest products

Some of the earliest forms of allergic alveolitis described involving fungal exposure were in sawmills. *Cryptostroma corticale* is a pathogen of *Acer* species in North America, also occurring on *A. pseudoplatanus* in Europe. The tree is infected while growing, but the fungus continues to grow under the bark after felling, producing abundant spores that are released when the bark is stripped in the mill. Up to 1.3×10^7 cfu m^{-3} air may be produced, causing maple bark pneumonitis in mill workers (Wenzel and Emanuel, 1967). *Aureobasidium pullulans* and *Graphium* species have similarly been implicated in sequoiosis, associated with redwood sawdust (Cohen *et al.*, 1967). *A. fumigatus* may be common in Swedish sawmills, with some isolates producing tremorgenic mycotoxins (Land *et al.*, 1987). These toxins may have a role in allergic alveolitis of wood-trimmers, although no toxin has yet been demonstrated in airborne spores.

In Canada and Scandinavia, wood cut into chips is stored in piles out of doors or in bulk under cover. Spontaneous heating and moulding of the chips can occur, to develop microfloras in which *A. fumigatus*, *A. niger*, *Mucor* species, *Paecilomyces variotii*, *Penicillium* species, *Rhizopus microsporus* var. *rhizopodiformis*, *Talaromyces* species, *Trichoderma* species, *Aureobasidium pullulans*, and *Phanerochaete chrysosporium* predominate (Thornquist and Lundstrom, 1982; van Assendelft *et al.*, 1985). *Rhizopus* has been implicated in allergic alveolitis; about 50% of Swedish wood-trimmers had precipitins, mostly to *Rhizopus*, and 10–20% suffered from allergic alveolitis (Belin, 1985).

The role of fungi in the allergic alveolitis of woodworkers is controversial. Late asthmatic responses have been found in patients exposed to wood dust alone (Pickering *et al.*, 1972), but Belin (1987) and Maier *et al.*, (1981) consider that such symptoms can be caused by moulds in a similar way to those that cause sequoiosis.

During processing, bales of cork are often submerged for several hours in boiling water, after which they are stacked in small humid warehouses. Here they become heavily colonized by fungi, including *Penicillium glabrum*, *P. glandicola*, *Aphanocladium album*, *Chrysonilia sitophila* and *Mucor plumbeus*. In a Portuguese cork factory, only *P. glabrum* was abundant throughout, and produced precipitins in the sera of cork workers with suberosis. Background concentrations reached 2.6×10^7 spores m^{-3} air in the small warehouses, and 10^6 to 10^7 m^{-3} within the factory. Workers were exposed to up to 9×10^7 spores m^{-3} air, the heaviest exposures occurring when moving the mouldy cork from the warehouse; about half the spores were of *Penicillium* species (Lacey, 1973; Avila and Lacey, 1974).

Processing of plant fibres

Sugar cane bagasse, the squashed, chopped fibre that remains after sugar is extracted from sugar cane, is a good substrate for microbial growth, leaving the factory with 50% water and up to 5% sucrose. It rapidly heats in bales to temperatures above 50°C which can be maintained for up to 2–3 months (Lacey, 1974). A characteristically thermophilic microflora develops, but many of the species differ from those in hay. Actinomycetes cause bagassosis, a form of allergic alveolitis, but *Aspergillus fumigatus* is often abundant and may present a risk of aspergillosis.

Cotton milling is associated with a range of occupational lung diseases, of which byssinosis is a classic example. However, its aetiology is still obscure, although bacterial endotoxins have recently received strong support (Rylander and Snella, 1976). Fungi and actinomycetes are common in cotton mill air, and numbered up to 4.0×10^5 and 1.2×10^7 m^{-3} air respectively, in a new mill in north-east England, and

1.2 × 10^6 and 9.8 × 10^6 m^{-3} respectively in an old Lancashire mill. The predominant fungi, *Aspergillus niger*, and *Cladosporium*, *Penicillium*, and *Eurotium* species, were all most common during the early stages of processing (Lacey and Lacey, 1987).

Exposure to fungal spores in industrial environments does not necessarily originate from the materials being handled, but may also come from air conditioning systems, especially if these are humidified. Water reservoirs and moist baffle plates provide environments for the growth of fungi, bacteria (including actinomycetes), protozoa, nematodes, and even aquatic mites. *Aspergillus fumigatus* colonized the air conditioning system of a factory producing synthetic fibres (Wolf, 1969), but there is usually a mixture of fungi present, including *Aureobasidium pullulans*, *Paecilomyces lilacinus*, *Phialophora* species (including *P. hoffmannii*), *Phoma*, *Penicillium*, *Fusarium*, and *Acremonium* species (Pickering *et al.*, 1976; Bernstein *et al.*, 1983; Austwick *et al.*, 1986). However, the roles of different microorganisms in humidifier fever is uncertain. Many have been implicated on different occasions (Liebert *et al.*, 1984; Banaszak *et al.*, 1970; Edwards *et al.*, 1976; Rylander and Haglind, 1984), but aerobiological studies of the numbers and types of microorganisms and antigens in the air are often lacking and subsequent studies have often failed to confirm earlier conclusions (Longbottom, 1986; McSharry *et al.*, 1987; Finnegan *et al.*, 1987). The cause still awaits identification (Edwards, 1986). Similarly ill-defined problems are associated with moulds in houses (Samson, 1985).

Municipal waste composting

Composting decreases the bulk of municipal waste and sewage and stabilizes putrescible material before it is disposed to landfill or used for other purposes. Heating to about 55°C occurs, with microbiological changes similar to those in mushroom composts. Actinomycetes are abundant and also fungi, of which 93% can be *Aspergillus fumigatus*, giving concentrations to 8.2 × 10^6 cfu m^{-3} air (Crook *et al.*, 1988*a*). In France, the air of four composting plants contained up to 7.8 × 10^4 cfu of mesophilic fungi m^{-3} air, and up to 6.1 × 10^4 cfu of thermophilic fungi m^{-3} air. *Penicillium* species were common in the raw garbage, while *A. fumigatus* increased during composting (Boutin *et al.*, 1987). Large numbers of *A. fumigatus* were also dispersed, at a rate of 4.6 × 10^6 cfu s^{-1}, from composted sewage sludge in the USA (Millner *et al.*, 1980).

Biotechnological processes

Biotechnological processes are being used increasingly to produce nutritional proteins, enzymes and other useful metabolites and to degrade

Table 8.4. Fungal metabolites implicated in occupational asthma.

Metabolite	Source fungus	Reference
penicillin	*Penicillium chrysogenum*	Davies *et al.* (1974)
'flaviastase' (protease)	*Aspergillus niger*	Pauwels *et al.* (1978)
'brinase' (protease)	*Aspergillus oryzae*	Forsbeck and Ekenvall (1978)
cellulase	*Aspergillus niger*	Losada *et al.* (1986)
amylase	*Aspergillus* sp.	Baur *et al.* (1986)
cephalosporin	*Acremonium strictum*	Butcher and Salvaggio (1986)

wastes and recalcitrant molecules. Aerosols are produced when handling culture mats and when liquid films break or fermenters leak. Allergic alveolitis has been found both to spores of *Aspergillus niger* and to contaminating *Aspergillus* and *Penicillium* species, dispersed from the surface of citric acid fermentations (Horejsi *et al.*, 1960; Topping *et al.*, 1985). *Fusarium* and yeasts are used to produce protein and *Aspergillus* species to produce enzymes. Although there is so far no evidence of disease caused by aerosols of these fungi, fungal enzymes and other metabolites have been implicated in similar disorders during their production and also during their use in bakeries (Table 8.4). With microbial insecticides, exposure to the fungus can occur both during production and also when they are applied, as when spraying *Verticillium lecanii* in glasshouses to control whitefly.

Conclusion

Fungi are common allergens and their importance is perhaps underestimated. They are common in outdoor air, and occupational exposure frequently leads to occupational asthma and allergic alveolitis. These diseases have a long history. In the early 18th century, Ramazzini (1713) described how threshers, sifters, and measurers of grain were plagued by the dust, especially from heated and crumbling grain, with symptoms closely resembling allergic alveolitis. Such occupational exposure occurs widely in agriculture and is now recognized in many other industries, especially where stored products are handled. In the future there will need to be greater awareness of the dangers of exposure and of occupational respiratory disease during biotechnological processes.

The use of adequate sampling equipment is important. The Durham slide and settle plates have produced much useful information, but they are subject to errors of interpretation resulting from the effects of differing terminal velocities of different spores and the erratic effects of wind speed and turbulence on deposition. As a consequence the importance of ascospores and basidiospores has been neglected. Even some newer samplers are subject to errors. The Reuter centrifugal sampler is convenient to use, but the sampling rate quoted by the manufacturers is open to question and patterns of deposition, particularly with larger particles, make interpretation difficult (Clark *et al.*, 1981). The suitability of different microbiological sampling methods has been reviewed by Crook *et al.*, (1988*b*), while Topping (1988) has described immunochemical methods for detecting airborne allergens. Such methods can be used to detect even unknown allergens in the atmosphere relevant to occupational lung disease.

No threshold limit values have yet been applied to fungal and actinomycete spores. These will also need adequate sampling methods for their determination and control, greater understanding of the dose-response relationships in occupational respiratory diseases, and information as to how such relationships are affected by constitutional factors in the exposed population.

The control of airborne spores is best achieved by the adequate drying of stored products, and also by preventing aerosols through good handling practice and the design of equipment. If the release of spores or aerosols is unavoidable, ventilation must be as close as possible to the point of release to prevent wide dispersal. Personal exposure may be limited by suitable respiratory protection efficient against spores down to 1 μm diam (Lacey *et al.*, 1982). Air conditioning systems present special problems. The accumulation of dirt and stagnant water must be avoided and the whole system cleaned to a high standard, with frequent steam cleaning or disinfection to prevent colonization.

References

Alavanja, M.C.R., Malker, H. and Hayes, R.B. (1987) Occupational cancer risk associated with the storage and bulk handling of agricultural foodstuff. *Journal of Toxicology and Environmental Health* 22, 247–254.

van Assendelft, A.H.W. (1985) Asthma due to contaminated humidifiers. *Thorax* 40, 560.

van Assendelft, A.H.W., Raitio, M. and Turkia, V. (1985) Fuel chip-induced hypersensitivity pneumonitis caused by *Penicillium* species. *Chest* 87, 394–396.

Austwick, P.K.C., Davies, P.S., Cook, C.P. and Pickering, C.A.C. (1986) Comparative microbiological studies in humidifier fever. *Colloque Institut National de la Santé et de Recherche Médicale, Paris* 135, 155–164.

Avila, R. and Lacey, J. (1974) The role of *Penicillium frequentans* in suberosis. *Clinical Allergy* 4, 109–117.

Bainbridge, A. and Legg, B.J. (1976) Release of barley-mildew conidia from shaken leaves. *Transactions of the British Mycological Society* 66, 495–498.

Banaszak, E.F., Thiede, W.H. and Fink, J.N. (1970) Hypersensitivity pneumonitis due to contamination of an air conditioner. *New England Journal of Medicine* 283, 271–276.

Baur, X., Fruhmann, G., Haug, B., Rasche, B., Reiher, W., and Weiss, W. (1986) Role of *Aspergillus* amylase in baker's asthma. *Lancet* 1, 43.

Belin, L. (1985) Health problems caused by actinomycetes and moulds in the industrial environment. *Allergy* 40, *Supplement* 3, 24–29.

Belin, L. (1987) Sawmill alveolitis in Sweden. *International Archives of Allergy and Applied Immunology* 82, 440–443.

Bernstein, R.S., Sorensen, W.G., Garabrant, D., Reaux, C. and Treitman, R.D. (1983) Exposure to respirable, airborne *Penicillium* from a contaminated ventilation system: clinical, environmental and epidemiological aspects. *American Industrial Hygiene Association Journal* 44, 161–169.

Boutin, P., Torre, M. and Moline, J. (1987) Bacterial and fungal atmospheric contamination at refuse composting plants: a preliminary study. In: de Bertoldi, M., Ferranti, M.P., L'Hermite, P. and Zucconi, F. (eds), *Compost: production, quality and use.* Elsevier, London, pp. 266–275.

Brown, H.M. and Jackson, F.A. (1978) Aerobiological studies based in Derbyshire. II. Simultaneous pollen and spore sampling at eight sites within a 60 km radius. *Clinical Allergy* 8, 599–609.

Burg, W.R. and Shotwell, O.L. (1984) Aflatoxin levels in airborne dust generated from contaminated corn during harvest and at an elevator in 1980. *Journal of the Association of Official Analytical Chemists* 67, 309–312.

Burge, H.A. (1985) Fungus allergens. *Clinical Reviews of Allergy* 3, 319–329.

Burge, H.A., Chatigny, M., Feeley, J., Kreiss, K., Morey, P., Otten, J. and Peterson, K. (1987) Guidelines for assessment and sampling of saprophytic bioaerosols in the indoor environment. *Applied Industrial Hygiene* 3, R10–R16.

Butcher, B.T. and Salvaggio, J.E. (1986) Occupational asthma. *Journal of Allergy and Clinical Immunology* 78, 547–559.

Cadham, F.T. (1924) Asthma due to grain rusts. *Journal of the American Medical Association* 83, 27.

Clark, S., Lach, V and Lidwell, O.M. (1981) The performance of the Biotest RCS centrifugal air sampler. *Journal of Hospital Infection* 2, 181–186.

de Closets, F., Nastorg, C., De Bièvre, C. and Roullier, A. (1985) Sensibilisations à *Aspergillus clavatus* découvert dans un élevage de lapins. *Bulletin de la Société Française de Mycologie Médicale* 14, 315–319.

Cohen, H.I., Merigan, T.C., Kosek, J.C. and Eldridge, F. (1967) Sequoiosis: a granulomatous pneumonitis associated with redwood sawdust inhalation. *American Journal of Medicine* 43, 785–794.

Cornillon, J., Touraine, J.-L., Bernard, J.-P., Lesterlin, P. and Touraine, R. (1975) Manifestations asthmatiques chez des ouvriers préparants des protéines alimentaires dérivées du pétrole (allergie à *Candida tropicalis*?). *Revue Française d'Allergologie et d'Immunologie Clinique* 16, 17–23.

Cox, A., Folgering, H.T.M. and van Grievesen, L.J.L.D. (1988) Extrinsic allergic alveolitis caused by spores of the oyster mushroom *Pleurotus ostreatus. European Respiratory Journal* 1, 466–468.

Crook, B., Bardos, P. and Lacey, J. (1988*a*) Domestic waste composting plants as sources of airborne micro-organisms. In: Griffiths, W.D. (ed.), *Aerosols: their Generation, Behaviour and Application, Aerosol Society Second Conference.*

Aerosol Society, London, pp. 63–68.

Crook, B., Griffin, P., Topping, M.D. and Lacey, J. (1988*b*) An appraisal of methods for sampling aerosols implicated as causes of work-related respiratory symptoms. In: Griffiths, W.D. (ed.), *Aerosols: their Generation, Behaviour and Application, Aerosol Society Second Conference.* Aerosol Society, London, pp. 327–333.

Crook, B., Venables, K.M., Lacey, J., Musk, A.W. and Newman Taylor, A.J. (1988*c*) Dust exposure and respiratory symptoms in a U.K. bakery. In: Griffiths, W.D. (ed.), *Aerosols: their Generation, Behaviour and Application, Aerosol Society Second Conference.* Aerosol Society, London, pp. 341–345.

Darke, C.S., Knowelden, J., Lacey, J. and Ward, A.M. (1976) Respiratory disease of workers harvesting grain. *Thorax* 31, 294–302.

Davies, P.D.O., Jacobs, R., Mullins, J. and Davies, B.H. (1988) Occupational asthma in tomato growers following an outbreak of the fungus *Verticillium albo-atrum* in the crop. *Journal of the Society of Occupational Medicine* 38, 13–17.

Davies, R.J., Hendrick, D.J. and Pepys, J. (1974) Asthma due to inhaled chemical agents: ampicillin, benzyl penicillin, 6-amino-penicillanic acid and related substances. *Clinical Allergy* 4, 227–247.

Deger, G.E. (1976) Aflatoxin – human colon carcinogenesis. *Annals of Internal Medicine* 85, 204.

Dennis, C.A.R. (1973) Health hazards of grain storage. In: Sinha, R.C. and Muir, W.E. (eds), *Grain storage – part of a system.* Avi Publishing, Westport, Connecticut, pp. 367–387.

Do Pico, G.A. (1986) Report on diseases. *American Journal of Industrial Medicine* 10, 261–265.

Duché, M.J. (1944) A propos du "Mal des Cannes de Provence". *Recueil de Travaux de l'Institut National d'Hygiène, Paris* 2, 242–252.

Dunner, L., Hermon, R. and Bagnall, D.J.T. (1946) Pneumoconiosis in dockers dealing with grain and seeds. *British Journal of Radiology* 19, 506–511.

Dvořáčková, I. (1986) Aflatoxin inhalation and alveolar cell carcinoma. *British Medical Journal* 1, 691.

Edmonds, R.L. [ed.] (1979) *Aerobiology: The ecological systems approach.* Dowden, Hutchinson and Ross, Stroudsburg, Pennsylvania.

Edwards, J.H. (1986) The contribution of different microbes to the development of humidifier fever antigens. *Colloque Institut National de la Santé et de Recherche Médicale, Paris* 135, 171–178.

Edwards, J.H., Griffiths, A.J. and Mullins, J. (1976) Protozoa as sources of antigen in 'humidifier fever'. *Nature, London* 264, 438–439.

Ehrlich, D.K., Lee, L.S., Ciegler, A. and Palmgren, M.S. (1982) Secalonic acid D: natural contaminant of corn dust. *Applied and Environmental Microbiology* 44, 1007–1008.

Emanuel, D.A., Lawton, B.R. and Wenzel, F.J. (1962) Maple-bark disease: pneumonitis due to *Coniosporium corticale. New England Journal of Medicine* 266, 333–337.

Finnegan, M.J., Pickering, C.A.C., Davies, P.S., Austwick, P.K.C. and Warhurst, D.C. (1987) Amoebae and humidifier fever. *Clinical Allergy* 17, 235–242.

Forsbeck, M. and Ekenvall, L. (1978) Respiratory hazards from proteolytic enzymes. *Lancet* 2, 524–525.

Ganderton, M.A. (1968) *Phoma* in the treatment of seasonal allergy. *Acta allergologica* 23, 173.
Gandy, D.G. (1955) *Stysanus stemonitis* in mushroom houses. *Mushroom Growers' Association Bulletin* 64, 550–552.
Garfield, A., Stone, C., Holmes, P. and Tai, E. (1984) *Serpula pinastri* – a wood decay fungus causing allergic alveolitis. *Australian and New Zealand Journal of Medicine* 14 (4, *Supplement* 2), 552–553.
Gari, M., Lavaud, F. and Pinon, J.M. (1983) Asthme a *Penicillium candidum* avec presence de precipitines chez un ouvrier fromager. *Bulletin de la Société Française de Mycologie Médicale* 12, 135–137.
Gerault, C., Morin, O., Dupas, D. and Vermeil, C. (1984) Les alveolites allergiques extrinseques dans l'industrie de fabrication des saucissons. *Archives des Maladies Professionelles de Médicine du Travail et de Securité Sociale* 45, 608–610.
Gerberick, G.F., Sorenson, W.G. and Lewis, D.M. (1984) The effects of T-2 toxin on alveolar macrophage function in vitro. *Environmental Research*, 33, 246–260.
Green, W.F., Harvey, H.P.B. and Blackburn, C.R.B. (1972) A shower curtain fungus (*Phoma violacea*) and allergic alveolitis. *Australian and New Zealand Journal of Medicine* 2, 310.
Gregory, P.H. (1973) *Microbiology of the Atmosphere.* 2nd edn. Leonard Hill, Aylesbury.
Hamilton, E.D. (1959) Studies in the air spora. *Acta allergologica* 13, 143–175.
Harries, M.G., Lacey, J., Tee, R.D., Cayley, G.R. and Newman Taylor, A.J. (1985) *Didymella exitialis* and late summer asthma. *Lancet* 1, 1063–1066.
Harris, L.H. (1939) Allergy to grain rusts and smuts. *Annals of Allergy* 10, 327–336.
Harvey, R. (1967) Air spora studies at Cardiff. I. *Cladosporium. Transactions of the British Mycological Society* 50, 479–495.
Hasnain, S.M., Newhook, F.J., Wilson, J.D. and Corbin, J.B. (1984) First report of *Ganoderma* allergenicity in New Zealand. *New Zealand Journal of Science* 27, 261–267.
Hasnain, S.M., Wilson, J.D., Newhook, F.J. and Segedin, B.P. (1985) Allergy to basidiospores: immunologic studies. *New Zealand Medical Journal* 98, 393–396.
Heinonen, O.P., Horsmanheimo, M., Vohlonen, I. and Terho, E.O. (1987) Prevalence of allergic symptoms in rural and urban populations. *European Journal of Respiratory Disease* 71, *Supplement* 152, 64–72.
Hoffman, D.R., Kozak, P.P., Gillman, S.A., Cummins, L.H. and Gallup, J. (1981) Isolation of spore specific allergens from *Alternaria. Annals of Allergy* 46, 310–315.
Horejsi, M., Sach, J., Tomsikova, A. and Mecl, A. (1960) A syndrome resembling farmer's lung in workers inhaling spores of aspergillus and penicillia moulds. *Thorax* 15, 212–217.
Hudson, H.J. (1969) Aspergilli in the air spora at Cambridge. *Transactions of the British Mycological Society* 52, 153–159.
Huuskonen, M.S., Husman, K., Jarvisalo, J., Korhonen, O., Kotimaa, M., Kuusela, T., Nordman, H., Zitting, A. and Mantyjarvi, R. (1984) Extrinsic allergic alveolitis in the tobacco industry. *British Journal Industrial Medicine* 41, 77–83.
Hyde, H.A. (1972) Atmospheric pollen and spores in relation to allergy. I. *Clinical Allergy* 2, 153–179.

Hyde, H.A. and Williams, D.A. (1953) The incidence of *Cladosporium herbarum* in outdoor air at Cardiff, 1949–50. *Transactions of the British Mycological Society* 36, 260–266.

Izrailet, L.I. and Feoktistova, R.P. (1970) Validation of sanitary zones for certain enzyme factories. *Gigiena i Sanitariya* 35, 412–414.

Jacobs, R.L., Thorner, R.E., Holcomb, J.R., Schwietz, L.A. and Jacobs, F.O. (1986) Hypersensitivity pneumonitis caused by *Cladosporium* in an enclosed hot-tub area. *Annals of Internal Medicine* 105, 204–206.

Jiminez-diaz, C., Lahoz, C. and Canto, G. (1947) The allergens of mill dust: asthma in millers, farmers and others. *Annals of Allergy* 5, 519–525.

Klaustermeyer, W.B., Bardana, E.J. and Hale, F.C. (1977) Pulmonary hypersensitivity to *Alternaria* and *Aspergillus* in baker's asthma. *Clinical Allergy* 7, 227–233.

Kondo, T. (1969) Case of bronchial asthma caused by the spores of *Lentinus edodes* (Berg.) Sing. *Japanese Journal of Allergy* 18, 81.

Kotimaa, M.H., Terho, E.O. and Husman, K.H. (1987) Airborne moulds and actinomycetes in the work environment of farmers. *European Journal of Respiratory Disease* 71, *Supplement* 152, 91–100.

Kroidl, R.F., Amthor, M., Durkes, U., Freitag, V. and Hain, E. (1984) Interstitielle Pneumonie durch Schimmelpilze aus einem Obstkuhlhaus. *Praxis und Klinik der Pneumologie* 38, 434–437.

Kroutil, L.A. and Bush, R.K. (1987) Detection of *Alternaria* allergens by Western blotting. *Journal of Immunology and Clinical Immunology* 80, 170–176.

Lacey, J. (1973) The air spora of a Portuguese cork factory. *Annals of Occupational Hygiene* 16, 223–230.

Lacey, J. (1974) Moulding of sugar-cane bagasse and its prevention. *Annals of Applied Biology* 76, 63–76.

Lacey, J. (1980) The microflora of grain dusts. In: Dosman, J.A. and Cotton, D.J. (eds), *Occupational Pulmonary Disease – Focus on Grain Dust and Health.* Academic Press, New York, pp. 417–440.

Lacey, J. (1981) The aerobiology of conidial fungi. In: Cole, G.T. and Kendrick, W.B. (eds), *The Biology of Conidial Fungi.* Vol. 1. Academic Press, New York, pp. 373–416.

Lacey, J. and Lacey, M.E. (1964) Spore concentrations in the air of farm buildings. *Transactions of the British Mycological Society* 47, 547–552.

Lacey, J. and Lacey, M.E. (1987) Micro-organisms in the air of cotton mills. *Annals of Occupational Hygiene* 31, 1–19.

Lacey, J., Nabb, S. and Webster, B.T. (1982) Retention of actinomycete spores by respirator filters. *Annals of Occupational Hygiene* 25, 351–363.

Lacey, M.E. (1962) The summer air spora of two contrasting adjacent rural sites. *Journal of General Microbiology* 29, 485–501.

Land, C.J., Hult, K., Fuchs, R., Hagelberg, S. and Lundstrom, H. (1987) Tremorgenic mycotoxins from *Aspergillus fumigatus* as a possible occupational health problem in sawmills. *Applied and Environmental Microbiology* 53, 787–790.

Lander, R., Jepsen, J.R. and Gravesen, S. (1985) Late asthmatic reactions caused by microfungi in the tobacco industry. *Ugeskrift for Læger* 147, 2918–2919.

Leskinen, L. and Klen, T. (1987) Storage mites in the work environment of farmers.

European Journal of Respiratory Disease 71, *Supplement* 152, 101–114.

Liebert, C.A., Hood, M.A., Deck, F.H., Bishop, K. and Flaherty, D.K. (1984) Isolation and characterization of a new *Cytophaga* species implicated in a work-related lung disease. *Applied and Environmental Microbiology* 936–943.

Longbottom, J.L. (1986) Antigen inter-relationships in outbreaks of humidifier fever. *Colloque Institut de la Santé et de Recherche Médicale, Paris* 135, 165–170.

Losada, E., Hinojosa, M., Moneo, I., Dominguez, J., Gomez, M.L.D. and Ibanez, M.D. (1986) Occupational asthma caused by cellulase. *Journal of Allergy and Clinical Immunology* 77, 635–639.

McSharry, C., Anderson, K. and Boyd, G. (1987) Serological and clinical investigation of humidifier fever. *Clinical Allergy* 17, 15–22.

Maier, A., Bronner, J., Orion, B., Wissler, J.C. and Hammann, M. (1981) Bronchopneumopathies des ouvriers du bois et sensibilisation aux moissisures. *Revue Française d'Allergologie* 21, 73–78.

Millner, P.D., Bassett, D.A. and Marsh, P.B. (1980) Dispersal of *Aspergillus fumigatus* from sewage sludge compost piles subjected to mechanical agitation in open air. *Applied and Environmental Microbiology* 39, 1000–1009.

Molina, C., Aiache, J.M., Brun, J. and Cheminat, J.C. (1971) Pneumopathies à precipitines. *Revue Française d'Allergologie* 11, 141–164.

Morin, O., Gerault, C., Gari, M., Vermeil, C. and Etrillard, S. (1984) Manifestations allergique a *Penicillium* dans une usine de fabrication de saucissons. Étude clinique, serologique, epidemiologique. *Bulletin de la Société Française de Mycologie Medicale* 13, 403–407.

Mullins, J., Harvey, R. and Seaton, A. (1976) Sources and incidence of airborne *Aspergillus fumigatus* (Fres.). *Clinical Allergy* 6, 209–217.

Newman Taylor, A.J. (1987) The lung and the work environment. In: Kay, A.B. and Goetzl, E.J. (eds), *Current Perspectives in the Immunology of Respiratory Disease.* Churchill Livingstone, Edinburgh, pp. 55–67.

Olivier, J.M., Pichot, E., Hocquet, P., Oury, M. and Durand, R. (1975) Étude de la microflore fongique des champignonnières en fonction des postes de travail. *Bulletin de la Société Française de Mycologie Médicale* 4, 65–67.

Packe, G.E. and Ayres, J.G. (1985) Asthma outbreak during a thunderstorm. *Lancet* 2, 199–204.

Palmgren, M.S. (1985) Microbial and toxic constituents of grain dust and their health implications. In: Lacey, J. (ed.), *Trichothecenes and other Mycotoxins.* John Wiley, Chichester, pp. 47–57.

Pascoe, I.G. (1990) *Fusarium* morphology II: Experiments on growing conditions and dispersal of mesoconidia. *Mycotaxon* 37, 161–172.

Patterson, R., Sommers, H. and Fink, J.N. (1974) Farmer's lung following inhalation of *Aspergillus flavus* growing in mouldy corn. *Clinical Allergy* 4, 79–86.

Patterson, R., Fink, J.N., Miles, W.B., Basich, J.E., Schleuter, D.B., Tinkelman, D.G. and Roberts, M. (1981) Hypersensitivity lung disease presumptively due to *Cephalosporium* in homes contaminated by sewage flooding or humidifier water. *Journal of Allergy and Clinical Immunology* 68, 128–132.

Pauwels, R., Devos, M., Callens, L. and van der Straeten, M. (1978) Respiratory hazards from proteolytic enzymes. *Lancet* 1, 669.

Pepys, J. (1986) Occupational allergic lung disease caused by organic agents. *Journal of Allergy and Clinical Immunology* 78, 1058–1062.

Pickering, C.A.C., Batten, J.C. and Pepys, J. (1972) Asthma due to inhaled wood dusts - Western red cedar and Iroko. *Clinical Allergy* 2, 213–218.
Pickering, C.A.C., Moore, W.K.S., Lacey, J., Holford-Strevens, V.C. and Pepys, J. (1976) Investigation of a respiratory disease associated with an air-conditioning system. *Clinical Allergy* 6, 109–118.
Pier, A.C. and McLoughlin, M.E. (1985) Mycotoxic suppression of immunity. In: Lacey, J. (ed.), *Trichothecenes and other Mycotoxins.* John Wiley, Chichester, pp. 507–519.
Ramazzini, B. (1713) *De morbis artificum diatriba.* [Translated by W.C. Wright.] Hafner, New York.
Rantio-Lehtimäki, A. (1977) Research on airborne fungus spores in Finland. *Grana* 16, 163–165.
Rati, E. and Ramalingam, A. (1965) Airborne aspergilli at Mysore. *Aspects of Allergy and Applied Immunology* 9, 139–149.
Revsbech, P. and Andersen, G. (1987) Storage mite allergy among grain elevator workers. *Allergy* 42, 423–429.
Revsbech, P. and Dueholm, M. (1990) Storage mite allergy among bakers. *Allergy* 45, 204–208.
Rhudy, J., Burrell, R.J. and Morgan, W.K.C. (1971) Yet another cause of allergic alveolitis. *Scandinavian Journal Respiratory Disease* 52, 177–180.
Richard, M. (1954) Atmospheric mold spores in and out of doors. *Journal of Allergy* 25, 429–439.
Riddle, H.F.V., Channell, S., Blyth, W., Weir, D.M., Lloyd, M., Amos, W.M.G. and Grant, I.W.B. (1968) Allergic alveolitis in a malt worker. *Thorax*, 23, 271–280.
Rylander, R. (1986) Lung diseases caused by organic dusts in the farm environment. *American Journal of Industrial Medicine* 10, 221–227.
Rylander, R. and Haglind, P. (1984) Airborne endotoxins and humidifier disease. *Clinical Allergy* 14, 109–112.
Rylander, R. and Snella, M.-C. (1976) Acute inhalation toxicity of cotton plant dusts. *British Journal of Industrial Medicine* 33, 175–180.
Rylander, R., Haglind, P., Lundholm, M., Mattsby, I. and Stenqvist, K. (1978) Humidifier fever and endotoxin exposure. *Clinical Allergy* 8, 511–516.
Sakula, A. (1967) Mushroom worker's lung. *British Medical Journal* 3, 708–710.
Samson, R.A. (1985) Occurrence of moulds in modern living and working environments. *European Journal of Epidemiology* 1, 54–61.
Santilli, J., Rockwell, W.J. and Collins, R.P. (1985) The significance of the spores of the basidiomycetes (mushrooms and their allies) in bronchial asthma and allergic rhinitis. *Annals of Allergy* 55, 469–471.
Schmitz-Schumann, M., Costabel, U., Krempl-Lambrecht, L. and Matthys, H. (1986) Exogen allergic alveolitis (EAA) to mouldy wood. *Abstracts of the 3rd International Conference on Aerobiology, Basel, Switzerland.* International Aerobiology Association, Basel, p. 95.
Schultz, K.H., Hausen, B.M. and Noster, U. (1974) Allergy to spores of *Pleurotus florida Lancet* 1, 625.
Shenoi, M.M. and Ramalingam, A. (1976) Air spora of a sorghum field at Mysore. *Journal of Palynology* 12, 43–54.
Solley, G.O. and Hyatt, R.E. (1980) Hypersensitivity pneumonitis induced by *Penicillium* species. *Journal of Allergy and Clinical Immunology* 65, 65–70.

Sorenson, W.G., Simpson, J.P., Peach, M.J., Thedell, T.D. and Olenchock, S.A. (1981) Aflatoxin in respirable corn dust particles. *Journal of Toxicology and Environmental Health* 7, 669–672.
Sreeramulu, T. and Ramalingam, A. (1966) A two-year study of the air spora of a paddy field near Visakhapatnam. *Indian Journal of Agricultural Science* 36, 111–132.
Stallybrass, F.C. (1961) A study of *Aspergillus* spores in the atmosphere of a modern mill. *British Journal of Industrial Medicine* 18, 41–46.
Symington, I.S., Kerr, J.W. and McLean, D.A. (1981) Type 1 allergy in mushroom soup processors. *Clinical Allergy* 11, 43–47.
Terho, E.O. and Lacey, J. (1979) Microbiological and serological studies of farmer's lung in Finland. *Clinical Allergy* 9, 43–52.
Terho, E.O., Husman, K. and Vohlonen, I. (1987) Prevalence and incidence of chronic bronchitis and farmer's lung with respect to age, sex, atopy and smoking. *European Journal of Respiratory Disease* 71, *Supplement* 152, 19–28.
Thomas, K.E., Trigg, C.J., Bennett, J.B., Baxter, P.J., Topping, M., Lacey, J., Crook, B., Whitehead, P. and Davies, R.J. (1991) Factors relating to the development of respiratory symptoms in coffee process workers. *British Journal of Industrial Medicine*, in press.
Thornqvist, T. and Lundstrom, H. (1982) Health hazards caused by fungi in stored wood chips. *Forest Products Journal* 32 (11/12), 29–32.
Tochigi, T., Nakazawa, T., Dobashi, K., Inazawa, M., Fueki, R. and Kobayashi, S. (1982) A case of hypersensitivity pneumonitis due to inhalation of spores of *Pholiota nameko*. *Japanese Journal of Thoracic Disease* 20, 1026–1031.
Topping, M.D. (1988) Detection and measurement of airborne allergens. In: Griffiths, W.D. (ed.), *Aerosols: Their generation, behaviour and application.* Aerosol Society, London, pp. 335–340.
Topping, M.D., Scarisbrick, D.A., Luczynska, C.M., Clarke, E.C. and Seaton, A. (1985) Clinical and immunological reactions to *Aspergillus niger* among workers at a biotechnology plant. *British Journal of Industrial Medicine* 42, 312–318.
Torok, M., de Weck, A.L. and Scherrer, M. (1981) Allergische Alveolitis infolge Verschimmelung der Schlafzimmerwand. *Schweizerische medizinische Wochenschrift* 111, 924–929.
Vallery-Radot, P. and Giroud, P. (1928) Sporomycose des pelleteurs de grains. *Bulletin et Mémoires de la Société Médicale des Hôpitaux de Paris* 52, 1632–1645.
Vijay, H.M., Young, N.M. Jackson, G.E.D., White, G.P. and Bernstein, I.L. (1985) Studies in *Alternaria* allergens. V. Comparative biochemical and immunological studies of three isolates of *Altenaria tenuis* cultured on synthetic media. *International Archives of Allergy and Applied Immunology* 78, 37–42.
Vincken, W. and Roels, P. (1984) Hypersensitivity pneumonitis due to *Aspergillus fumigatus* in compost. *Thorax* 39, 74–75.
de Weck, A.L., Gutersohn, J. and Bütikofer, E. (1969) La maladie des laveurs de fromage ("Käsewascherkrankheit"): une forme particulière du syndrome de poumon du fermier. *Schweizerische medizinische Wochenschrift* 99, 872–876.
Wenzel, F.J. and Emanuel, D.A. (1967) The epidemiology of maple bark disease. *Archives of Environmental Health* 14, 385–389.
Wolf, F.T. (1969) Observations on an outbreak of pulmonary aspergillosis.

Mycopathologia et Mycologia applicata 38, 359–361.

Yocum, M.W., Saltzman, A.R., Strong, D.M., Donaldson, J.C., Ward, G.W., Walsh, F.M., Cobb, O.M. and Elliott, R.C. (1976) Extrinsic allergic alveolitis after *Aspergillus fumigatus* inhalation: evidence of a Type IV immunologic pathogenesis. *American Journal of Medicine* 61, 939–945.

Yunginger, J.W., Jones, R.T., Nesheim, M.E. and Geller, M. (1980) Studies on *Alternaria* allergens. III. Isolation of the major allergenic fraction (ALT-1). *Journal of Allergy and Clinical Immunology* 66, 138–147.

Zadrazil, F. (1974) *Pleurotus*-sporen als Allergene. *Naturwissenschaften* 61, 456–457.

Zeitler, M.H. (1986) Staub-, Keim-, und Schad-gasgehalt in der Pferde-stallluft, unter besondere Berucksichtigung der FLH (Farmer's hay lung) -Antigene. *Tierärztliche Umschau* 41, 839–845.

Zennie, T.M. (1984) Identification of aflatoxin B_1 in grain elevator dusts in central Illinois. *Journal of Toxicology and Environmental Health*, 13, 589–593.

9

Lichens and Man

D.H.S. Richardson, *School of Botany, Trinity College, University of Dublin, Ireland.*

ABSTRACT Throughout the world, man's activities are affecting the lichen flora directly or indirectly. Gaseous air pollutants have led to major losses of sensitive lichen species although some recolonization has begun where clean-air legislation has been implemented to reduce pollution levels. The widespread destruction of woodland and wetland habitats by man has more than offset any opportunities provided by new substrata such as vulcanized rubber and plastics that have become available in recent years. The combination of habitat destruction and air pollution damage has resulted in a reduced diversity and abundance of lichens in many areas. This adversely affects the interesting micro- and macrofauna associated with these plants. An important reason for conserving lichens is that they are valuable for monitoring both radionuclides and a range of other substances in the environment. With respect to the direct effect of man on lichens, the major use of lichens, amounting to several thousand tons annually, is for the manufacture of extracts incorporated into perfumes, soaps and aftershave lotions. A large amount is also collected for use in wreaths and architects' models. At the moment these activities do not pose as serious a threat to lichen communities as did the collection of lichens for dyeing purposes in the past. The increasing awareness by politicians and legislators of the value of lichens for assessing environmental contamination leads to some hope that more effort will be made in future to conserve the habitats in which lichens thrive, and to ensure that air pollution is controlled.

Introduction

International Mycological Congresses provide the opportunity for mycologists to assess progress in areas of the subject, such as lichenology,

Frontiers in Mycology. Honorary and General Lectures from the Fourth International Mycological Congress, Regensburg 1990. Edited by D.L. Hawksworth.

which are often regarded as peripheral to mainstream mycology. It is pleasing therefore to report that lichenology is thriving, and in the past few years the number of available lichen text books, both elementary (e.g. Laundon, 1986), college-level (Hawksworth and Hill, 1984) and advanced (Lawrey, 1984; Kershaw, 1985; Galun, 1988) has increased. There are new identification manuals (e.g. Dobson, 1991; Galloway, 1985; Swinscow and Krog, 1988), and a new guide to the lichen floras of the world (Hawksworth and Ahti, 1990). Finally, the standard of illustration in A-4-sized chartlets and field guides has dramatically improved (Dalby, 1981, 1987; Jahns, 1983; Vitt *et al.*, 1988; Wirth, 1987). While symposia and keynote lectures on lichens were included in early International Mycological Congresses (Smith, 1978), the organizers of the present Congress have allocated lichenology a substantial segment of the programme. The main lichen symposia at this Congress are: lichen taxonomy and systematics; lichenicolous and fungicolous fungi; morphogenesis in ascomycetes (including lichenized taxa); distribution of lichens and fungi in the Southern Hemisphere; the ecological role of lichens. Lichenologogical papers also form part of the symposia on chemotaxonomy II and uptake of heavy metals and radionuclides. The absence of a session on the physiology of lichens reflects the fact that research funding in this area is difficult to obtain, often being regarded as of 'low national priority'. Much still needs to be learned about the lichen symbiosis which would assist in interpreting data from the more applied aspects of lichenology such as pollution studies and lichenometry (Lock *et al.*, 1979; Nash and Wirth, 1988).

My aim is not to summarize the presentations within the above symposia, but rather to provide information on the topic of lichens and man which may be used to hold student interest during the lectures on lichens that round out courses in mycology or plant pathology.

Traditional uses of lichens by man

Perhaps I should first remind you that lichens grow from extreme low-tide level as on the seashore in western England to the top of the highest mountains as in Norway, and from the Canadian high arctic to the tropics. Within particular human populations, lichens have been used from the cradle for the treatment of thrush (Smith, 1921) through adult life as decoration, as in New Guinea (Fig. 9.1), to the grave, as in central Europe (Fig. 9.2). In the latter context, lichens have the advantage that unlike flowers they last for many weeks on the graves. Over 2000 tonnes of *Cladonia stellaris* are picked annually in Scandinavia and exported to Central Europe for this purpose. They are used elsewhere as trees on architects' models (Kauppi, 1979). Lichens have also featured in poetry,

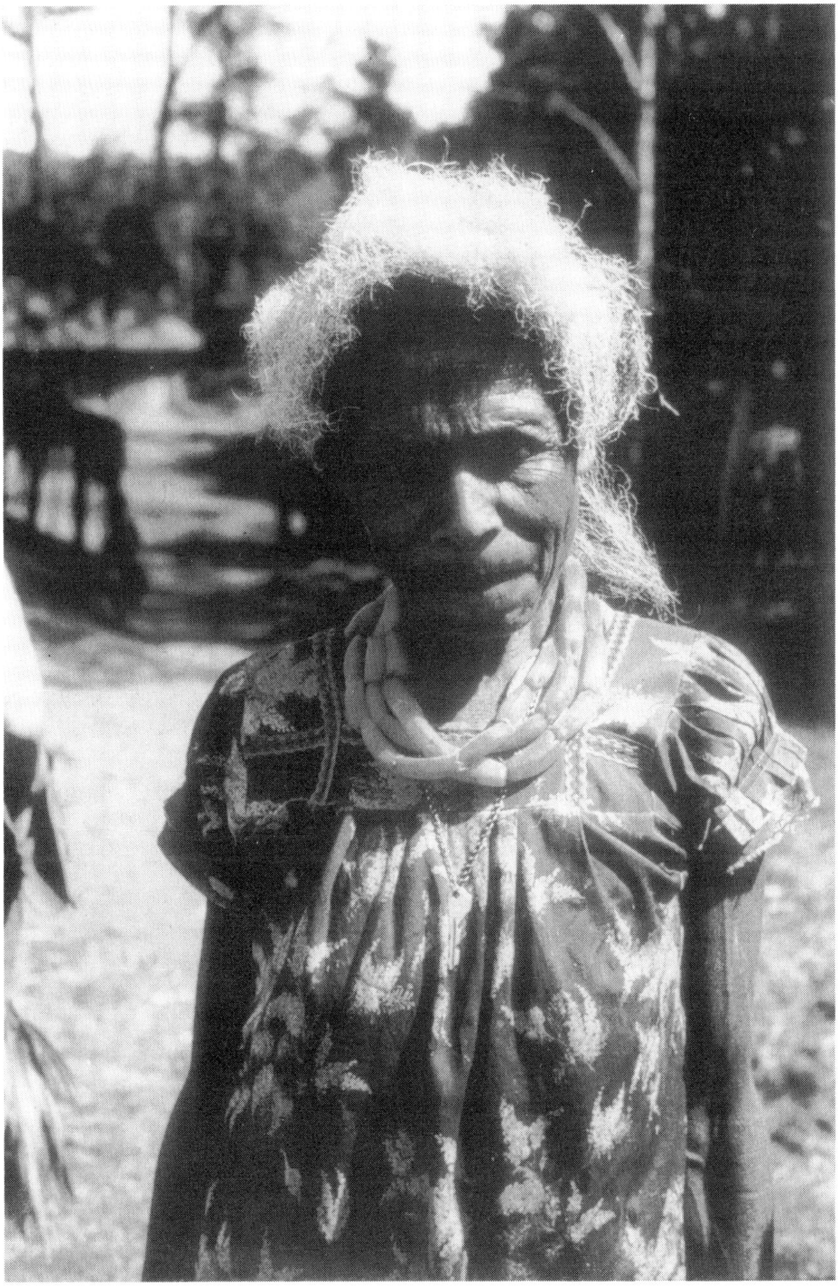

Figure 9.1. Old woman from Enga Province, Papua New Guinea using *Usnea* for personal decoration. Photo: P.W. Lambley.

literature (Wyndham, 1963), and even art as in the "Waterfall" by Maurits Escher.

In times of hardship, lichens have been eaten by many native peoples (Smith, 1921; Richardson, 1975, 1988*a*). For example, in the past Scandinavians have produced flour from *Cetraria islandica* and the North American Indians prepared a soup from *Umbilicaria muhlenbergii.* As far as I know, only the Japanese, and people from the Indian subcontinent still eat lichens in any quantity. The Japanese consume several hundred kilograms annually of *Umbilicaria esculenta* as a delicacy in soups, or deep fried (Richardson and Young, 1977). In India, Kubal garam masala (Fig. 9.3), a curry additive, includes a high proportion of various Parmeliaceae (especially *Parmotrema* and *Cetrariastrum* species) as well as *Ramalina* and *Usnea.* Garam masalas containing lichen can be found on sale as far afield as Saudi Arabia or London, being imported from northern India and Nepal. In addition, the above lichens are also sold loose and added to curry as a bulking agent with mild preservative properties. The amount of material collected for these purposes places a heavy burden on the diminishing lichen flora of the Indian subcontinent (M.R.D. Seaward, pers. comm.).

In the past, man has used lichens for dyeing wool and silk yellow, brown, or purple (Richardson, 1975). By far the largest use was for the production of orchil, the purple dye from *Roccella* species, which was collected by the shipload from about 1450 to 1850 A.D.. The main sources of supply were regions of Mediterranean climate such as the Canaries, the Cape Verde Islands, Madagascar, Mexico and parts of South America (Perkins, 1986). Today, apart from the production of a small amount of litmus, the other red dye from this lichen, in the Netherlands (as an acid/ base indicator for school laboratories), the use of dyes from lichens has

Table 9.1. The amount of *Evernia prunastri* (oakmoss) and *Pseudevernia furfuracea* (treemoss and cedarmoss) used in the manufacture of fixatives for the perfumery industry (from Moxham, 1986).

Countries of collection	Tonnes
Yugoslavia	4900–5200
France	2000–3000
Morocco	900–1000
India	120
Nepal	800
Unknown, but processed in India	200
Total	8920–10320

Note: No figures are available for Eastern Europe, Russia, or the United States.

Figure 9.2. An Austrian gravestone with a wreath incorporating *Cladonia stellaris*. Photo: M.R.D. Seaward.

Figure 9.3. A packet of Kubal garam masala, a curry additive, together with the lichens used as an ingredient; these lichens are also sold loose and form a bulking agent for curry in India. Photo: M.R.D. Seaward.

largely ceased. Indeed their use by craft dyers is actively discouraged for conservation reasons by most of the lichen societies around the world. The same general history applies to the use of lichens for medicinal purposes. There was a long tradition of the value of lichen-derived medicines which led to a spate of interest after the discovery of penicillin and to small-scale commercial production of antibiotics like "Usno" from reindeer lichens (*Cladonia* species) (Richardson, 1988*a*). This waned, largely because of the difficulty of ensuring a sustained supply of raw material at a reasonable price. However, it should be mentioned that the International Mycological Institute, Nottingham University, and two biotechnology companies are currently searching for novel compounds with biological activity in more than a thousand isolated fungal components of lichens (Crittenden, 1990). Significant amounts of Iceland moss *Cetraria islandica* are still sold annually in European pharmacies for the home concoction of herbal tonics and laxatives. A number of companies produce pastilles and pills for sore throats made from this lichen, or from the beard lichen *Usnea* (Richardson 1988*a*).

The major current commercial use of lichens is in the perfumery industry and involves oakmoss *Evernia prunastri*, and to a lesser extent treemoss *Pseudevernia furfuracea*. The bulk of the raw lichen, collected mainly in France, Yugoslavia, and Morocco, is remarkable (Table 9.1)

Figure 9.4. Piles of oakmoss, *Evernia prunastri* being pitchforked into the processing vats in France. Photo: T.H. Moxham.

considering the low dry weight of lichens. In Yugoslavia, *Evernia* is collected with the help of scrapers from trunks and branches of the oak forests in the mountains, where cloud seems to result in rapid lichen growth. In France, *Pseudevernia* is collected in the pine forests of the central massif, where lichens are picked from fallen and lower branches of the trees. Consequently the product contains more twigs. Traditionally, a distinction has been made between treemoss from France and cedarmoss from Morocco but the lichen is the same and the difference in fragrance is probably attributable to aromatic compounds provided by the incorporated twigs. The rate of pay of the lichen pickers and profit margin of the men who bale and supply the lichen to the perfume manufacturers is low (Fig. 9.4). Thus lichen picking merely boosts the income of those who are involved. Bales of lichens, each of 50–90 kg, are transported to Grasse in southern France where oakmoss and treemoss extracts are produced using a series of organic solvents. The extracts are used as a fixative in perfumes, soaps and aftershave lotions as well as giving a mossy fragrance. In spite of the large quantity of lichen processed, its abundance in the preferred collecting areas and reasonably rapid growth seem to indicate that conservation measures are not necessary at the moment though trans-boundary air pollution may be a long-term threat (Moxham, 1986).

Man's influence on lichen communities

New substrata

In the past, executions or battles resulted in human bones being available for lichen colonization, mainly by the yellow and grey genera *Caloplaca*, *Xanthoria* and *Physcia.* When the doctrine of signatures was the basis for medical treatment, lichen growing on human skulls could be worth its weight in gold as a cure for epilepsy, especially if the lichen was from the skull of a person executed for a criminal offence (Smith, 1921)! Another early substratum associated with the growing human population was thatch on which a diverse *Cladonia* flora develops. Since the industrial revolution a whole new range of man-made substrata have become available to lichens including stained glass, building materials, vulcanized rubber and plastic (Brightman and Seaward, 1977). In New Zealand, the growth of lichens on glasshouses can be sufficient to reduce the light below (Green and Snelgar, 1977; Fig. 9.5). Even in Ireland, at Carlow, lichens have developed on the upper surfaces of the shelves in open-top chambers within three years of their erection. The chambers are used for studying the effects of ambient air pollution levels on crop growth as part of a European network. Carlow is a rural site in extreme western Europe and except for occasional ozone episodes has extremely low levels of air pollutants.

Plastic and rubber are initially a very inhospitable substratum for lichens, but as they degrade from the effects of ultraviolet light, become cracked, and get a fine cover of wind-blown dust, they eventually become colonized. An example is *Candelariella vitellina* on the mouldings around windows of rusting cars at Clare Island in Co. Mayo.

Habitat loss

I do not need to stress to mycologists the seriousness of the loss of tropical forests, as they are fully aware that their fungal and lichen floras have even now received very little study. In European terms, the loss of our swamp woodlands, peatlands and deciduous forests is still occurring although at a slower rate than in past decades. These vascular plant communities provide major lichen habitats and I will limit myself to two examples from Ireland to illustrate the sort of devastation that can ensue from exploitation. The first is Slish wood, an oak woodland in Co. Sligo, where A.L. Smith recorded many rare species, which has been devastated by the widespread felling of oaks and underplanting with conifers (Alexander *et al.*, 1989). Second, a view of a raised bog in the Irish midlands following peat extraction shows the catastrophic impact of this activity on the vegetation including lichens (Fig. 9.6A). Almost all of the many large raised bogs in this region

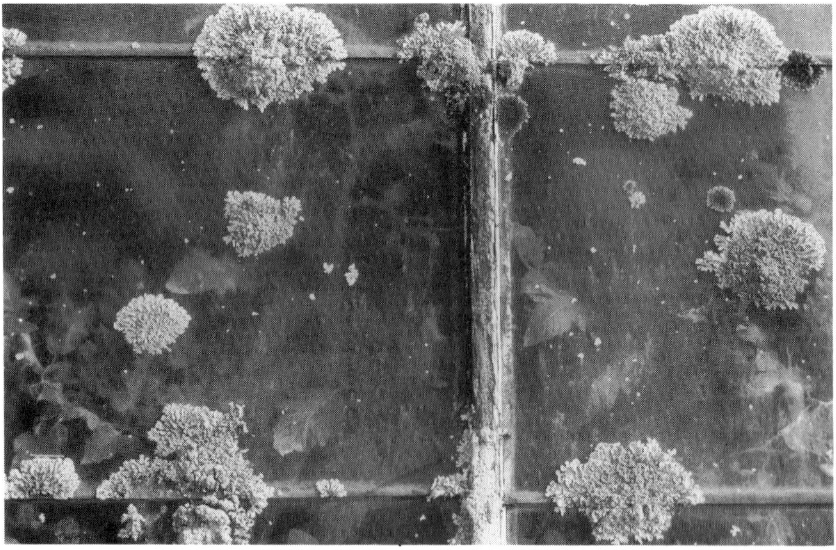

Figure 9.5. Growths of foliose lichens, *Parmelia* sp. and others on greenhouse glass near Hamilton, New Zealand. Photo: D.H.S. Richardson.

Figure 9.6. A, Mongon Bog, one of the few intact large raised bogs in central Ireland which has remained unburned for over 20 years and harbours rich lichen communities. B, Cut-away raised bog near Clogham, Ireland, showing the effect of exploitation. Following drainage, the peat has been scraped from the central area and piled into large windrows for removal. Photos: D.L. Kelly.

have been exploited, but Clara bog is one of the few where conservation is being attempted (Fig. 9.6B). On uncut bogs there is a rich *Cladonia* flora on the hummocks and many other lichens on the older heather stems (McCarthy *et al.*, 1985).

Harmful effects of lichens

Buildings and works of art

Public display of ancient buildings, statues and frescos following excavation provides a great tourist attraction in Mediterranean and Central American countries, such as Guatemala. Unfortunately, lichens often colonize the freshly exposed surfaces (Fig. 9.7). There are a few particularly aggressive species that have done serious damage (Seaward *et al.*, 1989). For example, at the Museo Nationale Romano, thalli of *Lecanora muralis* less than 15 years old are causing damage to excavated earthenware pots. Crowding of fruit-bodies in the thallus centre creates a central blister and pulls away fragments of the substratum from the remainder of the terracotta pot. Another example is the Palazzo Farnese, built in 1547–49, where frescos are being attacked by another aggressive lichen, *Dirina massiliensis* f. *sorediata.* Lichens are now being laboriously removed with a scalpel and each area repainted. Liquid chemicals cannot be used due to the water-based nature of the fresco paints. The growth of lichens even on modern buildings can make them unsightly, at least to the architects, and the performance of power line insulators covered with lichen can be affected (Naito *et al.*, 1990).

Humans

Contact dermatitis among forestry and horticultural workers has been known for over 60 years and is known as Woodcutters' Eczema or cedar poisoning. The offending allergens are secondary products of lichens and liverworts (Thune, 1987; Goncalo, 1987). Among the many lichen substances implicated are usnic, evernic, fumaroprotocetraric, and stictic acids, and atranorin (Richardson, 1988*a*). The increasing use of perfumes and aftershave lotions in modern society suggests that this type of dermatitis will become more common. Many women, while not developing classical dermatitis, know that they develop an allergic response to one or all perfumes. In such cases it is frequently the lichen extract used as a fixative which is the allergen (Richardson, 1988*a*).

Figure 9.7. Modern calcareous statue heavily colonized by lichens, mainly *Peltula*, outside Faculty of Medicine building, University of Caracas, Venezuela. Photo: M. Tabasso.

Lichens and pollution

Gaseous air pollutants

This topic could occupy the whole of this contribution. I propose merely to draw attention to a few facets of a subject where the literature is now extensive and readily available (e.g. Henderson, 1990).

The first point is that the distribution of lichens has been, and is, changing in response to prevailing pollution climates throughout the industrialized world. In the latter half of the last and first half of this century, high sulphur dioxide and smoke levels affected relatively limited areas around the larger and more industrialized cities. Exhaust from the internal combustion engine, and high stack emission policies in industry, have led to acid rain, ozone, and peroxyacetyl nitrates (PAN) becoming pollutants of significant importance over large areas of the developed world.

Lichens containing cyanobacteria as photobionts appear to be damaged by acid rain and the more sensitive species seem to be disappearing from many regions. This has been well-documented for *Lobaria scrobiculata* in southern Sweden where the species was formerly recorded from more than 300 localities. Recently, it could only be confirmed as still present at just two of 50 thoroughly investigated old sites (Hallinback, 1989). It may be that even weak acid rain, from transboundary sources, progressively overcomes the buffering capacity of the tree bark to the point where the surface pH of the bark falls significantly (Nieboer *et al.*, 1984). This may prevent the germination of *Lobaria* ascospores or the growth of the symbiotic cyanobacterium and thus prevent regeneration of thalli (Richardson, 1988*b*). *Lobaria* is now so scarce in most parts of Britain that considerable efforts are directed at its conservation. Lord Lonsdale and the Nature Conservancy Council were very concerned over the future of *Lobaria amplissima* in Lowther Park, Cumbria. A programme of transplanting the lichen was instituted when one of the old trees had to be felled as it was unsound and a danger to the public. Ten of the 14 thalli transplanted continue to thrive after ten years (Gilbert, 1991). It is interesting that in Oregon, small (2–5 cm) pieces of *L. oregana* and *L. pulmonaria* with a 2 mm hole punched in them have been cultured by being threaded on 1.7 mm nylon monofilament from a strimmer (weed cutter) with plastic beads between the lichen pieces. When suspended in the forest, the pieces of *L. oregana* showed a 16% weight gain per year and the other species half of this over a three year period (Denison, 1988).

The changing distribution of *Usnea* species in the British Isles is remarkable. This lichen was widespread until about 1800 but it then disappeared from an area covering some 70000 km^2 as a result of increased pollution, especially of sulphur dioxide (Seaward, 1987, 1989). However, as sulphur dioxide levels have fallen, *Usnea* has re-established itself, at first on trees such as *Fraxinus* and *Salix* (Fig. 9.8). These trees have bark with a higher pH relative to most other trees (Farmer *et al.*, 1990).

We have some understanding as to how gaseous air pollutants affect lichens but our knowledge is far from complete (Richardson 1988*b*). The ability of lichens to accumulate anions (including dissolved gaseous pollutants such as sulphur dioxide) actively and rapidly as well as the lack of a protective cuticle and poikilohydrous water relations all contribute to their sensitivity. Of the various metabolic processes in lichens, laboratory studies have generally shown the following to be the sequence of sensitivity with respect to sulphur dioxide damage: nitrogen fixation < photosynthesis, respiration < pigment status < potassium efflux. The loss of ATP is another parameter and this seems to be a more sensitive indicator of air pollution damage than pigment breakdown. For example, at Abou-Kabir in Israel, chlorophyll a levels fell by 8% from July to March in transplanted thalli of *Ramalina duriaei* while the ATP levels dropped by 50% (Kardish *et al.*, 1987). The likely sulphur dioxide damage mechanisms include:

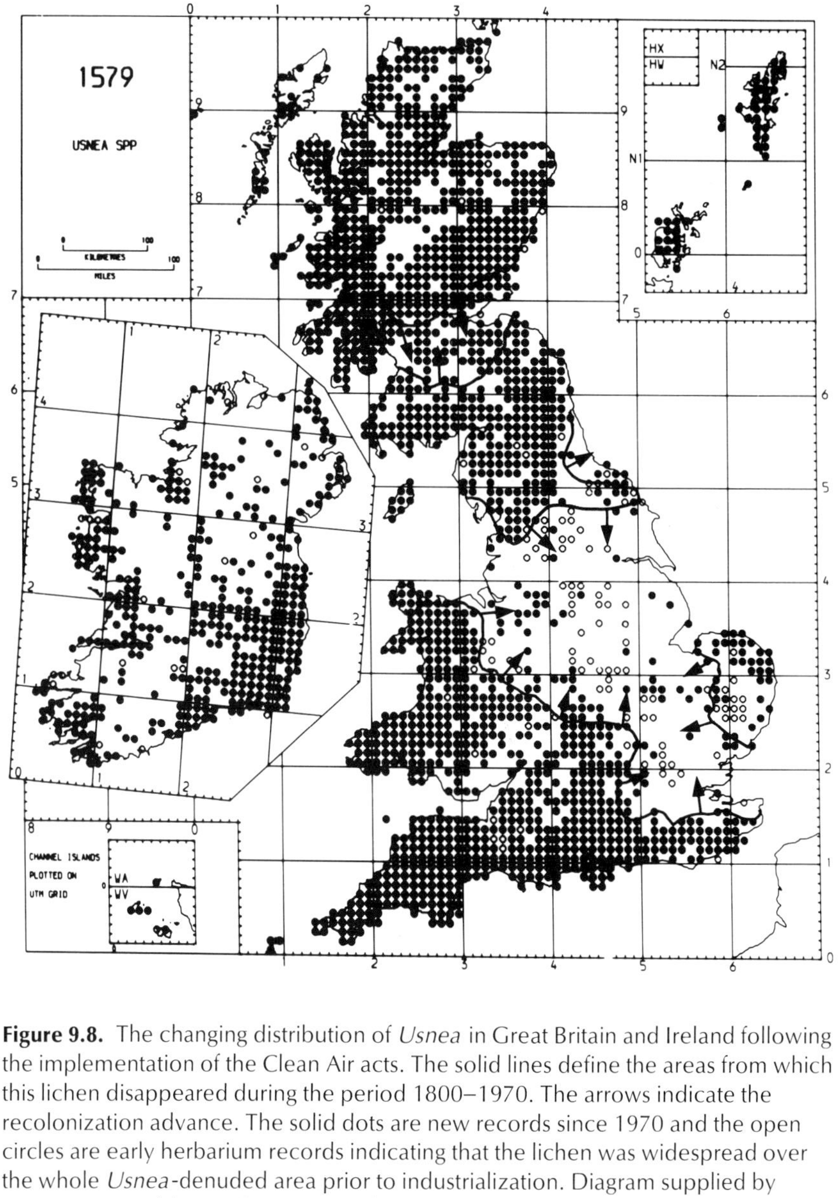

Figure 9.8. The changing distribution of *Usnea* in Great Britain and Ireland following the implementation of the Clean Air acts. The solid lines define the areas from which this lichen disappeared during the period 1800–1970. The arrows indicate the recolonization advance. The solid dots are new records since 1970 and the open circles are early herbarium records indicating that the lichen was widespread over the whole *Usnea*-denuded area prior to industrialization. Diagram supplied by M.R.D. Seaward from information in the British Lichen Society's Mapping Scheme Database at Bradford University.

(1) enzyme deactivation; (2) enzyme stimulation; (3) interaction with reactive biomolecules; (4) formation of free radicals (Richardson and Nieboer, 1983). Ultrastructural changes in photobiont mitochondria and chloroplasts are among the earliest consequences of sulphur dioxide fumigation in lichens. Deformation of chloroplast and thylakoid membranes and mitochondrial swelling is seen to be associated with slight or intermediate injury. As damage progresses, it is reflected by changes in the fluorescence of the algal cells which appear orange or white rather than red under the UV microscope (Holopainen and Kauppi, 1989). Recent studies in Switzerland indicate that the long-term effects of pollution on lichen metabolism may be different from that following briefer laboratory exposures or transplantation to urban sites. Thus investigations on thalli of *Parmelia sulcata* growing naturally in different urban pollution zones have shown that the overall protein content, dark respiration, gross and net photosynthesis did not change between lichens from urban and suburban habitats. However, lichens from the former habitat exhibited up to seven times less growth, and the amount of assimilate released by the algae was 15 times less in lichens from the centre versus the periphery of the city. The higher chlorophyll content in lichen samples from the central city areas is thought to be related to the stimulatory effects of nitrogen oxides from traffic emissions on chlorophyll synthesis (von Arb and Brunold, 1989).

Another recent threat to lichens is ammonia emissions from intensive cattle rearing, which, in parts of The Netherlands and Denmark, are injuring populations of *Cladonia portentosa*, *C. arbuscula* and *C. ciliata* on dunes and heathlands (Søchting, 1987).

As urban smoke and sulphur dioxide levels fall, a group of pollution-tolerant crustose lichens, including *Lecanora muralis* and *Xanthoria elegans*, have invaded stonework in what used to be “lichen deserts” (Henderson-Sellers and Seaward, 1979). There have also been recent records of epiphytic foliose and fruticose species such as *Ramalina farinacea* and *Xanthoria polycarpa* spreading into suburbs and even into central London (Hawksworth and McManus, 1989). In some cases nutrient-rich dust seems to be promoting the growth of these lichens.

There have been, as mentioned above, many studies on the harmful effects of sulphur dioxide on lichens. In spite of this research, the reasons why different species of the same lichen genus, containing apparently closely related algal symbionts, vary in their pollution sensitivity has still to be revealed. However, that different lichen species do exhibit a marked variation in pollution sensitivity has meant that studies on lichen distribution can yield useful information on air quality. The result has been that lichens are now the most popular and widely accepted bioindicators of air quality (Hawksworth and Rose, 1976; Henderson, 1990; Nash and Wirth, 1988).

Metallic pollutants

The ability of lichens to accumulate high levels of various metals in a distance-related way has led to them being excellent monitors for atmospheric fallout around smelters and other industrial plants. Typically high levels are found in lichens collected close to the emission source and these levels fall off rapidly at first and then more slowly, as for example around the Sudbury nickel smelter in Canada (Nieboer and Richardson, 1980). We know that lichens can accumulate metallic elements by trapping insoluble particulates (usually oxides, sulphates, and sulphides) and taking up metal ions by extracellular ion exchange processes. The latter may be accompanied by a slower uptake into the cells (Brown 1985). The efficiency of these uptake mechanisms has resulted in lichens even being packed into porous PVC tubes and suspended in rivers for two weeks to monitor for a wide variety of metal contaminants (Beck and Ramelow, 1990), or being used to construct lichen-modified carbon paste electrodes to produce electrochemical biosensors (Connor *et al.*, 1991).

In general, lichens can accumulate high levels of metallic elements without apparent harm (Nash, 1989). It is the nature and the form of the accumulated element that dictates the response. Where metals are presented in a soluble form, are present in excess, or are by their nature highly toxic (Nieboer and Richardson, 1980), then the lichen is adversely affected. I could show you a complex table based on laboratory experiments to demonstrate this (Nieboer *et al.*, 1978; Richardson *et al.*, 1978) but instead will give a few graphic examples based on the empirical observation of lichens growing in their natural habitat which will make the same point. Firstly, in Haiti King Cristof (1807–1820) built a French-style palace at sea level and an impregnable castle on a nearby mountain top (792 m above sea level) in case the French re-invaded. They never came with the result that all the cannon and cannon balls are still in place and available for colonization by lichens. The iron cannon balls are covered by a wide range of crustose and foliose lichens including species of *Heterodermia*, *Ramalina*, *Teloschistes*, *Parmelia*, and *Chiodecton*. These are growing on the iron surfaces without apparent harm (Richardson, 1978). In contrast, the effects of zinc, and particularly lead and copper, are much more damaging to lichens. The effects of zinc can be seen on tree trunks below the points at which barbed wire is attached (Seaward, 1974) or on lichen-covered roofs in rural areas below galvanized supports of television aerials. The effects of copper telephone wires can also be seen in such situations. To illustrate this, let me give as an example a telegraph pole from southwestern Australia; it has lichens on the ends of the cross-poles and on the central pole, but not beneath the wires where dissolved copper ions drip onto the wood and inhibit lichen growth. Even more dramatic is the effect of soluble lead ions. A gravestone in western Ireland has been inlaid with lead so that the

Figure 9.9. A gravestone in County Sligo, Ireland, colonized by lichens except where the inscription has been inlaid with lead from which toxic lead ions have been leached by rainwater. Photo: D.H.S. Richardson.

inscription is more permanently visible (Fig. 9.9). Lichens have colonized the whole stone except beneath the lead where leached ions have prevented growth. In places where the lead has fallen out, lichens have been able to re-invade.

Radionuclides

The efficiency with which lichens absorb radionuclides from the air and the importance of the lichen–reindeer–man food-chain has been known since the 1960s following the time when atmospheric nuclear bomb testing was widespread (Åberg and Hungate, 1969). Studies revealed that male Lapps had higher ^{137}Cs levels than female Lapps, and that the peak was in early summer. Male Lapps consumed more contaminated meat from the reindeer which had been feeding on lichen as a major winter food. In contrast, female Lapps had higher levels of ^{55}Fe, another isotope released by atmospheric nuclear testing, even though they eat less meat. This could be accounted for by the higher uptake efficiency of the female gut for iron which has to be replaced as a result of menstruation (Richardson, 1975). However, it was the Chernobyl accident in April 1986 which has had more serious economic consequences to the Lapp reindeer herders in Scandinavia due to the unacceptably high levels of radioactivity in the reindeer meat (over 10 000 Bq of ^{137}Cs kg^{-1} in many animals, with a legal limit for sale of only 300 Bq kg^{-1}) that has brought further focus to the subject (Mackenzie, 1986; O'Clery, 1986). Scientists around the world now conclude that lichens are excellent collectors of atmospheric aerosols owing to their high surface-to-mass ratio, slow growth rates, and the fact that they derive nutrients from atmospheric sources. These scavenging properties make them cost-effective monitoring devices for radionuclides (Smith and Ellis, 1990). The Chernobyl plume arrived in Canada 11 days after the accident and its movement across Canada provided a unique opportunity to validate a model describing radionuclide uptake by lichens. A deposition velocity of 1.1 cm sec^{-1} gave good agreement between the model and experimental results. Using this result, combined with atmospheric residence time data, it was possible to calculate that the height of the radioactive cloud as it passed over Canada was 10 000 m (Smith and Ellis, 1990). Thalli of *Umbilicaria* in southwest Poland exhibited higher levels of radioactivity when collected above 800 m. The lichen showed up to 165 times the radioactive levels recorded before the Chernobyl accident, with maximum values in excess of 35 000 Bq kg^{-1} (Seaward *et al.*, 1988). Similar results have been recorded for Austria (Turk, 1988).

While lichens provide a major item of diet for reindeer and caribou in winter, the White-tailed deer of eastern America also use lichen in winter and consumed 63% of *Usnea* and 56% of *Evernia* offered at feeding

Figure 9.10. Red deer grazing on lichen-covered rocks near the sea on islands off Western Ireland; above the reach of the deer the *Ramalina siliquosa* is over 5 cm tall whereas it is cropped almost to the base below this level. Photo: S. Ryan.

stations (Hodgman and Bower, 1985). In addition, many other deer including mule deer, the Columbian black-tailed deer, the black-tailed deer of Western America, the Roosevelt elk, the musk deer of central Asia and roe deer will each lichens when other fodder is in short supply and can thereby accumulate radioactive caesium (Book *et al.*, 1972; Richardson and Young, 1977). Even red deer can occasionally be seen grazing lichens as on islands off the coast of western Ireland where they feed on species such as *Ramalina siliquosa* growing on rocks near the sea (Fig. 9.10). There is a browse line above which the lichens are over 5 cm long but below it the thalli are grazed almost to the base. The consumption of lichens by deer, and that they also often feed on upland ericaceous vegetation which has mycorrhizal roots that conserve and recycle ^{137}Cs, accounts for the significantly higher radioactivity level in these animals *vis-a-vis* cattle. Enhanced levels of radioactivity has led to a reluctance of consumers to purchase deer meat with economic consequences to upland landowners and deer farmers.

The future of lichen communities and their associated fauna

In this final section of my contribution, I would like briefly to discuss the many interesting and close lichen/fauna interactions. The relevance of this in terms of man and lichens will I trust become evident later, but I will first mention the generally overlooked protozoa and other microinvertebrate associates of epiphytic lichens. Studies in Belfast have shown that micro-invertebrates increase in abundance and diversity as the lichen flora improves from the centre of the city to rural areas (Roberts and Zimmer, 1990).

Throughout the world, there are caterpillars and other invertebrates that feed on lichens, as well as a variety of insects and some small vertebrates that exhibit lichen crypsis (Richardson, 1975; Gerson and Seaward, 1977; Roberts and Zimmer, 1990). I would like to mention a few of these to indicate their variety and beauty. First a bush-cricket from the tropical forest of Africa; the background tree leaves are covered with folicolous lichens, and the bush-cricket wings mimic the leaf and its venations, with also mimicked patches of folicolous lichens. Second is a praying mantis from Australia, where the body segments resemble foliose lichens. Finally, a North American tree frog whose grey colour and mottling enable it to be almost invisible against a lichen-covered branch. In Europe too, there are many invertebrate examples of lichen crypsis, of which the most famous is the Peppered Moth *Biston betularia.* The concept of industrial melanism was first developed as a result of observations on this moth. The common "*typica*" form of this moth evolved to resemble, and thus be cryptic with lichen-covered bark. The moth also throws up the occasional dark mutant,

Figure 9.11. Two Peppered moths in copulation showing lichen crypsis and the typical resting pose on twigs; Pollagh, Co. Offaly, Ireland. Photo: D.L. Kelly and P. Fay.

"*carbonaria*" which became predominant around Manchester and many industrial areas of England by the end of the nineteenth century. Based on bird-predation studies, it was later demonstrated that the dark "*carbonaria*" form survived better on the trunks of bare, blackened, trees in industrial areas while the pale "*typica*" form was more advantaged in rural areas where the trees were lichen covered (Kettlewell, 1955, 1956). A recent re-examination of melanism in this moth has revealed that the situation is more complex and interesting than originally thought, with the "*carbonaria*" form varying widely in coloration and twigs rather than trunks being the main resting sites (Fig. 9.11; Majerus, 1989). Recent studies on the Peppered Moth in Britain following the implementation of the Clean Air Act have revealed that the proportion of dark "*carbonaria*" forms to pale forms trapped has dropped from over 90% to less than 40% in Cambridge (Majerus, 1989). In northern Britain, across a transect from urban Manchester to the Welsh border, an eastward shift of the decline in frequency of the dark form has been explained in terms of the spread of the pollution-tolerent lichen *Lecanora conizaeoides* on trees within the former lichen desert at the eastern end and by an increase in lichen diversity in the central portion of the transect (Cook *et al.*, 1990). Other moths which have evolved lichen crypsis and have melanic forms or exhibit polymorphisms have enchanting names such as the Dotted Border, the Spring Usher, the Brindled Beauty, the Earth Thorn, the Green Brindled Crescent, and the Mottled Umber. Research on these could be scientifically very rewarding (Majerus, 1989).

It is clear from my comments above that lichen communities face serious direct or indirect threats from man as a consequence of air pollution and habitat destruction. The latter can be due, for example, to drainage, tree felling, or the replacement of native tree species with exotic conifers or

Eucalyptus species. In addition, fungicides and artificial fertilizers used in large amounts in modern agricultural practices can damage many macrolichen species (James, 1973). Within the European Economic Community, politicians, planners and legislators are increasingly paying heed to information derived from lichens with respect to air quality and are requiring baseline lichen studies as part of environmental impact statements prior to industrial developments in rural areas. However, there is as yet little willingness to allocate funds for habitat conservation or for research as to the best way to manage forests in order to retain the rare lichen species typical of ancient woodlands. If only man can develop and implement effective air pollution abatement and conservation techniques for the rich lichen communities around the world, he will also protect the fascinating associated fauna and fulfil the rallying call of many environmentalists "Air fit for lichens and water fit for trout".

Acknowledgements

I am most grateful to Dr C. Giacobini, Dr D.L. Kelly, Mr T.H. Moxham, Mr P.W. Lambley, Mr S. Ryan, and Professor M.R.D. Seaward for the loan of a number of slides for my lecture at the Fourth International Mycological Congress at Regensburg, Germany, which was based on the above manuscript. I would also like to thank Professor Seaward for making the unpublished map derived from the British Lichen Society's Mapping Scheme Database showing the latest change in *Usnea* distribution in Britain and Ireland available. I would also like to thank Professors M.R.D. Seaward and D.L. Hawksworth for helpful comments on the manuscript.

References

Åberg, B. and Hungate, F.P. (1969) *Radioecological Concentration Processes.* Pergamon Press, Oxford.

Alexander, R.W., Richardson, D.H.S., Cotton, D. and Seaward, M.R.D. (1989) Field Meeting to Sligo and Connemara National Park. *Lichenologist* 21, 159–168.

Arb, C. von and Brunold, C. (1989) Lichen physiology and air pollution. I. Physiological responses of *in situ Parmelia sulcata* among air pollution zones within Biel, Switzerland. *Canadian Journal of Botany* 68, 35–42.

Beck, J.N. and Ramelow, G.J. (1990) Use of lichen biomass to monitor dissolved metals in natural waters. *Bulletin of Environmental Contamination and Toxicology* 44, 302–308.

Book, S.A., Connolly, G.E. and Longhurst, W.M. (1972) ^{137}Cs accumulation in two adjacent populations of northern California deer. *Health Physics* 22, 379–385.

Brightman, F.H. and Seaward, M.R.D. (1977) Lichens of man-made substrates. In:

Seaward, M.R.D. (ed.), *Lichen Ecology.* Academic Press, London, pp. 253–293.

Brown, D.H. [ed.] (1985) *Recent Advances in Lichen Physiology.* Plenum Press, London.

Connor, M., Dempsey, E., Smyth, M.R. and Richardson, D.H.S. (1991) Determination of some metal ions using lichen-modified carbon paste electrodes. *Electroanalysis* 3, in press.

Cook, L.M., Rigby, K.D. and Seaward, M.R.D. (1991) Melanic moths and changes in epiphytic vegetation in north-west England and north Wales. *Biological Journal of the Linnean Society* 39, 343–354.

Crittenden, P. (1990) Research news and notes. *International Lichenological Newsletter* 23, 23–24.

Dalby, C. (1981) *Lichens and Air Pollution.* [Wallchart and A-4 sized chartlets.] British Museum (Natural History), London.

Dalby, C. (1987) *Lichens on Rocky Seashores.* [Wallchart and A-4 sized chartlets.] British Museum (Natural History), London.

Denison, W.C. (1988) Culturing the lichens *Lobaria oregana* and *L. pulmonaria* on nylon monofilament. *Mycologia* 80, 811–814.

Dobson, F. (1991) *Lichens. An Illustrated Guide.* 3rd edn. Richmond Publishing, Burnham.

Farmer, A.M., Bates, J.W. and Bell, N.J.B. (1990) A comparison of methods for the measurement of bark pH. *Lichenologist* 22, 191–194.

Galloway, D.J. (1985) *Flora of New Zealand Lichens.* Government Printer, Wellington.

Galun, M. [ed.] (1988) *CRC Handbook of Lichenology.* 3 vols. CRC Press, Boca Raton.

Gerson, U. and Seaward, M.R.D. (1977) Lichen-invertebrate associations. In: Seaward, M.R.D. (ed.), *Lichen Ecology.* Academic Press, London, pp. 69–119.

Gilbert, O.L. (1991) A successful transplant operation involving *Lobaria amplissima. Lichenologist* 23, 73–76.

Goncalo, S. (1987) Contact sensitivity to lichens, Compositae in Frullania dermatitis. *Contact Dermatitis* 16, 84–86.

Green, T.G.A. and Snelgar, W.P. (1977) *Parmelia scabrosa* on glass in New Zealand. *Lichenologist* 9, 170–171.

Hallinback, T. (1989) Occurrence and ecology of the lichen *Lobaria scrobiculata* in southern Sweden. *Lichenologist* 21, 331–342.

Hawksworth, D.L. and Ahti, T. (1990) A bibliographic guide to the lichen floras of the world (second edition). *Lichenologist* 22, 1–78.

Hawksworth, D.L. and Hill, D.J. (1984) *The Lichen-forming Fungi.* Blackie, Glasgow.

Hawksworth, D.L. and McManus, P.M. (1989) Lichen recolonization in London under conditions of rapidly falling sulphur dioxide levels, and the concept of zone-skipping. *Botanical Journal of the Linnean Society* 100, 99–109.

Hawksworth, D.L. and Rose, F. (1976) *Lichens as Pollution Monitors.* Studies in Biology no. 66. Edward Arnold, London.

Henderson, A. (1990) Literature on air pollution and lichens XXXI. *Lichenologist* 22, 173–182.

Henderson-Sellers, A. and Seaward, M.R.D. (1979) Monitoring lichen reinvasion of ameliorating environments. *Environmental Pollution* 19, 207–213.

Hodgman, T.P. and Bowyer, R.T. (1985) Winter use of arboreal lichens, ascomycetes, by white-tailed deer, *Odocoileus virginianus*, in Maine. *Canadian Field Naturalist* 99, 313–316.
Holopainen, T. and Kauppi, M. (1989) A comparison of light, fluorescence and electron microscopic observations in assessing the SO_2 injury of lichens under different moisture conditions. *Lichenologist* 21, 119–134.
Jahns, H.M. (1983) *Collins Guide to Ferns, Mosses and Lichens of Britain and North and Central Europe.* Collins, London.
James, P.W. (1973) The effect of air pollutants, other than hydrogen fluoride and sulphur dioxide on lichens. In: Ferry, B.W., Baddeley, M.S. and Hawksworth, D.L. (eds), *Air Pollution and Lichens.* Athlone Press of the University of London, London, pp. 143–175.
Kardish, N., Ronen, R., Bubrick, P. and Garty, J. (1987) The influence of air pollution on the concentration of ATP and on chlorophyll degradation in the lichen *Ramalina duriaei* (De Not.) Bagl. *New Phytologist* 106, 697–706.
Kettlewell, H.B.D. (1955) Selection experiments on industrial melanism in the Lepidoptera. *Heredity* 9, 323–342.
Kettlewell, H.B.D. (1956) Further selection experiments on industrial melanism in the Lepidoptera. *Heredity* 10, 287–301.
Kershaw, K.A. (1985) *Physiological Ecology of Lichens.* Cambridge University Press, Cambridge.
Kauppi, M. (1979) The exploitation of *Cladonia stellaris* in Finland. *Lichenologist* 11, 85–89.
Laundon, J.R. (1986) *Lichens.* Shire Publications, Aylesbury.
Lawrey, J.D. (1984) *The Biology of Lichenized Fungi.* Praeger, New York.
Lock, W.W., Andrews, J.T. and Webber, P.J. (1979) *A Manual for Lichenometry.* British Geomorphological Research Group Technical Bulletin No. 26. Institute of British Geographers, London.
McCarthy, P.M., Mitchell, M.E. and Schouten, M.G.C. (1985) Lichens epiphytic on *Calluna vulgaris* (L.) Hull in Ireland. *Nova Hedwigia* 42, 91–98.
Mackenzie, D. (1986) The rad-dosed reindeer. *New Scientist* 112 (1539), 37–40.
Majerus, M.E.N. (1989) Melanic polmorphism in the Peppered moth, *Biston betularia* and in other Lepidoptera. *Journal of Biological Education* 23, 267–284.
Moxham, T.H. (1986) The commercial exploitation of lichens for the perfume industry. In: Brunke, E.J. (ed.), *Progress in Essential Oil Research.* Walter de Gruyte, Berlin, pp. 491–503.
Naito, K., Matsuoka, R. and Sakanish, K. (1990) Investigations of the insulator performance of the insulator covered with lichen. *IEEE Transactions on Power Delivery (Japan)* 5, 1634–1640.
Nash, T.H. III (1989) Metal tolerance in lichens. In: Shaw, J. (ed.), *Heavy Metal tolerance in Plants: Evolutionary aspects.* CRC Press, Boca Raton, pp. 119–132.
Nash, T.H. III and Wirth, V. [eds] (1988) Lichens, bryophytes and air quality. *Bibliotheca lichenologica* 30, 1–297.
Nieboer, E., Richardson, D.H.S., Lavoie, P. and Padovan, D. (1978) The role of metal ion binding in modifying the toxic effects of sulphur dioxide on the lichen *Umbilicaria muhlenbergii.* I. Potassium efflux studies. *New Phytologist* 82, 621–632.

Nieboer, E., McFarlane, J.D. and Richardson, D.H.S. (1984) Modifications of plant cell buffering capacities by gaseous air pollutants. In: Koziol, M. and Whatley, F.R. (eds), *Gaseous Air Pollutants and Plant Metabolism.* Butterworths, London, pp. 313–330.
Nieboer, E. and Richardson, D.H.S. (1980) The replacement of the nondescript term heavy metals by a biologically and chemically significant classification of metal ions. *Environmental Pollution,* B, 1, 3–26.
O'Clery, C. (1986) The Chernobyl fallout: How Lapland is paying the price. *Irish Times* [weekend supplement] 13 September.
Perkins, P. (1986) Ecology, beauty, profits: Trade in lichen-based dyestuffs through western history. *Journal of the Society of Dyers and Colourists* 102, 221–227.
Richardson, D.H.S. (1975) *The Vanishing Lichens: Their history, biology and importance.* David and Charles, Newton Abbot.
Richardson, D.H.S. (1978) Lichens on iron cannon balls. *Lichenologist* 10, 233–234.
Richardson, D.H.S. (1988*a*) Medicinal and other economic aspects of lichens. In Galun, M. (ed.), *CRC Handbook of Lichenology.* CRC Press, Boca Raton, vol. 3, pp. 93–108.
Richardson, D.H.S. (1988*b*) Understanding the pollution sensitivity of lichens. *Botanical Journal of the Linnean Society* 96, 31–43.
Richardson, D.H.S. and Nieboer, E. (1983) Ecophysiological responses of lichens to sulphur dioxide. *Journal of the Hattori Botanical Laboratory* 54, 331–351.
Richardson, D.H.S. and Young, C.M. (1977) Lichens and vertebrates. In: Seaward, M.R.D. (ed.), *Lichen Ecology.* Academic Press, London, pp. 121–144.
Richardson, D.H.S., Nieboer, E., Lavoie, P. and Padovan, D. (1978) The role of metal ion binding in modifying the toxic effects of sulphur dioxide on the lichen *Umbilicaria muhlenbergii* II. ^{14}C-fixation studies. *New Phytologist* 82, 633–643.
Roberts, D. and Zimmer, D. (1990) Microfaunal communities associated with epiphytic lichens in Belfast. *Lichenologist* 22, 163–172.
Seaward, M.R.D. (1974) Some observations on heavy metal toxicity and tolerance in lichens. *Lichenologist* 6, 158–164.
Seaward, M.R.D. (1987) Effects of quantitative and qualitative changes in air pollution on the ecological and geographical performance of lichens. *NATO ASci Series* G16, 439–450.
Seaward, M.R.D., Heslop, J.A., Green, D. and Bylinska, E.A. (1988) Recent levels of radionuclides in lichens from southwest Poland with particular reference to ^{134}Cs and ^{137}Cs. *Journal of Environmental Radioactivity* 7, 123–129.
Seaward, M.R.D. (1989) Lichens as pollution monitors: adapting to modern problems. In: Ozturk, M.A. (ed.), *Plants and Pollutants in Developed and Developing Countries.* University of Bornova, pp. 307–319.
Seaward, M.R.D., Giacobini, C., Giuliani, M.R. and Roccardi, A. (1989) The role of lichens in the biodeterioration of ancient monuments with particular reference to central Italy. *International Biodeterioration* 25, 49–55.
Smith, A.L. (1921) *Lichens.* Cambridge University Press, Cambridge.
Smith, D.C. (1978) What can lichens tell us about real fungi? *Mycologia* 70, 915–934.
Smith, J.N. and Ellis, K.M. (1990) Time dependent transport of Chernobyl radioactivity between atmospheric and lichen phases in eastern Canada. *Journal*

of Environmental Radioactivity 11, 151–168.

Søchting, U. (1987) Injured reindeer lichens in Danish lichen heaths. *Graphis Scripta* 1, 103–106.

Swinscow, T.D.V. and Krog, H. (1988) *Macrolichens of East Africa.* British Museum (Natural History), London.

Türk, R. (1988) Bioindikation von Luftverunreinigungen mittels Flechten. In: Grill, D. and Guttenberger, H. (eds), *Okophysiologische Probleme durch Luftverunreinigungen.* Karl-Franzens Universität, Graz, pp. 13–27.

Thune, P. (1987) Lichens, Compositae and photosensitivity. *Photodermatology* 4, 1–4.

Vitt, D.H., Marsh, J.E. and Bovey, R.B. (1988) *Mosses, Lichens and Ferns of Northwest America.* Lone Pine Publishing, Edmonton.

Wirth, V. (1987) *Die Flechten Baden-Württembergs.* Eugen Ulmer, Stuttgart.

Wyndham, J. (1963) *Trouble with Lichen.* Penguin Books, Harmondsworth.

10

Modified Amatoxins and Phallotoxins for Biochemical, Biological and Medical Research

H. Faulstich, *Max-Planck-Institut für medizinische Forschung, Postfach 10 38 20, D-6900, Heidelberg 1, Germany.*

ABSTRACT The phallotoxins, a group of cyclic peptides produced by *Amanita phalloides*, form complexes with filamentous actin and by this inhibit depolymerization. A second group of toxic cyclopeptides found in this mushroom, the amatoxins, bind to eukaryotic RNA polymerase II, thus blocking the transcription of hnRNA (mRNA). Because of their highly specific bioactivities, both these kinds of natural toxins are valuable tools for cell biologists. This review describes phallotoxins and amatoxins that have been chemically modified for special biological or medical purposes.

1. Amatoxins conjugated with proteins were studied for their specific cytotoxicity for phagocytosing cells. Monoclonal and polyclonal antibodies prepared from such conjugates were used for immunoassays, for an immunotherapy of *Amanita* mushroom poisoning, and for studying the physical interactions in a peptide-protein complex.
2. Amatoxins bound to a macromoleculr carrier and labelled with a ligand (epidermal growth factor) were shown to be internalized into certain tumor cells possessing EGF receptors, by receptor-mediated endocytosis.
3. An amatoxin conjugated with a fluorescent moiety was used for visualizing RNA polymerase II in resting and mitotic cells.
4. Cytoskeletal structures containing actin were identified with phallotoxins bound to fluorescein (green), tetramethylrhodamine (red), or coumarine (blue). Using coumarine-labelled phalloidin, actin stress fibres could be identified side by side with two other cytoskeletal components in one and the same cell preparation (triple staining).
5. On the level of electron microscopy, phalloidin linked to biotin *via* a 12 atom spacer was used for detecting actin filaments after incubation with gold particle-dotted streptavidin.

Frontiers in Mycology. Honorary and General Lectures from the Fourth International Mycological Congress, Regensburg 1990. Edited by D.L. Hawksworth.

Introduction

The green death cap, *Amanita phalloides*, produces two families of extremely toxic peptides, phallotoxins and amatoxins (Wieland and Faulstich, 1978). Due to the specific transporting abilities of hepatocytes, both toxins exhibit their toxic activities mainly in the liver, although in quite different ways. In laboratory animals parenteral application of, for example phalloidin, the main component of the phallotoxin family (Fig. 10.1), impairs the cortical web associated with the plasma membrane, leading to death within 3–5 hours from haemorrhagic necrosis of the liver. Phallotoxins, however, do not contribute to human *Amanita* poisoning. The typical human *Amanita* mushroom intoxication is solely due to the amatoxins, which, after parenteral, and in some animal species also enteral application, cause necrosis and fatty degeneration of the liver. Laboratory animals die after 5–8 days from combined hepatic and renal failure. The main component of the amatoxins is α-amanitin (Fig. 10.2).

Phallotoxins form strong complexes with filamentous actin. Although formation of this complex does not disturb the interaction of filaments with myosin, it is one of the toxic effects induced by phalloidin in non-muscle cells. Conjugates of phalloidin with reporter groups such as fluorescein or biotin are nowadays widely used for the visualization of microfilaments. Such compounds and their application in cell biology are discussed below.

Amatoxins inhibit the transcription process in all eukaryotic cells. Enzymes of mammalian origin form particularly strong complexes, having K_D values around 0.01 μM. Organisms such as amphibians, nematodes, and insects possess enzymes with K_D values of *c.* 0.05 μM, as do vascular plants. Much less sensitive are the RNA polymerases II of yeasts and filamentous fungi such as *Saccharomyces cerevisiae, Physarum polycephalum*, and *Mucor rouxii*, exhibiting K_D values of about 1 μM. The edible mushroom *Agaricus bisporus* was reported to have an RNA polymerase II with a K_D value of 7 μM. As expected, amatoxin-accumulating mushrooms such as *Amanita phalloides* or *A. suballiacea* have the most insensitive RNA polymerase II enzymes with K_D values of *c.* 100 μM, or higher (see Faulstich, 1980).

Because of their high specificity for RNA polymerase II (RNA polymerase I is not inhibited at all, and RNA polymerase III only at 10^3 times higher concentrations of amanitin) the amatoxins constitute excellent tools for assaying whether complex biological processes including hormone action, virus replication, or the development of certain stages of an oocyte, depend on *de novo* syntheses of mRNA (hnRNA). The importance of α-amanitin as a specific inhibitor of transcription in biological research is comparable to that of cycloheximide as an inhibitor of translation. This review, however, is not of the native toxins, but rather their derivates or conjugates prepared for specific biological or medical purposes.

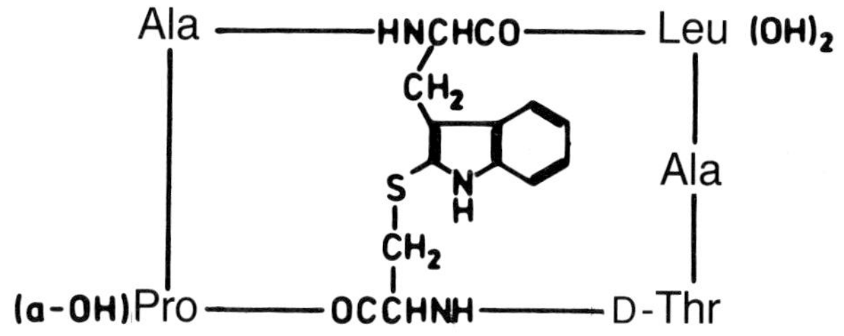

Figure 10.1. Structure of phalloidin.

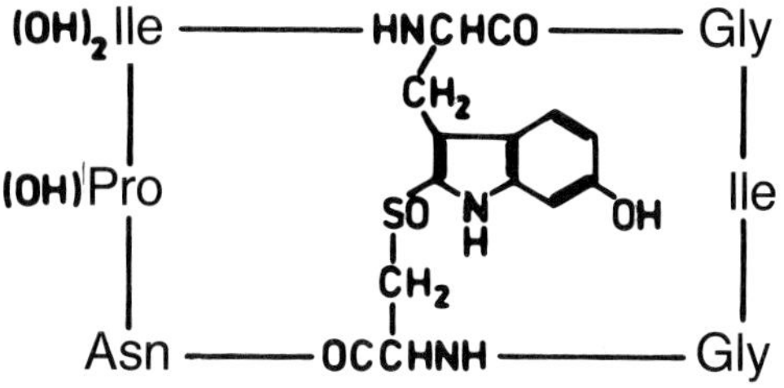

Figure 10.2. Structure of α-amanitin.

Amatoxin conjugates of proteins

One of the earliest approaches in this field was the preparation of antibodies. Amatoxin-specific antibodies were desirable for several purposes, such as the establishment of immunoassays for the detection of amatoxins, the possible use as an antiserum in immunotherapy in human *Amanita* poisonings, or the use of immunoglobulins as a model for studying the physical interaction of the cyclopeptides with proteins. Because of their low molecular weight (*c.* 900 D), amatoxins are not immunogenic by themselves but have to be coupled to protein carriers. For this, the acidic component of the amatoxins, β-amanitin, which contains an aspartic acid instead of asparagine, was employed. Later, we introduced a linker moiety into α-amanitin *via* an ether bridge, which in turn could react with lysine residues in proteins (Fig. 10.3; see Faulstich and Fiume, 1985). But it soon became apparent that amatoxins, when part of such protein conjugates, exhibit a toxicity many times higher than that of the native toxins. On average, the increase in *in vivo* toxicity is 5–10 fold, but may be as high as 50–100 fold where the pharmacodynamics of the toxin conjugate were changed, for example by the introduction of positively charged molecules

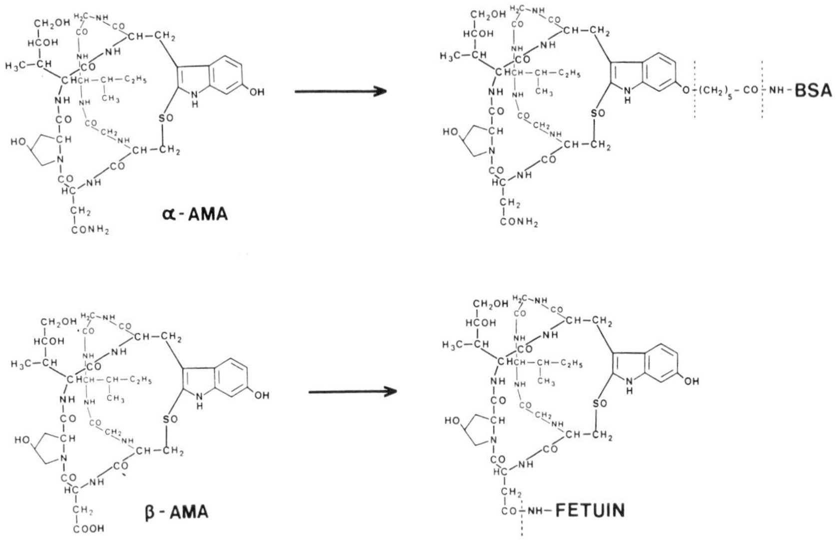

Figure 10.3 Protein derivatives of α- and β-amanitin.

of the coupling reagent into the protein as with the water-soluble carbodiimide ECDI, or, when fetuin is treated with neuraminidase, producing a glycoprotein with terminal galactose moieties. The *in vivo* toxicities of some amatoxin conjugates in comparison to the free toxins are shown in Table 10.1.

Fiume and co-workers have shown that the enhancement of amatoxin toxicity was due to a preferential uptake of the protein conjugates by phagocytosing cells, particularly RES-cells of the liver or cells of the proximal tubules of the kidney, while the normal target cells of the amatoxins, the hepatocytes, remained almost intact (Derenzini *et al.*, 1973).

Table 10.1. Toxicity of amatoxins in the free state and when attached to proteins (Faulstich and Fiume, 1985).

Amatoxin	Toxicity
α-amanitin	1
β-amanitin	1
bovine serum albumin-β-amanitin	6–100*
asialo-fetuin-β-amanitin	9
asialo-fetuin-α-amanitin	70

α-amanitin: LD_{50} = 0.35–0.75 mg/kg white mouse.
*Depends on the coupling method.

These workers also showed that macrophages, another type of phagocytosing cells, are highly sensitive to protein-bound amatoxins. For example, the IC_{25} value, i.e. the concentration required to kill 25% of macrophages in culture, was *c.* 50 times lower (0.05 $\mu g\ ml^{-1}$) for an albumin-β-amanitin derivate than for free amanitin (2.5 $\mu g\ ml^{-1}$; Barbanti-Brodano and Fiume, 1973). In contrast, non-phagocytosing cells such as rat sarcoma cells showed nearly equal sensitivity to amanitin in the free and bound states (IC_{25} values of 4.2 and 5.0 $\mu g\ ml^{-1}$ respectively). These data suggest that the endocytosis of amanitin bound to macromolecules followed by proteolytic digestion of the carrier can indeed provide a way for allowing amatoxins to act in cells other than hepatocytes. On the other hand, it is obvious that destruction of protein-clearing cells in an organism is undesirable, and that the killing of some types of phagocytosing cells contributes to the high *in vivo* toxicity of such toxin conjugates.

Amatoxin-specific antibodies

By the choice of an amanitin-resistant animal such as the rat, or by use of an appropriate protein carrier such as the glycoprotein fetuin, it was possible to raise polyclonal antibodies against amatoxins that in the past were widely employed in radioimmunoassays for measuring amanitin concentrations in urine, blood or gastric juice of *Amanita*-poisoned patients. One of the β-amanitin-fetuin conjugates was used also to prime rat lymphocytes that could be fused with mouse myeloma cells. A clone of these hybridomas was isolated that produced an IgG2a antibody with high affinity for amatoxins (K_D *c.* 0.1 – 1 nM). When injected intravenously into mice poisoned with α-amanitin by intra-peritoneal administration, the antibody not only failed to prevent the deaths of these animals, but even enhanced the toxicity of α-amanitin by a factor of two. The Fab fragment prepared from this monoclonal antibody by papain digestion and applied in the same way enhanced the amanitin toxicity even by a factor of 50. The explanation for this tremendous increase of toxic activity was found by histological examination of the animals, which revealed damage to the proximal tubule cells in the kidneys. Obviously, complexes of α-amanitin with the two immunoproteins were filtered to some extent in the kidney and taken up by phagocytosing endothelial cells (Faulstich *et al.*, 1988*a*). It appears that such immunocomplexes of amatoxins exhibit a kind of cytotoxicity to phagocytosing cells that is very similar to that of covalently bound protein conjugates, but differs from the latter by a preference for phagocytosing cells of the kidney.

It appeared that complexes composed of α-amanitin and this Fab fragment would represent a valuable model for studying the physical interactions of peptides with proteins, for example by X-ray analysis. Such studies would identify all the hydrogen bonds, hydrophobic, and van der Waals interactions that must cooperate in such a complex in order to explain apparent dissociation constants (K_D) in the range of 0.1 − 1.0 nM. In joint work with R. Huber's group, we were able to crystallize the Fab fragment, and fortunately found that the protein crystals could be soaked with an iodinated amatoxin without destroying them. Work is currently in progress to match the X-ray reflections of the Fab fragment with and without the toxin. We hope that this study will provide an answer to the question of whether the structural features of α-amanitin recognized as essential for inhibiting RNA polymerase II are similar, or identical to, those required for binding the peptide to the immunoglobulin.

Amatoxins internalized by receptor-mediated endocytosis

The lesions caused in phagocytosing cells by amatoxins bound to proteins were of the same kind as those caused by the free amatoxins in hepatocytes. This suggests that amatoxins can exhibit their toxic activities in all mammalian cells, provided that the cells have the ability to internalize the toxin. In most mammalian cells, uptake of natural amatoxins is poor. Hepatocytes, however, are very sensitive to amatoxins due to the presence of a suitable transporting system. Cultured hepatocytes, for example, suffer lethal poisoning with only 0.3 μM of α-amanitin (IC_{50}), while the corresponding value for cultured lymphocytes is at least 10 times higher (*c.* 5 μM). Conjugation of amatoxins to macromolecules considerably enhances their toxic activity for cells exhibiting phagocytosis. From experiments with macrophages, it can be estimated that for these cells

Table 10.2. Number of toxin molecules (*α*- or *β*-amanitin) covalently coupled to macromolecules.

Toxin	Carrier	Size of Macromolecule (kD)	Ratio toxin : protein
β	bovine serum albumin	67 000	1–3
β	human *γ*-globulin	160 000	1
β	concanavalin A	55 000	2
β	fetuin	48 000	2–4
α	bovine serum albumin	67 000	1–8
α	human *γ*-globulin	160 000	3
α	egg albumin	45 000	1–4
α	fetuin	48 000	4
β	poly-L-ornithine HBr	32 000	5–19

amanitin bound to proteins is more toxic than is the same amount of free toxin for hepatocytes. With this in mind, we anticipated that the attachment to the toxin vehicle of a homing molecule, for example, a hormone that binds to a receptor on the cell surface with high affinity and that is internalized together with the receptor, would lead to a special type of phagocytosis, the so-called receptor-mediated endocytosis, able to further increase the toxic activities of macromolecular bound amatoxins.

Proteins are limited in their capacity to serve as toxin carriers by a small number of lysine residues accessible for substitution. As shown in Table 10.2, the number of amatoxin molecules attached per protein molecule was in most cases between one and four, and thus could be expected to be too low to produce an amatoxin concentration inside the cell sufficient to inhibit transcription. We therefore designed a toxin vehicle composed of poly-L-ornithine (32 kD) as carrier, spiked with as many β-amanitin molecules as possible, which were linked to the side-chains of the polymer *via* carboxamide groups. The highest load achieved so far was *c.* 20 toxin molecules per polymer chain, corresponding to a toxin mass of about one third of the total. *Via* a disulfide-containing linker moiety, we attached to this toxin vehicle hr EGF_L, human recombinant (Leu-21) epidermal growth factor, at a ratio of 1-2 hormone molecules per toxin package (Fig. 10.4).

With this hormone-labelled and polymer-supported amatoxin, it was shown that very low concentrations of β-amanitin indeed suffice to inhibit the growth of a tumor cell line (A 431), which possesses on its surface an increased number of EGF receptors. The IC_{50} value was found to be 0.028 μM, and thus *c.* 20 times lower than for cells with nonspecific protein phagocytosis, such as macrophages or some endothelial cells types of liver and kidney. In a second cell line, SV-80 transformed human fibroblasts bearing a lower but still increased number of EGF receptors, the effect of the hormone-mediated endocytosis was less pronounced, but still present. With an IC_{50} value of 0.08 μM it was still significantly (6×) lower than the IC_{50} value expected for cells exhibiting unspecific endocytosis (Bermbach and Faulstich, 1990). Further experiments are needed to show whether susceptibility of a cell to the EGF-labelled toxin conjugate in all cases correlates with the number of EGF receptors present on the cell surface. If so, the correlation may provide a rationale for the therapy of certain tumor cells, such as lung or breast cancer cells, which have been reported to express on their surfaces EGF receptors in numbers distinctly higher than normal cells.

The high cytotoxicity of β-amanitin in the hormone-labelled conjugate allows several conclusions to be drawn concerning internalization, transport and processing of the toxin vehicle:

1. Internalization must have occurred, because amatoxins cannot exhibit toxic activities from the exterior of a cell.

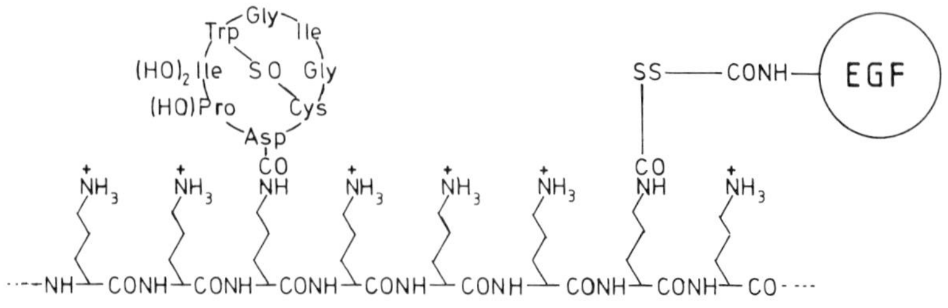

Figure 10.4. Structure of poly-L-ornithine spiked with β-amanitin and labelled with human recombinant (Leu-21) epidermal growth factor (EGF) via a disulphide-containing linker moiety.

2. The polymeric carrier must have been degraded, probably in a lysosomal compartment, because amatoxins bound to polymers have been shown to possess only *c.* 1% of the inhibitory activity to RNA-polymerase II compared to the free state.
3. The amatoxin molecule obviously had resisted the proteolytic attack, as was to be expected since no mammalian enzyme is known able to inactivate the cyclopeptides.
4. The toxin after its release from the carrier must have been transported from the lysosomal compartment to its target in the nucleus, either by diffusion through the cytoplasm or by vesicular transport.

Amatoxins bound to polymers and conjugated to certain hormones may also become of interest to cell biologists as tools for investigating the intricate sequence of events that follows the binding of some hormones to their receptors. The internalization and the following processing of hormone-labelled macromolecules that includes the present knowledge on this process is shown diagrammatically in Figure 10.5. Since amatoxins withstand proteolytic digestion, it should be possible to localize amatoxins in each stage of the processing, either free or bound, by the use of amatoxin-specific antibodies combined with an antibody-detecting system. As a preliminary basis for future studies, we measured the period of time between the addition of the hormone-labelled toxin and the first decrease observed in [^{3}H]U-incorporation into the RNA of A 431 cells. The earliest significant inhibition of RNA synthesis was found after 15 minutes, indicating that the whole processing, and thus each of the single steps, had taken not more than this time. A detailed electron-microscopic study of this kind of receptor-mediated endocytosis is in progress.

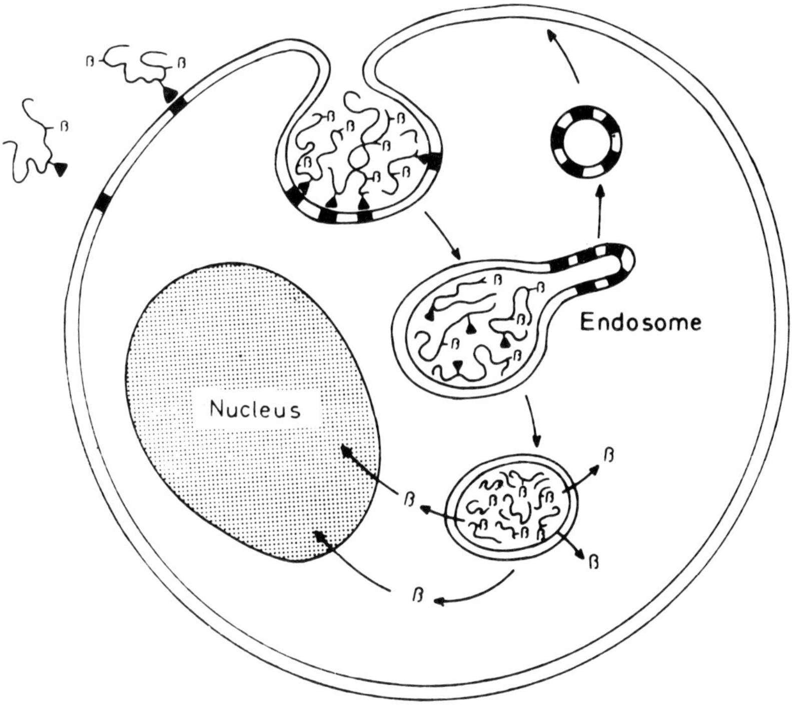

Figure 10.5. Scheme for the proposed internalization into a cell of a carrier-bound amanitin (β) labelled with a ligand *via* receptor-mediated endocytosis. Endosomal or lysosomal degradation may release the toxin which makes its way into the nucleus.

Fluorescent amatoxins

Besides the derivatization of amatoxins with macromolecules like proteins or polyaminoacids, cyclopeptides were submitted to chemical modifications of the side chains in a way that would allow their coupling to reporter groups, such as fluorescent moieties. Based upon the specific interaction of amatoxins with RNA-polymerase II, such amanitin derivatives should allow the localization of the enzyme in cells to be studied by light microscopy, provided substitution with the fluorescent moiety is not deleterious to the complex formation. From studies of the structure-activity relationship of amatoxins (Wieland and Faulstich, 1983) we knew that the hydroxy groups of hydroxy-L-proline and hydroxy-L-isoleucine are involved in binding to the enzyme and therefore should be left intact. We further knew that the aspartic acid/asparagine moiety would not tolerate a substitution with bulky residues without significant decrease of activity. The only functional group identified as not essential for binding was the 6′-hydroxy group of the indol nucleus.

By etherification of this OH group we introduced a linker moiety which subsequently was reacted with fluorescein isothiocyanate (FITC) (Fig. 10.6). In collaboration with Friedlinde Bautz's group, we were able to show that in cultured rat kangaroo (PtK 1) cells RNA polymerase II is indeed localized in the nuclei when the cells are at rest, but becomes diffusely distributed over the whole cell body when the cells are in mitosis. A bright fluorescence of the spindle apparatus and the centrioles could be attributed to an amanitin-binding protein different from RNA polymerase II. This protein is of unknown structure and function. A corresponding α-amanitin derivative linked to tetramethylrhodamine, a red fluorescent moiety with a higher stability against bleaching, proved useless because of high background staining, probably due to nonspecific binding of the cationic rhodamine moiety to DNA.

Fluorescent phallotoxins

Fluorescent derivates were also prepared from phalloidin. The cytoskeleton of cells is under investigation in many laboratories, and fluorescent phallotoxins have become useful for identifying microfilaments or other cell structures containing filamentous actin. Here too, the choice of the side-chain to be substituted by the bulky fluorescent groups required consideration. Studies of the structure-activity relationship of phallotoxins had revealed that the dihydroxylated leucine side-chain is probably located juxtaposed to the actin binding site (Wieland and Faulstich, 1983). A functional amino group was consequently introduced into this side-chain and reacted with various fluorescent reagents (Fig. 10.7). The fluorescein derivative of phalloidin was the first tool for visualizing stress fibres

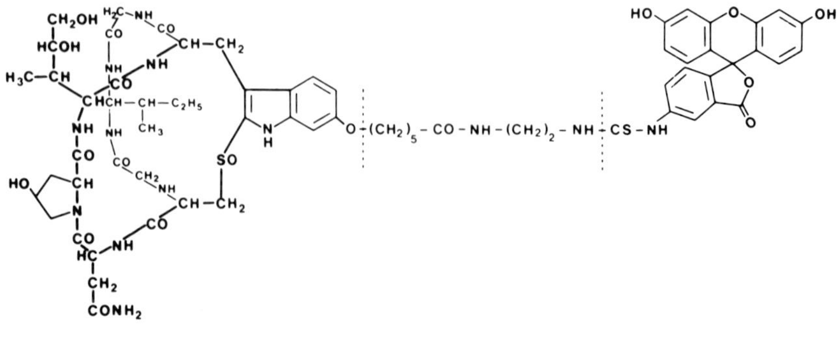

Figure 10.6. Structure of fluorescein-labelled α-amanitin.

(bundled actin filaments) in mammalian cells independent from the double immuno-technique (Wulf *et al.*, 1979). Because of its higher stability under irradiation, however, many laboratories today prefer the use of phalloidin labelled with the tetramethylrhodaminyl residue (Faulstich *et al.*, 1988*b*).

One of the major interests in cytoskeletal studies is to reveal interactions that may exist between the various components of the cytoskeleton. Interactions can be expected if morphological studies show that different cytoskeletal components are co-localized. In order to detect co-localization, it is necessary for two or three cytoskeletal components to be visualized in one and the same cell preparation. Double staining with, for example, a fluorescein-labelled antibody for one component and a rhodamine-labelled antibody for a second, are regarded as routine. Less usual is the triple staining of three cytoskeletal components, which requires that the spectra of the three fluorescent stains used show no or only minimal overlap. In

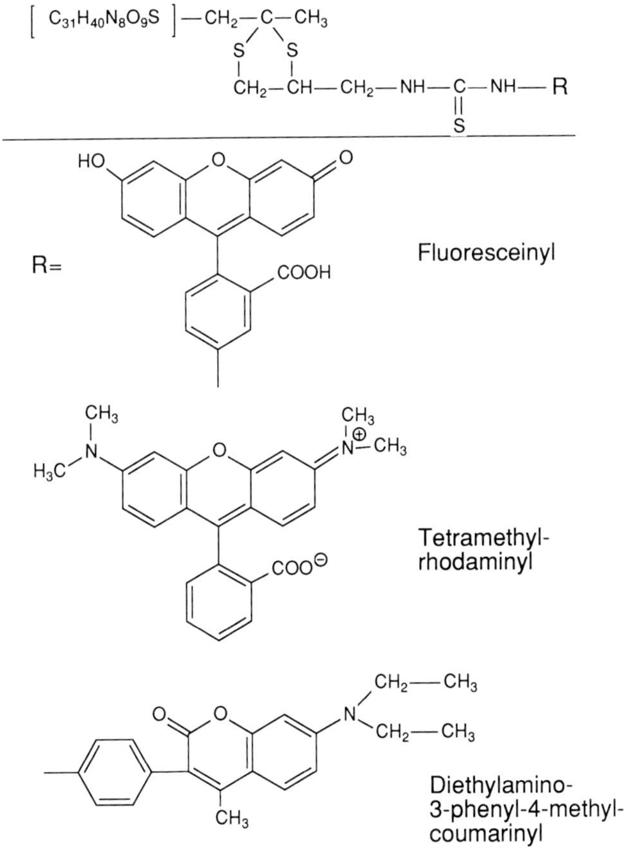

Figure 10.7. Structures of various fluorescin-labelled phallotoxins.

collaboration with Victor Small's group (Small *et al.*, 1988) we established that a coumarine-labelled phalloidin, because of its blue fluorescence, can provide a useful tool for making visible actin-containing structures in fibroblasts (Fig. 10.7), side by side with, for example, tubulin and vinculin, if the latter are reacted with their specific antibodies and made visible with other antibodies labelled with tetramethylrhodamine and fluorescein.

Biotinylated phalloidin

Studies on the organization of the cytoskeleton in non-muscle cells are also performed at the level of electron microscopy. Here, the detection of filaments would require their decoration with standardized gold particles. It was our plan to design a phalloidin-biotin conjugate, which could be sandwiched between actin and streptavidin dotted with, for example 10 nm gold particles. The compound shown in Figure 10.8 met these requirements, particularly by containing a 12-atom spacer moiety between aminodithiolano-phalloidin and biotin. Spacers that were less extended, for example glycyl or diglycyl, did not work. The latter compounds were able to form complexes with filamentous actin but not to bind the streptavidin component at the same time.

In collaboration with Brigitte Jokusch's group (Faulstich *et al.*, 1989), we were able to prove that the biotinylated phalloidin shown (Fig. 10.8) was indeed able to locate gold particles bound to streptavidin specifically at places where negative staining revealed the presence of a thin filament. This was shown for a preparation of filamentous muscle actin as well as for the cortical web of fixed mammalian cells. Although the number of gold particles in close proximity with filaments was much lower than expected from the number of actin units constituting the filaments, the decoration was significant because the number of gold particles in the background or in controls was extremely low.

We believe that biotinyl phalloidin will be a useful tool for identifying

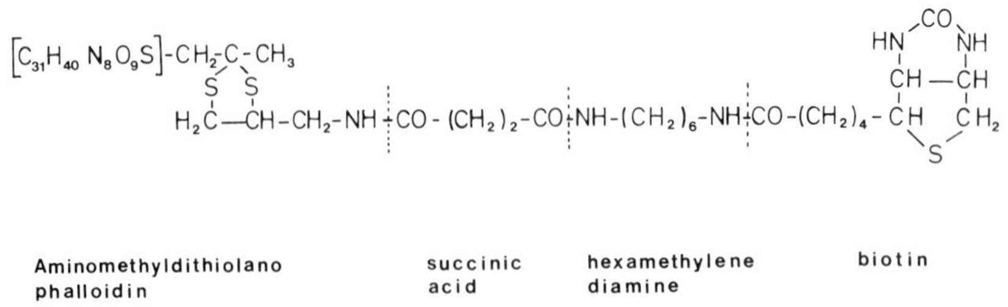

Figure 10.8. Structure of a biotinylated phallotoxin with a 12-atom spacer moiety.

actin in complex structures in which the presence of filaments is not obvious, or, alternatively, for distinguishing real actin filaments from actin-like filaments of other composition. For a chemist it is of course a challenge to further improve the structure of the biotinylated phallotoxins until actin decoration in electron microscopic pictures attains its optimum.

References

Barbanti-Brodano, G. and Fiume, L. (1973) Selective killing of macrophages by amanitin-albumin conjugates. *Nature, London* 243, 281–283.

Bermbach, U. and Faulstich, H. (1990) Epidermal growth factor labeled β-amanitin-poly-L-ornithine: preparation and evidence for specific cytotoxicity. *Biochemistry* 29, 6839–6845.

Derenzini, M., Fiume, L., Marinozzi, V., Mattioli, A., Montanero, L. and Sperti, S. (1973) Pathogenesis of liver necrosis produced by amanitin-albumin-conjugates. *Laboratory Investigation* 29, 150–158.

Faulstich, H. (1980) The amatoxins. In: Hahn, F., Kersten, H., Kersten, W. and Szybalski, W. (eds), *Progress in Molecular and Subcellular Biology.* Springer, Berlin, pp. 88–122.

Faulstich, H. and Fiume, L. (1985) Protein conjugates of fungal toxins. In: Widder, K.J. and Green, R. (eds), *Methods in Enzymology.* Vol. 112. *Drug and Enzyme Targeting.* Part A. Academic Press, New York, pp. 225–237.

Faulstich, H., Kirchner, K. and Derenzini, M. (1988*a*) Strongly enhanced cytotoxicity of the mushroom toxin α-amanitin by an amatoxin-specific Fab or monoclonal antibody. *Toxicon* 26, 491–499.

Faulstich, H., Zobeley, S., Rinnerthaler, G. and Small, J.V. (1988*b*) Fluorescent phallotoxins as probes for filamentous actin. *Journal of Muscle Research and Cell Motility* 9, 370–383.

Faulstich, H., Zobeley, S., Bentrup, V. and Jockusch, B.M. (1989) Biotinylphallotoxins: preparation and use as actin probes. *Journal of Histochemistry and Cytochemistry* 37, 1035–1045.

Small, J.V., Zobeley, S., Rinnerthaler, G. and Faulstich, H. (1988) Coumarine-phalloidin: a new actin probe permitting triple immunofluorescence microscopy of the cytoskeleton. *Journal of Cell Science* 89, 21–24.

Wieland, T. and Faulstich, H. (1978) Amatoxins, phallotoxins and antamanide: the biologically active components of poisonous *Amanita* mushrooms. *Critical Reviews in Biochemistry* 5, 185–260.

Wieland, T. and Faulstich, H. (1983) Peptide toxins from *Amanita.* In: Keeler, R.F. and Tu, A. (eds), *Handbook of Natural Toxins.* Vol. 1. *Plant and Fungal Toxins.* Marcel Dekker, New York, pp. 585–635.

Wulf, E., Deboben, A., Bautz, F.A., Faulstich, H. and Wieland, T. (1979) Fluorescent phallotoxins, a new tool for the visualization of cellular actin. *Proceedings of the National Academy of Sciences, USA* 76, 4498–4502.

Wulf, E., Bautz, F.A., Faulstich, H. and Wieland, T. (1980) Distribution of fluorescent α-amanitin (FAMA) during mitosis in cultured rat kangaroo (PtK1) cells. *Experimental Cell Research* 130, 415–420.

11

Mycology, Mycologists and Biotechnology

J.D. Miller, *Plant Research Centre, Agriculture Canada, Ottawa, Ontario K1A OC6, Canada.*

ABSTRACT The majority of mycologists are taxonomically and/or ecologically oriented. The following discussion presents the case that these scientists can play a critical role in the most economically-important sectors of industrial biotechnology. Specialized knowledge relating to the ecology of some species of *Aspergillus* have recently led to the discovery of some potent natural insecticides. The collection of isolates of higher marine fungi for taxonomic studies by academic researchers has been exploited by a number of pharmaceutical companies. Reliable taxonomy has been deemed essential by regulatory authorities with respect to the registration process of fungi in novel food technology and biorational control strategies. In contrast to other industries, biotechnology companies have emerged forming strong linkages to university and government-based researchers. Mycologists can and should take advantage of industrial interest in their skills. This will result in useful technologies and more funding for basic and applied research on fungi.

Introduction

Most members of the Mycology Society of America or the British Mycology Society list themselves as ecology or taxonomically oriented and there is a modest number of physiologists. It is obvious that the latter group as well as the growing number of molecular biologists that use filamentous fungi in their work are key to the "new" biotechnology. The potential for other mycologists to contribute to the economic and social benefits of biotechnology is less obvious and hence underestimated. In many OECD countries, academic funding agencies are directing more resources towards

Frontiers in Mycology. Honorary and General Lectures from the Fourth International Mycological Congress, Regensburg 1990. Edited by D.L. Hawksworth.

Table 11.1. International comparison of biotechnology companies. Adapted from Anon. (1989b).

	Canada	UK	USA	Japan
Health	30%	28%	57%	26%
Agriculture	18%	22%	13%	23%
Suppliers	5%	28%	19%	26%
Other	47%[1]	22%	11%	25%[2]

[1]Environmental and food biotechnologies.
[2]Food biotechnology.

the development of industry-university linkages. Research partnerships between industry and government laboratories have also been emphasized in recent years in Canada, USA, and various EC countries. The pool of resources for academic research has been held constant for about 10 years in many countries. The average university or government researcher must be content with little funding or learn to collaborate with industry.

With the increased emphasis on industry-research organization linkages, the biotechnology sector has been targeted for special attention. Canada's National Biotechnology Strategy emphasizes industrial linkages though a number of programmes that leverage industrial research and development (R&D). These include the Industrial Research Assistance Program and Natural Science and Engineering Research Council-Industry Program, among others. Basic research is funded concerning "industrial microorganisms for the chemical and food industries" and "microbial ecology in relation to fermentation and waste treatment systems and to predict behaviour in environmental and agricultural applications" (Anon., 1989*a*). This kind of research offers many opportunities for mycologists.

Studies of biotechnology companies in Canada, USA, UK and Japan have been published in 1989 and 1990. Approximately 2,100 companies were identified as engaged in biotechnology research. Between the four countries, the product areas of these companies were quite similar (Table 11.1). More of the Canadian companies are in the natural resource sector, environment and food. The USA has a higher proportion of companies involved in health care products. An important feature of the organization of biotechnology companies in the countries studied was their propensity to form alliances. Alliances were reported with universities, other research companies, government laboratories additional to financial interactions. About 75% of the companies examined had at least one research alliance. Canadian biotechnology companies report technical advice comes about equally from universities and their own staff (Anon., 1989*b*). Biotechnology companies in these countries are engaged in R&D in health care, agri-

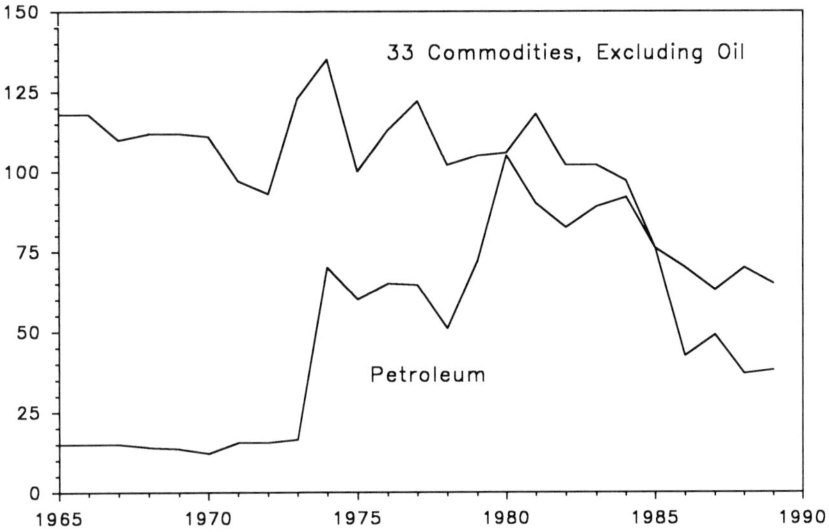

Figure 11.1. "Real" commodity prices, 1965–1989. Index 1979–1981 = 100; adapted from World Bank data by K.A. Newton, Economic Council of Canada.

culture/forestry, speciality chemicals, environment and food technologies. These organizations have adopted as a principal strategy the formation of alliances with university and government researchers.

At least three economic factors will drive commercial R&D for the next decade: economic growth of developing countries, free trade and stability of commodity prices. There is a general convergence of real gross domestic product per capita of many countries towards that of the USA and Canada. Assuming a continued liberalization of trade, competitive advantage will be scrutinized for each product or commodity. Commodity prices have been declining in real terms for the past 15 years and this trend is largely expected to continue (Fig. 11.1). These factors will have a major impact on commercial R&D. Technology that increases value added and/or improves the return on a product will be essential to continued economic growth. This general principle is especially important in agriculture and forestry.

A model to illustrate the opportunities for increasing profit in a time of low commodity prices can be found in agriculture (or forestry). Crop yield is a function of many variables (Fig. 11.2). There is an absolute constraint on yield, namely the genetic potential of the plant. The yield that should accrue is that based on inputs of management, energy, fertilizer, protection as appropriate. This is the "affordable" yield. In practice, the "actual" yield is less than the "affordable" yield because of biological stress. This includes chronic infections by pathogens including viruses and even the use of

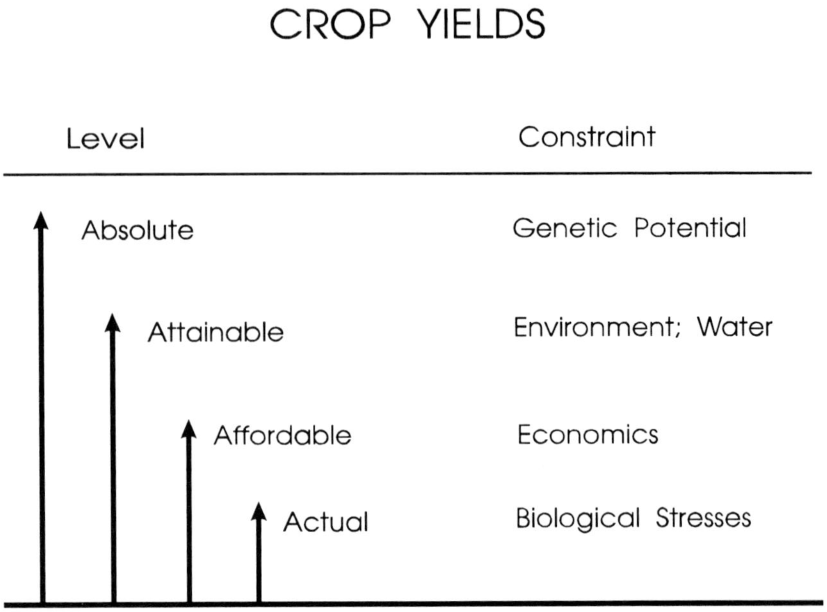

Figure 11.2. Factors relating to crop yield. Adapted from a slide from a presentation by Dr R.J. Cook (USDA-ARS, Pullman, Washington, USA).

photosynthate by phylloplane microbes (Smeddegaard-Petersen and Tolstrup, 1986). Technologies that reduce the difference between the affordable and actual yields of any crop will be attractive investments.

Additional to these economic factors, the concept of "sustainable economic growth" will continue to attract more attention in society. In general terms this refers to products and technologies that for example "reduce, reuse and recycle" and reduce or eliminate toxic hazards and wastes. Technologies that replace environmentally-damaging products in a cost-effective fashion will have markets.

The purpose of this contribution is to consider the economic and social framework for industrial fungal biotechnology in relation to mycology and mycologists. There are many useful reviews of the technology part of the fungal biotechnology equation (e.g. Leong and Berka, 1990; May *et al.*, 1989). These have largely concentrated on aspects of the so-called "new" biotechnology, that is technologies employing some molecular biology. This would include for example, the use of a fungus to produce a human protein. These products are not yet very important in industry; the "old" biotechnologies, particularly industrial fermentations and enzyme production account for the vast majority of economic activity directly involving fungi.

As the biotechnology industry matures, research and development will focus on the development of products. Market opportunities are mostly driven by economic and social factors and less by technology (albeit with some spectacular exceptions). Research that may be interesting may not be commercial. However, the nature of the emerging biotechnology industry is that it seeks collaborations. The activities of mycologists are most important to these companies. The remaining part of this discussion will focus on some examples in agriculture and forestry to put the above issues into a mycological context. The case could similarly be made for the many opportunities for mycologists in other "biotechnologies" such as environmental pollution control and remediation technology.

Secondary metabolites

Investigations of fungi for secondary metabolites other than mycotoxins are largely confined to pharmaceutical companies. The production, isolation and characterization of secondary metabolites requires considerable infrastructure. Modern organic chemistry involves extensive use of high resolution nuclear magnetic resonance and mass spectroscopy, both costly instruments. In the past, the large drug companies have employed mass screening techniques in the discovery of secondary metabolites. This practice was very productive until recently. The largest companies are now placing greater emphasis on ecological information to guide the collection and metabolite screening process (Dreyfuss, 1990; Geissbuler *et al.*, 1987).

This practice has two broad directions: (1) the investigation of previously overlooked ecological groups; and (2) ecological studies of fungi that suggest the presence of compounds with biological activity. Since the mid 1970s, marine higher fungi have been investigated for bioactive compounds by a number of companies. This has opened up opportunities for several marine mycologists to collect and provide strains of this group of fungi for screening (Miller, 1986, 1991; Newman and Clement, 1991). Another example are recent studies of endophytes of various plants. Endophytes of grasses, trees and even seaweeds are known to produce bioactive compounds (Clark *et al.*, 1989; Clay, 1987, 1988). Some of the research on endophytes has been supported by pharmaceutical companies (Dreyfuss, 1990). In both of these examples, fungal taxonomists and ecologists have been supported and allowed to publish the taxonomic and ecological aspects of the industrial collaborations (e.g. Jones, 1988; Petrini *et al.*, 1990).

Studies of interference competition of fungi have also been a useful tool for identification of strains with biological activity (Demain, 1989; Wicklow, 1981, 1989). This has been used to select isolates of species that appear to be competitive in nature. Some lignicolous marine fungi were found to restrict the growth of other species on small wood blocks that had been placed in the sea (Miller *et al.*, 1985). Further studies of some of these strains led to the identification of antifungal compounds (Strongman *et al.*, 1987). Studies of lignicolous freshwater fungi have similarly provided evidence of the production of antifungal compounds (Shearer and Zare-Maivan, 1988; Shearer and Bartolata, 1990).

A stunning example of the value of fungal ecology as a tool for the discovery of bioactive compounds has come from studies of *Aspergillus flavus.* The recognition that *A. flavus* produces sclerotia that contain various indole metabolites not found in the mycelium or conidia led to investigations of the significance of this observation (Wicklow and Cole, 1982). Nitidulid beetles overwinter on maize colonized by *A. flavus.* Feeding trials demonstrated that *Carpophilus hemipterus* consumed mycelium and conidia of the fungus but avoided the sclerotia. This was explained by the presence of a number of insecticidal aflavanines found only in the sclerotia (Gloer, *et al.*, 1988; Wicklow *et al.*, 1988). Other, new, insecticidal compounds have been since found uniquely in the sclerotia of the related species *A. nomius* (Gloer *et al.*, 1989) and from *A. tubingensis* (TePaske *et al.*, 1989). These findings were considered in relation to the chemical ecology of the fungus (Wicklow and Cole, 1982). More recently, it has been shown that there is a complex relationship between the spread of *A. flavus* to maize, nitidulid beetles and other insects (Lussenhop and Wicklow, 1990). These studies of the chemical ecology of sclerotial fungi have resulted in the discovery of a number of natural insecticides with potential commercial value.

Biorational pesticides

Particularly in North America, there has been increasing concern with the use of chemical pesticides. In California, public concern has recently been translated into action. The "Safe water and toxic enforcement act of 1986", an initiative measure (Proposition 65), requires among other things that business is prohibited from "knowingly or unknowingly exposing any individual to a chemical known to the State to cause cancer or reproductive toxicity without giving clear and reasonable warning" (Anon., 1986). A consultation proposal of the Pesticide Registration Review of Canada contained a minority report from the Canadian Labour Congress calling for the phasing out of "all pesticide products with a potential for harm to human health by 1988" (Anon., 1990*a*). A review of the politics of pesticides noted that "chemical makers no longer dismiss environmentalists as hopeless modern Luddites" (Bosso, 1988).

Of the approximately \$19 $\times$ 10^9 US worth of pesticides that were sold in 1989, only 0.5% were biological products. A number of useful products were registered in the 1970s including *Bacillus thuringiensis* and polyhedrosis virus. However, these successes did not prompt further product development. This is largely because chemicals were cheap, effective and there was less public concern about the environment. The development of a new chemical pesticide requires \$15–30 $\times$ 10^6 US of investment in an 8–10 year development cycle (Carlton, 1990). Guidelines for experimental release and registration of naturally-occurring microbes in crop protection and production are available in North America. For many products, registration costs of \$0.5–1.0 $\times$ 10^6 can be expected (Anon., 1990*b*; Popiel and Olkowski, 1990). Pesticide-resistant strains of insects and plant pathogens have risen from virtually nil to > 450 from 1950–80 (Cairncross, 1989). The discovery phase of biorational pesticides will come largely from collaborations of large and small companies and university and government laboratories (Carlton, 1990).

The use of so-called "hard chemicals" in North America is declining mainly with improved application technology and increased potency. However, the above-noted social and economic factors appear to be stimulating interest in biorational products. There are many fungi that are "pests of pests" and careful studies of these may justify commercial investment. However, in North America, the registration of products requires an understanding of the general mechanism and the product must past the "tier 1" toxicology package (Table 11.2). This point seems to have eluded many investigators. This includes those investigating the use of toxigenic moulds such as *Trichoderma* and *Gliocladium* in biological control (Kommedahl and Windels, 1981). These particular moulds produce many dangerous mycotoxins (Taylor, 1986). However, one laboratory is attempting to produce strains of *Trichoderma* that retain their useful properties without producing mycotoxins (A. Bottalico, pers. comm.).

The chemical structure of one or more or the biochemicals that confer the biological activity of interest can be of greater commercial interest than the organism itself. The use of mouldy bread to treat bacterial infections preceded the large-scale production and use of penicillin. It is obvious that the latter mode of chemotherapy is more useful (Wainwright, 1989). For many years, efforts have been made to develop mycoherbicides, fungal pathogens of weeds (TeBeest and Templeton, 1985; Templeton *et al.*, 1986). Despite much effort, there are still only a few products registered in the USA and none as yet in Canada. There is increasing interest in mycoherbicides in Europe (Ayres and Paul, 1990). One of the major problems is that mycoherbicide products depend on the successful development and spread of an "epidemic" in the field. This relates to environmental conditions, particularly the availability of moisture just after application, and virulence.

Where phytotoxins are involved in the pathology of the weed disease, the opportunity exists to identify strains that produce more phytotoxin thus

Table 11.2. Generalized tier 1 requirements for US/Canada registration of microbials in crop production. Adapted from Anon. (1987, 1990).

Human health risk	acute oral toxicity acute dermal toxicity acute pulmonary acute intravenous irritation hypersensitivity/immunological
Nontarget testing	avian oral wild mammal freshwater fish nontarget insects nontarget plants honeybee

increasing virulence. In recent years, a number of groups have examined the phytotoxins from weed pathogens. Several of these have been found to be host-specific and not to be mycotoxins (Kenfield *et al.*, 1988, 1989). More investigations of this type are needed (see Duke and Lydon, 1987). The attitude of North American regulatory authorities towards unmodified host-specific phytotoxins may be similar to that towards the *Bacillus thuringiensis* toxin. Killed cell preparations of "Bt" are treated as bacteria by the guidelines, not as chemicals. Herbicidal compounds from soil microorganisms (Cutler, 1988) would probably be considered as chemicals.

Both of these areas of research are appealing to the makers of chemical herbicides because these natural compounds can be produced in fermentation and chemically modified to increase their potency/host specificity (Fischer and Bellus, 1984). Some argue that this approach would overcome efficacy problems associated with mycoherbicides. Presumably enzymes capable of degrading microbial phytotoxins exist (Duke and Lydon, 1987; Kenfield *et al.*, 1988). Additionally, the genetic information for the production of some kinds of phytotoxins (e.g. peptides) offers another alternative. A phylloplane-adapted bacterium with the ability to produce a fungal phytotoxin may be a convenient application vehicle. This allows the production of the "toxin" *in situ* possibly under a broader range of environmental conditions.

Information on the biochemical basis of a biological control which one wishes to sell has three essential merits: (1) facilitates or, in some cases, enables registration; (2) the "toxin" (or siderophore, etc.) may be more useful in itself than the fungus; and (3) more effective patent protection can be obtained. Strain "X" of a naturally-occurring fungus may well have a useful biological activity. However, other strains could be isolated and registered with the same activity described in a process patent. If the mode of action is associated with particular biochemicals and this is part of the patent, better protection can probably be obtained.

Biorational control strategies may emerge in a fairly direct fashion such as those from studies of weed pathogens. Alternatively, such technologies could come from systems with less obvious potential. Poisonings of domestic animals from ingestion of grasses have been known in the southwestern USA, Germany, New Zealand and India for many years (Clay, 1987). Most of these occurrences have now been explained in terms of the production of toxins by various ascomycetes living inside leaf tissue. It has been demonstrated that this association results in reduced herbivory by insect pests. A number of other benefits to the plant have been shown or speculated upon. These include improved disease resistance, nematode resistance, drought tolerance and increased vegetative reproduction (Clay, 1987, 1989).

In the case of the clavicipitaceous endophytes of grasses, reduced herbivory has been explained by the production of ergot alkaloid-like toxins. In the case of pasture or rangeland grasses for cattle, these endophytes are a problem in terms of animal health and productivity. However, endophyte-infected grasses are considerably more productive than otherwise. Recent studies have demonstrated that there is a compatibility system between endophyte strains and their hosts (Leuchtmann and Clay, 1989). It may be possible to develop strains of endophytes that produce insecticidal toxins but with modest toxicity to grazing animals (Clay, 1989). Endophyte-based biorational control products (infected seeds) are already marketed for lawns (Kaplan, 1990). Similar possibilities exist with respect to some tree species. Some conifer needle endophytes have been demonstrated to produce insecticidal toxins (Clark *et al.*, 1989).

The research that has led to the proposition that endophytes may be useful in biorational control came from research directed to preventing animal poisonings. The work that followed was focused on the toxins produced and on the ecological significance of the phenomenon. Finally, the identification of the presence of compatibility systems allows the meaningful possibility of the development of useful commercial products. The properties of such products would include low mammalian toxicity of the grass, host specificity, and zero or negligible spread to uninoculated (host) plants.

There are a number of other fungal systems where detailed study of the ecology mechanism of action could lead to useful biorational control products. This would include fungal diseases of insects and fungus-plant-insect associations. Among other things, this would identify other toxins that have useful properties. There is general recognition of the notion that natural molecules evolved to be "at their best". Although there are exceptions, natural molecules are potent and, quite often, functionally specific. The new recognition that fungi make mixtures of toxins and "synergizers" also offers interesting possibilities for biorational control strategies.

This was first clearly demonstrated with the metabolites of *Fusarium graminearum*. This fungus produces many metabolites including several trichothecenes. Trichothecenes are among the most potent inhibitors of protein synthesis known (Feinberg and McLaughlin, 1989). *F. graminearum* produces several dozen other metabolites from several biogenic origins (Miller *et al.*, 1991). This includes perhaps seven that accumulate in quantity in naturally-contaminated grain (Foster *et al.*, 1986). Studies of the effects of mixing some of the latter metabolites have been conducted on caterpillars. Compounds that were non-toxic in themselves were potent synergizers of the trichothecene tested (deoxynivalenol; Dowd *et al.*, 1989). This same phenomenon has been found with metabolites of some *Aspergillus* and *Penicillium* species (Dowd, 1988, 1989).

No insect (or animal) has been found that is resistant to trichothecenes (see Strongman *et al.*, 1990). This remarkable fact may be true because the trichothecene-producing *Fusarium* and *Myrothecium* species produce mixtures of compounds that affect a number of biochemical/organ systems in the "target" animal. This would minimize the possibility of resistance to the principal toxin. The discovery of synergizers to allow both the reduction of the amount of chemical pesticide applied and prolong the "life" of chemicals is a fertile area of research (Gressel and Shaaltiel, 1988). Natural compound synergizers may well be of interest. As these molecules are found, it has been cogently argued that they should not be "squandered". For example, the techniques of molecular biology have been employed to cause the Bt toxin to be produced by plants. Already, insect resistance to Bt preparations is a problem (Marrone, 1990; McGaughey, 1990).

Plant disease

Traditionally, plant pathology and mycology have had various interactions. The advent of molecular techniques in plant pathology has opened some new possibilities for the management of plant diseases. These studies involve a range of expertise including molecular biology but also fungal biochemistry and enzymology. In principle, the latter have been incorporated in studies of plant pathogens for many years. In practice, such efforts have not been directed at one host-pathogen interaction for long enough to acquire a detailed understanding. Workers at the Central Research and Development Department of DuPont in Delaware have studied the rice blast fungus/rice interaction in some detail.

This group has been able to identify that the so-called Mendelian genes for resistance to this disease in rice are in fact clusters of "molecular genes". Race-specific resistance mechanisms have been found by empirical means that are both "complete" and "partial" (Bonman *et al.*, 1989). Resistance is comprised of genes coding for a variety of biochemical systems including the formation of fungal metabolites (Chumley and Valent, 1990). Many of these genes have not been identified prior to the use of molecular techniques. These studies are costly, but it is reasonable to conjecture that they will allow the development of safer crop protection strategies. In addition, disease-resistance with increased durability can be expected by virtue of combining the most effective resistance mechanisms (Valent, 1990). Such plants may not suffer an inordinate yield penalty because genes that code for marginal components of resistance can be excluded.

Taxonomy

One of the critical areas of fungal biotechnogy that is habitually overlooked and underestimated is the importance of reliable identifications and culture collections. Knowledge of the identity and hence of related species of fungi found to produce valuable metabolites is very useful. The regulation of either natural or genetically-altered fungi in food processes or for release as crop protection or production technologies *depends* on taxonomic placement. If organisms are in a toxigenic or pathogenic family or group, certain tests are indicated.

The US Federal Insecticide, Fungicide and Rodenticide Act requires information on the identity, method of identification and limit of detection. Provisional Canadian requirements for organisms with pesticidal or fertilizer claims include "taxonomic position, serotype, strain, phage type, forma speciales, drug resistance, biochemical and plasmid profile (as relevant), geographical distribution, known or suspected relationship to a known human, avian, mammalian or fish pathogen, toxin production". UK requirements include "nature of organism or agent i.e. species, host range, pathogenicity to man, animals, plants or microbes". OECD, EC, Australian and New Zealand requirements are not substantively different from these (Anon., 1987).

A number of research projects for the production of food-grade materials have not been commercialized because the mould used produced mycotoxins. Similarly, cases have been reported where microbial inoculants were incorrectly identified. The correct placements sometimes moved strains in or out of toxigenic genera. It is clear, that taxonomic position is not an infallible indicator of the pathogenic or toxigenic potential of a fungus. None the less, correct placement is an extraordinarily useful indicator of these properties. Resources for taxonomy and taxonomists are in a decline. This situation is very undesirable for the development of certain kinds of biotechnology. Corporate support for the maintenance of taxonomic expertise on a world-wide basis would seem important.

Summary

Biotechnology involving fungi can be much more than molecular biology. There are a great many opportunities for ecologically-, physiologically- or taxonomically-oriented mycologists to participate in industrial research. A feature of the biotechnology industry, perhaps unlike other commercial organizations, is the willingness and need for collaborative ventures with university and government researchers. This allows the strengths of each to be combined in interdisciplinary research. The most basic of mycological skills, the identification of fungi, is one of the most useful contributions that we can make to biotechnology. Appropriate arrangements with companies interested in the properties of fungi will allow research on the ecology and diversity of fungi to continue and be published while contributing to sustainable economic development.

At the Fourth International Mycological Congress (Regensburg), there were symposia concerning aquatic fungi, plant breeding, secondary metabolites, phytotoxins, pathogens/pests, endophytes, biological control, pollution, and biological processing among others. Many of the individual papers contained information useful to the emerging biotechnology industry.

Acknowledgements

I would like to thank the organizing committee of IMC4 for allowing me the opportunity to speak on this subject and Drs D. Wicklow, E.B.G. Jones, R.J. Cook and R. Greenhalgh for sharing their insight. Opportunities for me to understand Canadian R&D afforded by Agriculture Canada, the National Research Council (IRAP) and the Natural Science and Engineering Research Council are appreciated. Collaborations with a number of companies in Europe, USA, and Canada have also influenced my thinking.

References

Anon. (1986) *California Code of Regulations Division* 2 *Chapter* 3. *Safe Drinking Water and Toxic Enforcement Act of 1986.*

Anon. (1987) *Review of International Regulations.* Pesticides Directorate, Agriculture Canada, Ottawa.

Anon. (1989*a*) *Annual Report: National Biotechnology Advisory Committee.* Industry, Science and Technology Canada, Ottawa.

Anon. (1989*b*) *Canadian Biotech '89: on the Threshold.* Industry, Science and Technology Canada, Ottawa.

Anon. (1990*a*) *A Proposal for a Revised Federal Pest Management Review.*

Secretariat, Federal Pesticide Registration Review, Ottawa.
Anon. (1990*b*) *Proceedings: Workshop on Microbial Pest Control Agents.* Pesticides Directorate, Agriculture Canada, Ottawa.
Ayres, P. and Paul, N. (1990) Weeding with fungi. *New Scientist* 127(1732), 36–39.
Bosso, C.J. (1988) Transforming adversaries into collaborators: interest groups and the regulation of chemical pesticides. *Policy Sciences* 21, 3–22.
Bonman, J.M., Bandong, J.M., Lee, Y.H., Lee, E.J., Valent, B. (1989) Race-specific partial resistance to blast in temperate Japonica rice cultivars. *Plant Disease* 73, 496–499.
Cairncross, F. (1989) The environment. *The Economist* 312(7618), S1–18.
Carlton, B.C. (1990) Economic considerations in marketing and application of biocontrol agents. In: Baker, R. and Dunn, P. (eds), *New Directions in Biological Control: alternatives for suppressing agricultural pests and diseases.* Alan R. Liss, New York, pp. 419–434.
Chumley, F.G. and Valent, B. (1990) Genetic analysis of melanin-deficient, nonpathogenic mutants of *Magnaporthe grisea. Molecular Plant-Microbe Interactions* 3, 135–143.
Clark, C.L., Miller, J.D., Whitney, N.J. (1989) Toxicity of conifer needle endophytes to spruce budworm. *Mycological Research* 93, 508–512.
Clay, K. (1987) The effect of fungi on the interaction between host plants and their herbivores. *Canadian Journal of Plant Pathology* 9, 380–388.
Clay, K. (1988) Fungal endophytes of grasses: a defensive mutualism between plants and fungi. *Ecology* 69, 10–16.
Clay, K. (1989) Clavicipitaceous endophytes of grasses: their potential as biocontrol agents. *Mycological Research* 92, 1–12.
Cutler, H.G. (1988) Perspectives on discovery of microbial phytotoxins with herbicidal activity. *Weed Technology* 2, 525–532.
Demain, A.L. (1989) Functions of secondary metabolites. In: Hershberger, C.L., Queener, S.W. and Hegeman, G. (eds), *Genetics and Molecular Biology of Industrial Microorganisms.* American Society for Microbiology, Washington, D.C., pp. 1–11.
Dowd, P.F. (1988) Synergism of aflatoxin B toxicity with the co-occurring fungal metabolite kojic acid to two caterpillars. *Entomology Experimental and Applied* 47, 69–71.
Dowd, P.F. (1989) Toxicity of naturally occurring levels of the *Penicillium* mycotoxins citrinin, ochratoxin A and penicillic acid to the corn ear worm, *Heliothis zea* and the fall army worm, *Spodoptera frugiperda* (Lepidoptera: Noctuidae). *Environmental Entomology* 18, 24–29.
Dowd, P.F., Miller, J.D. and Greenhalgh, R. (1989) Toxicity and interactions of some *Fusarium graminearum* metabolites to caterpillars. *Mycologia* 81, 646–650.
Dreyfuss, M.M. (1990) Fungal ecology as a tool for screening secondary metabolites. In: *Abstracts, Fourth International Mycological Congress, Regensburg.* Fourth International Mycological Congress, Regensburg, p. 117.
Duke, S.O. and Lydon, J. (1987) Herbicides from natural compounds. *Weed Technology* 1, 122–128.
Feinberg, B. and McLaughlin, C.S. (1989) Biochemical mechanism of action of trichothecene mycotoxins. In: Beasley, V.R. (ed.), *Trichothecene Mycotoxicosis:*

pathophysiological effects. Vol. I. CRC Press, Boca Raton, pp. 27–36.

Fischer, H.P. and Bellus, D. (1984) Phytotoxins from microorganisms and related compounds. *Pesticide Science* 14, 334–346.

Foster, B.C., Neish, G.A., Lauren, D.L., Trenholm, H.L., Prelusky, D.B. and Hamilton, R.M.G. (1986) Fungal and mycotoxin content of slashed corn. *Microbiology Aliments Nutrition* 4, 199–203.

Geissbuler, H., d'Hondt, C., Kunz, E., Nyfeler, R. and Pfister, K. (1987) Reflections on the future of chemical plant protection research. In: Greenhalgh, R. and Roberts, T.R. (eds), *Pesticide Science and Biotechnology*. Blackwell Scientific Publications, Oxford, pp. 3–14.

Gloer, J.B., TePaske, M.R., Sima, J.S., Wicklow, D.T. and Dowd, P.F. (1988) Antiinsectan aflavanine derivatives from the sclerotia of *Aspergillus flavus*. *Journal of Organic Chemistry* 53, 5457–5460.

Gloer, J.B., Rinderknecht, B.L., Wicklow, D.T. and Dowd, P.F. (1989) Nominine: a new insecticidal indole diterpene from the sclerotia of *Aspergillus nomius*. *Journal of Organic Chemistry* 54, 2530–2532.

Gressel, J. and Shaaltiel, Y. (1988) Biorational herbicide synergists. In: Hedin, P.A., Menn, J.J. and Hollingworth, R.M. (eds), *Biotechnology for Crop Protection*. American Chemical Society, Washington, D.C., pp. 4–24.

Jones, E.B.G. (1988) Do fungi occur in the sea? *The Mycologist* 2, 150–157.

Kaplan, J.J. (1990) Grass-fungus symbiosis saves lawns. *Agricultural Research* 38, 11.

Kenfield, D., Bunkers, G., Strobel, G.A. and Sugawara, F. (1988) Potential new herbicides – phytotoxins from plant pathogens. *Weed Technology* 2, 519–524.

Kenfield, D., Bunkers, G., Strobel, G. and Sugawara, F. (1989) Fungal phytotoxins – potential new herbicides. In: Graniti, A., Durbin, R.D. and Ballio, A. (eds), *Phytotoxins and Plant Pathogenesis*. Springer-Verlag, Berlin, pp. 319–336.

Kommedahl, T. and Windels, C.E. (1981) Introduction of microbial antagonists to specific courts of infection: seeds, seedlings and wounds. In: Papavizas, G.C. (ed.), *Biological Control in Crop Production*. Osmun, Allanheld, pp. 227–248.

Leong, S.A. and Berka, R.M. [eds] (1990) *Molecular Industrial Mycology*. Marcel Dekker, New York.

Leuchtmann, A. and Clay, K. (1989) Experimental evidence for genetic variation in compatibility between the fungus *Atkinsonella hypoxyon* and its three host grasses. *Evolution* 43, 825–834.

Lussenhop, J. and Wicklow, D.T. (1990) Nitidulid beetles (Nitidulidae: Coleoptera) as vectors of *Aspergillus flavus* in pre-harvest maize. *Transactions of the Mycological Society of Japan*, 31, 63–74.

McGauhey, W.H. (1990) Insect resistance to *Bacillus thuringiensis* endotoxin. In: Baker, R. and Dunn, P. (eds), *New Directions in Biological Control: alternatives for suppressing agricultural pests and diseases*. Alan R. Liss, New York, pp. 583–597.

Marrone, P. (1990). Comment. In: Baker, R. and Dunn, P. (eds), *New Directions in Biological Control: alternatives for suppressing agricultural pests and diseases*. Alan R. Liss, New York, p. 642.

May, G.S., Waring, R.B., Osmani, S.A., Morris, N.R. and Denison, S.H. (1989) The coming of age of molecular biology in *Aspergillus nidulans*. In: Nevalainen, H.

and Penttila, M. (eds), *Molecular Biology of Filamentous Fungi.* Foundation for Biotechnical and Industrial Fermentation Research, pp. 11–20.

Miller, J.D. (1986) Secondary metabolites in lignicolous marine fungi. In: Moss, S.T. (ed.), *The Biology of Marine Fungi.* Cambridge University Press, Cambridge, pp. 61–67.

Miller, J.D. (1991) Screening marine fungi for bioactive molecules. In: Jones, E.B.G. (ed.), *The Isolation and Study of Marine Fungi.* John Wiley & Sons, Chichester, in press.

Miller, J.D., Jones, E.B.G., Moharir, Y.E. and Findlay, J.A. (1985) Colonization of wood blocks in Langstone Harbour. *Botanica Marina* 28, 251–257.

Miller, J.D., Greenhalgh, R., Wang, Y.Z., Lu, M. (1991) Trichothecene chemotypes of three *Fusarium* species. *Mycologia* 83, in press.

Newman, D.J. and Clement, J.J. (1991) Production of pharmacologically active materials by some higher marine fungi. *Canadian Journal of Botany,* submitted.

Petrini, O., Hake, U., and Dreyfuss, M.M. (1990) An analysis of fungal communities isolated from fruticose lichens. *Mycologia* 82, 444–451.

Popiel, I. and Olkowski, W. (1990) Biological control of pests and vectors: pros and cons. *Parasitology Today* 6, 205–207.

Shearer, C.A. and Bartolata, M. (1990) Experimental determination of *in situ* competitive interactions among aquatic lignicolous fungi. *Abstracts, Fourth International Mycological Congress, Regensburg.* Fourth International Mycological Congress, Regensburg, p. 156.

Shearer, C.A. and Zare-Maivan, H. (1988) *In vitro* hyphal interactions among wood- and leaf-inhabiting ascomycetes and fungi imperfecti from freshwater habitats. *Mycologia* 80, 31–37.

Smedegaard-Petersen, V. and Tolstrup, K. (1986) Yield-reducing effect of saprophytic leaf fungi in barley crops. In: Fokkema, N.J. and van Den Heuvel, J. (eds), *Microbiology of the Phyllosphere.* Cambridge University Press, Cambridge, pp. 160–174.

Strongman, D.B., Miller, J.D., Calhoun, L., Findlay, J.A. and Whitney, N.J. (1987) The biochemical basis for interference competition among some lignicolous marine fungi. *Botanica Marina* 30, 21–26.

Strongman, D.B., Strunz, G.M. and Yu, C.M. (1990) Trichothecene mycotoxins produced by *Fusarium sporotrichioides* DAOM 197255 and their effects on spruce budworm *Choristoneura fumiferana. Journal of Chemical Ecology* 16, 1605–1609.

Taylor, A. (1986) Some aspects of the chemistry and biology of the genus *Hypocrea* and its anamorphs *Trichoderma* and *Gliocladium. Proceedings of the Nova Scotia Institute of Science* 36, 27–58.

TeBeest, D.P. and Templeton, G.E. (1985) Mycoherbicides: progress in the biological control of weeds. *Plant Disease* 69, 6–10.

Templeton, G.E., Smith, R.J. and TeBeest, D.O. (1986) Progress and potential of weed control with mycoherbicides. *Reviews in Weed Science* 1, 1–14.

TePaske, M., Gloer, J.B., Wicklow, D.T. and Dowd, P.F. (1989) Three new aflavanines from the sclerotia of *Aspergillus tubingensis. Tetrahedron* 45, 4961–4968.

Valent, B. (1990) Rice blast as a model system for plant pathology. *Phytopathology* 80, 33–36.

Wainwright, M. (1989) Moulds in ancient and more recent medicine. *The Mycologist* 3, 21–23.

Wicklow, D.T. (1981) Interference competition and the organization of fungal communities. In: Wicklow, D.T. and Carroll, G.C. (eds), *The Fungal Community.* Marcel Dekker, New York, pp. 351–375.

Wicklow, D.T. (1989) Fungal succession: technology transfer of ecological data. In: Hattori, T., Ishida, Y., Maruyama, Y., Morita, R.Y. and Uchida, A. (eds), *Recent Advances in Microbial Ecology.* Japan Scientific Societies Press, Tokyo, pp. 280–284.

Wicklow, D.T. and Cole, D.T. (1982) Tremorgenic indole metabolites and aflatoxins in sclerotia of *Aspergillus flavus* Link: an evolutionary perspective. *Canadian Journal of Botany* 60: 525–528.

Wicklow, D.T., Dowd, P.F., TePaske, M.R. and Gloer, J.B. (1988) Sclerotial metabolites of *Aspergillus flavus* toxic to a detritivorous maize insect (*Carpophilus hemipterus*, Nitidulidae). *Transactions of the British Mycological Society* 91, 433–438.

Conservation and Education

12

Mycologists and Nature Conservation

E. Arnolds, *Biological Station, Landbouwuniversiteit Wageningen, Kampsweg 27, 9418 PD Wijster, The Netherlands.*

Abstract Only in the last decade have mycologists started to pay attention to fungi as objects of nature conservation. The motives for such efforts are briefly mentioned. Data on fungi can provide important additional information for nature conservation. One of the main reasons to pay more attention to the conservation of fungi is the increasing information on drastic changes in the mycoflora in large parts of Europe. The methods required to study such changes are concisely outlined. A selection of data is provided on declining species diversity in general, and on decreasing saprobic fungi in grasslands and ectomycorrhizal fungi in forests in particular. The importance of natural forests for many wood-inhabiting fungi is stressed. The use of Red Lists, containing surveys of threatened fungi in particular areas, is discussed. The effects of various management practices on different groups of the mycoflora are described with special reference to forests and nutrient-poor grasslands. Finally, suggestions for future research in relation to the conservation of fungi are provided.

Introduction

Mycology and nature conservation have developed separately for a long time. Mycologists were not interested in conservation aspects of fungi because they were not aware of possible threats to the mycoflora or they simply were so busy with their own research that they ignored the signs around them. Nature conservationists, on the other hand, hardly paid any attention to fungi, mainly because of lack of knowledge, but maybe also because they thought that special measures for this group were unnecessary. This pattern has changed in recent years, at least in Western and Central Europe, where mycologists were confronted with alarming changes in the

Frontiers in Mycology. Honorary and General Lectures from the Fourth International Mycological Congress, Regensburg 1990. Edited by D.L. Hawksworth. © C · A · B International. 1991.

mycoflora. So far the studies and efforts have been mainly focused on macrofungi and little attention has been paid to microfungi.

The significance of fungi for nature conservation

Motives for the conservation of fungal species and populations are not fundamentally different from motives to conserve other organisms (Winterhoff, 1978; Winterhoff and Krieglsteiner, 1984):

- *Ecological importance*: Fungi are essential components of biocoenoses by their functions as decomposers of organic matter, parasites of other organisms and mutualistic symbionts in mycorrhizas. This is also true for ecosystems of economic value, in particular forests.
- *Value as indicator organisms*: Fungi can be excellent bioindicators in view of their functions, niche-differentiation and species diversity. For instance, many species of ectomycorrhizal fungi are indicators for the degree and kind of air pollution (e.g. Fellner, 1988, 1989; Arnolds, 1990); wood-inhabiting fungi for intensity of forestry practice, and grassland fungi for soil conditions, type of management, and degree of disturbance (Arnolds, 1982; Nitare, 1988).
- *Economic importance*: Fungi are important (potential) sources of food and medicines, and they can be used for selective delignification of wood and straw as well as the degradation of xenobiotics (e.g. Bumphus and Aust, 1987).
- *Importance for science*: Conservation of the gene pools of fungi is needed in order to extend our understanding of evolutionary processes and the resulting diversity in taxa, morphological structures, and ecological strategies.
- *Value for recreation and education*: Collecting of edible sporocarps of wild fungi is an important form of healthy recreation in many regions; the study of fungi is a hobby of a growing number of naturalists (Kreisel, 1960).
- *Esthetic value*: Sporocarps of many macrofungi are appreciated by many people as interesting and beautiful components of our environment and as a source of joy and creative activities, for instance photography.
- *Ethical motive*: Mankind is responsible for the continued existence of the variety of life-forms (including fungi), developed during evolution.

In my opinion the latter motive on its own is sufficient justification for efforts in the field of fungal conservation.

It may be argued that macrofungi are automatically protected by the good conservation and management of ecosystems, adjusted to vascular plants and animals (which can often more easily be recognized). However, special attention to fungi is justified because:

- Some habitats are very important for fungi, but to a much lesser degree for vascular plants and animals, for instance coniferous and deciduous forests on very poor soils (e.g. Jansen, 1984; Sammler, 1988; Wöldecke and Wöldecke, 1990), some types of old pastures (e.g. Rald, 1985; Nitare, 1988) and roadsides with old trees (P.J. Keizer, pers. comm.). These differences are also reflected in figures on species diversity of phanerogams and macrofungi in various plant communities, which are often not correlated (Arnolds, 1981; Senn-Irlet, 1986).
- A decrease of fungi may not coincide in some habitats with a decrease of other groups of organisms, and different factors may be responsible. This is for instance the case with ectomycorrhizal fungi (see below).
- Some fungal assemblages require special measures in order to maintain them (see below).
- The ecological consequences of changes in the mycoflora are different from those of primary producers and consumers. For instance, a reduction of mycorrhizal fungi may lead to lower vitality or even death of the host plants (e.g. Meyer, 1984; Fellner, 1988); a decrease of saprobic fungi may lead to litter accumulation and disturbed nutrient cycles (e.g. Kuyper and de Vries, 1990).

Data on the mycoflora should be used more often as (additional) arguments for the conservation of certain areas by the establishment of nature reserves or by planning measures. This is still exceptional nowadays, but nature conservation organizations and environmental planners are often willing to consider such data (e.g. Hille and Scholz, 1988; Wöldecke and Wöldecke, 1988, 1990).

Finally, all means and arguments must be mobilized to prevent further destruction of the ecosystems on this planet. Mycologists have their own responsibility in this respect.

Recent changes in the mycoflora

Methods to study changes in the mycoflora

The composition of the mycoflora is, like that of vascular plants and the fauna, subject to evolutionary and geological processes and the natural succession of plant communities, and is therefore constantly changing. Most of these processes are so slow that the resulting changes cannot be established during a human lifetime. In the last millenium the mycoflora has also been strongly influenced by local man-made changes in the landscape, such as deforestation and afforestation, reclamation of bogs, etc. Only in the last decades have changes been noticed which are not directly related to such interferences, but instead to unwanted side-effects of human activities, such as environmental pollution.

Quantitative and semiquantitative data on changes in the macrofungal flora are scarce in comparison with data on phanerogams and larger animals. This is due to some fundamental and practical methodological problems. The former concern: (1) the dependence on the above ground appearance of reproductive structures, sporocarps, which abundance may not closely reflect the below ground occurrence of vegetative structures which are difficult to study and attribute to relevant fungi – in addition some sporocarps remain below the soil surface, and such hypogeous fungi may be quantitatively important in some ecosystems (e.g. Fogel, 1976); (2) the short lifespan of sporocarps which may therefore be missed when spot observations are made (e.g. Richardson, 1970); (3) the pronounced periodicity of most species (e.g. Krieglsteiner, 1977; Arnolds, 1985); (4) fluctuations from year to year attributable to weather conditions (e.g. Thoen, 1976; Agerer, 1985); and (5) which are coincidental with successional changes in plant cover and environment (e.g. Arnolds, 1988*b*).

The numbers of sporocarps of a species may vary from year to year by factors of 10 to 100, while the intervals between the successive appearance of the same fungus on a certain spot may extend from several years to decades. Practical problems involved in long-term studies include: (1) the scarcity of reliable older, and often recent, data; (2) difficulties related to taxonomy and nomenclature. Together these problems necessitate careful attention to methods and the reliability of different interpretations (Arnolds, 1985, 1988*a*; Derbsch and Schmitt, 1984, 1987).

Suitable methods, subject to the above restrictions are: (1) comparison of foray reports of different periods from a limited area; (2) comparison of distribution maps of well-investigated species (e.g. Arnolds, 1985; Parent and Thoen, 1986); (3) comparison of observations in selected plots or stands over the years (e.g. Nitare, 1988; Arnolds, 1990); and (4) data on the supply of edible macrofungi to markets (Derbsch and Schmitt, 1987).

In the following sections special attention is paid to changes in species diversity of macrofungi in general, and to changes in species composition of the floras of ectomycorrhizal fungi, saprobic fungi in grasslands, and wood-inhabiting fungi. The data are restricted to Northwest and Central Europe, since hardly any data from other parts of the world are available.

Changes in species diversity and functional groups of macrofungi

The most reliable data on the decreasing species diversity of macrofungi is that of Derbsch and Schmitt (1987). These authors compared the annual

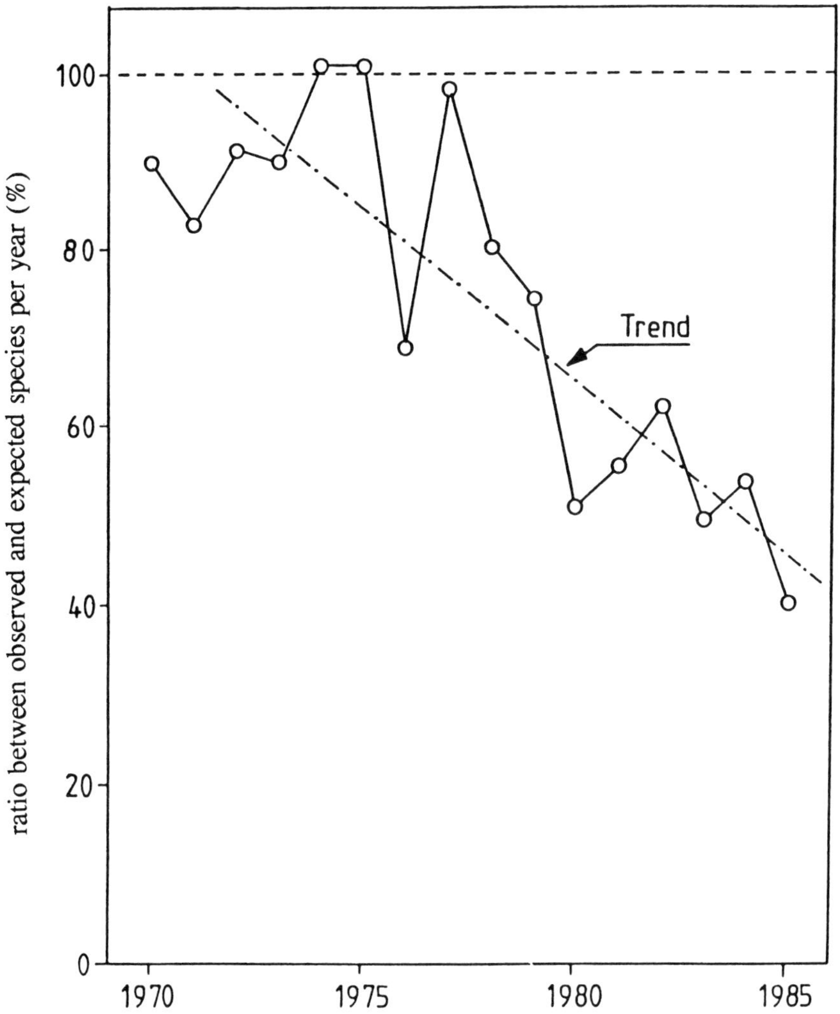

Figure 12.1. The ratio between observed and expected species of macrofungi per year in the period 1970–85 in Saarland (Germany), corrected for the annual numbers of excursions (modified from Derbsch and Schmitt, 1987).

numbers of species observed between 1970 and 1985 during more than 8000 excursions in Saarland, Germany. They demonstrated a decrease of annual species diversity of almost 60% during these years (Fig. 12.1). These developments can neither be attributed to unfavourable weather conditions during more recent years, not to changes in methodology. In a more detailed study restricted to one forest, Völklinger Kreuzberg, which they visited more than 3500 times between 1950 and 1985, Derbsch and

Schmitt (1987) recorded 852 species of macromycetes. Among these species 81 (9.5%) were recorded for the last time between 1950 and 1959, 180 (21.1%) between 1960 and 1969, and 194 (22.8%) between 1970 and 1979. They estimated that at least 40% of the local mycoflora has become extinct in 35 years.

Similar observations have been made by Rücker and Peer (1988) on the Hellbrunner Berg, near Salzburg. They compared data collected by Leischner-Siska in 1937, Friedrich in 1953 (very incomplete) and themselves in 1987. Although the methods applied by these authors were different, it seems probable that the number of species of macrofungi decreased from 172 in 1937 to 95 in 1987, a reduction of 45%. Grosse-Brauckmann and Grosse-Brauckmann (1978) compared the occurrence of sporocarps in the Darmstadt area between 1970 and 1976 with data collected by Kallenbach between 1918 and 1942. Of the 236 species listed by the latter, only 137 (60%) species had been found in recent years. A less pronounced decrease of species diversity per excursion has been reported from the Netherlands (Arnolds, 1985).

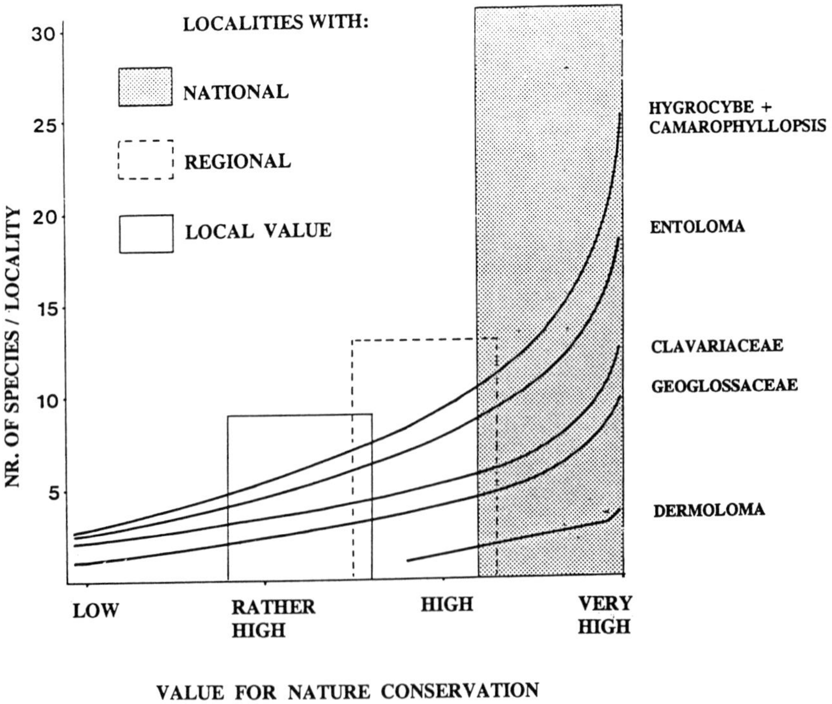

Figure 12.2. A scheme for assigning the value for nature conservation to seminatural grasslands in Sweden, based on the number of species of grassland fungi (modified from Nitare, 1988).

In Saarland both saprobic, soil-inhabiting, and ectomycorrhizal fungi living in mutualistic symbioses with trees and shrubs, are strongly decreasing (Derbsch and Schmitt, 1987). In the Netherlands a decrease is only prominent in ectomycorrhizal species and saprobic fungi from grasslands, whereas saprobic and necrotrophic wood-inhabiting fungi are significantly increasing (Arnolds, 1985, 1988*a*).

Saprobic fungi in grasslands

It has been calculated that in the Netherlands 365 species of macrofungi are predominantly found in grasslands (Arnolds and de Vries, 1989). While this is only 11% of the entire flora of macrofungi, the number of fungal species is higher than that of phanerogams in this habitat. At least 75% of these fungi are restricted to poor pastures or hayfields, often rich in mosses, on soils with low contents of nitrogen and phosphorus. Characteristic fungi include many species of the genera *Hygrocybe* (including *Camarophyllus*), *Camarophyllopsis, Clavaria, Dermoloma, Entoloma* (mainly subgenera *Leptonia* and *Nolanea*), and *Geoglossum* (Arnolds, 1980; Nitare, 1988). Many species of these groups have become rare in large parts of Europe (e.g. Winterhoff and Krieglsteiner, 1984; Rald, 1985). In The Netherlands (Arnolds, 1980) and Denmark (Rald, 1985) species of *Hygrocybe* s.lat. were used as indicator species for such grasslands with a rich mycoflora. Of the thousands of square kilometres of grasslands present in The Netherlands around the beginning of this century, only 50 relics with at least five species of *Hygrocybe* remain, covering together less than 10 km^2 (Arnolds, 1988*c*). The situation in Denmark, where Boertmann (1985) and Rald (1986) investigated *Hygrocybe* grasslands, is almost as dreary. Nitare (1988) restudied 157 grasslands in central and northern Sweden between 1982 and 1987, where Hakelier had collected Geoglossaceae between 1961 and 1965. In 85% of the localities the plant cover appeared to have drastically changed and the total number of records of Geoglossaceae in the 1980s was only 6.5% of that in the 1960s. Nitare (1988) introduced a model to evaluate the mycoflora in grasslands, which is probably also valid for other temperate parts of Europe (Fig. 12.2).

The mycoflora of grasslands is threatened by a number of factors, including the application of dung and artificial fertilizers (Arnolds, 1989*c*), ploughing and resowing, abandonment followed by spontaneous forest development (Nitare, 1988), afforestation (Nitare, 1988), and suboptimal management, also in nature reserves (see below).

Ectomycorrhizal fungi

A review of data concerning the decrease of ectomycorrhizal fungi in

Table 12.1. Changing species composition and abundance of above-ground sporocarps of mycorrhizal fungi between 1972 and 1989 in three replicate plots (each 1000 m^2) in mature oak forest (*Dicrano-Quercetum*) on dry, acidic sand in Drenthe, north-east Netherlands (from Arnolds, 1990).

	Average number of species per plot			Average maximum number of sporocarps per 1000 m^2		
	1972–73	1976–79	1988–89	1972–73	1976–79	1988–89
Coltricia	0.3	–	–	0	–	–
Hydnellum (incl. *Sarcodon*)	1.3	0.3	–	3	0	–
Tricholoma	2.7	1.0	–	19	8	–
Cantharellus	1.0	1.0	–	233	55	–
Hebeloma	0.7	1.3	–	37	10	–
Thelephora	1.0	1.0	–	1	2	–
Cortinarius	9.0	7.0	0.3	287	70	1
Inocybe	3.7	3.0	0.7	18	27	2
Boletus (incl. *Xerocomus*)	3.3	1.7	1.0	12	26	2
Amanita	2.7	2.0	1.3	341	41	3
Russula	4.0	6.0	2.0	717	52	11
Lactarius	3.7	4.0	3.0	1347	599	196
Paxillus	1.0	1.0	0.7	467	73	10
Laccaria	2.0	2.3	1.7	873	67	17
Scleroderma	1.0	1.0	1.3	370	77	128
total mycorrhizal fungi	37	32	12	4720	1110	370
saprobes on soil	21	33	22	2430	920	350
saprobes and parasites colonizing wood	15	15	26	180	330	680
total macromycetes	73	81	60	7340	2350	1400

Europe has been published by Arnolds (1990). Consequently only a selection of the most important points is presented here.

In The Netherlands complete lists of records of 15 forays made between 1912 and 1954 were compared with those of 15 forays made between 1973 and 1982 (Arnolds, 1985, 1988*a*). The average number of ectomycorrhizal species per foray decreased from 71 in the first period to 38 in the second period, or from 47% to 30% of the total number of species. Recent calculations, based on all available data (over 85000 records) confirmed this trend: the proportion of ectomycorrhizal species found on a foray was constant (46–48%) in the periods 1900–49, 1950–59 and 1960–69, but decreased sharply to 35% in the decade 1970–79 and only 26% in the

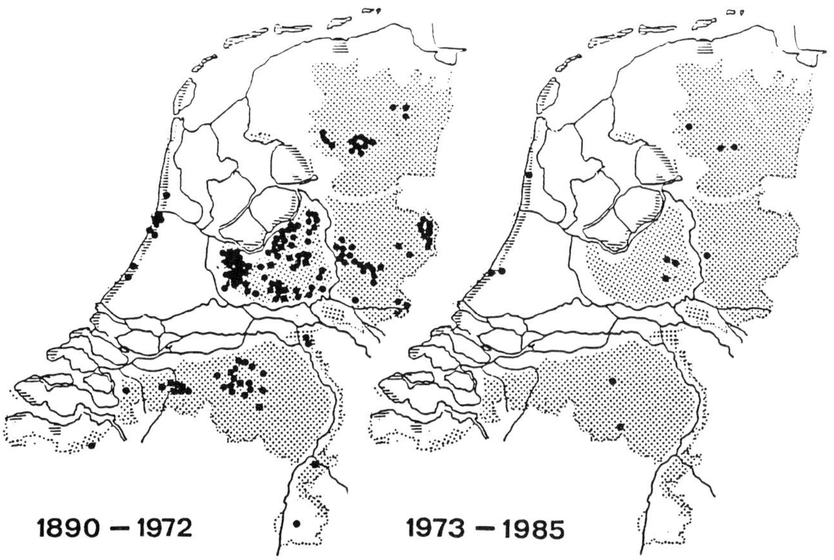

Figure 12.3. Occurrences (•), recorded between 1890 and 1972, and 1973 and 1985 in the Netherlands of sporocarps of four species of the ectomycorrhizal genus *Phellodon* (after Arnolds 1989a). Habitat characters: dotted = acid, pleistocene sands; hatched = calcareous, holocene sands.

decade 1980–89 (E. Arnolds and E. Jansen, unpubl. data).

Between 1972 and today, detailed information has been obtained on macrofungal communities in forests of *Quercus robur* on dry, nutrient-poor, acidic sand in the northeast Netherlands. Sporocarps were recorded in three replicate plots (each 1000 m^2) between 1972 and 1973, 1976 and 1979 (Jansen, 1984) and again in 1988 and 1989 (Table 12.1). The total number of macrofungi has only slightly decreased, but the number of mycorrhizal species has greatly decreased from 37 per plot in 1972–73, to 32 in 1976–79, and to only 12 in 1988–89. On the other hand, the number of soil-inhabiting saprobic species has not significantly changed and the number of lignicolous species has increased markedly from 15 to 26 per plot. The numbers of ectomycorrhizal sporocarps decreased by 77% between 1972–73 and 1976–79, and again by 67% between 1976–79 and 1988–89; now only 8% of the sporocarps found in the early 1970s are being produced. Decrease of species diversity was observed in most taxonomic groups, in particular in *Cortinarius, Inocybe, Tricholoma,* boletes, and hydnoid fungi (Table 12.1). The reduction of sporocarp production was more than 80% in all genera of mycorrhizal fungi except *Scleroderma.*

A substantial decrease in the number of records was demonstrated in The Netherlands for a large number of mycorrhizal species, for instance all hydnoid fungi (Fig. 12.3; Arnolds, 1989*a*) and *Cantharellus cibarius*

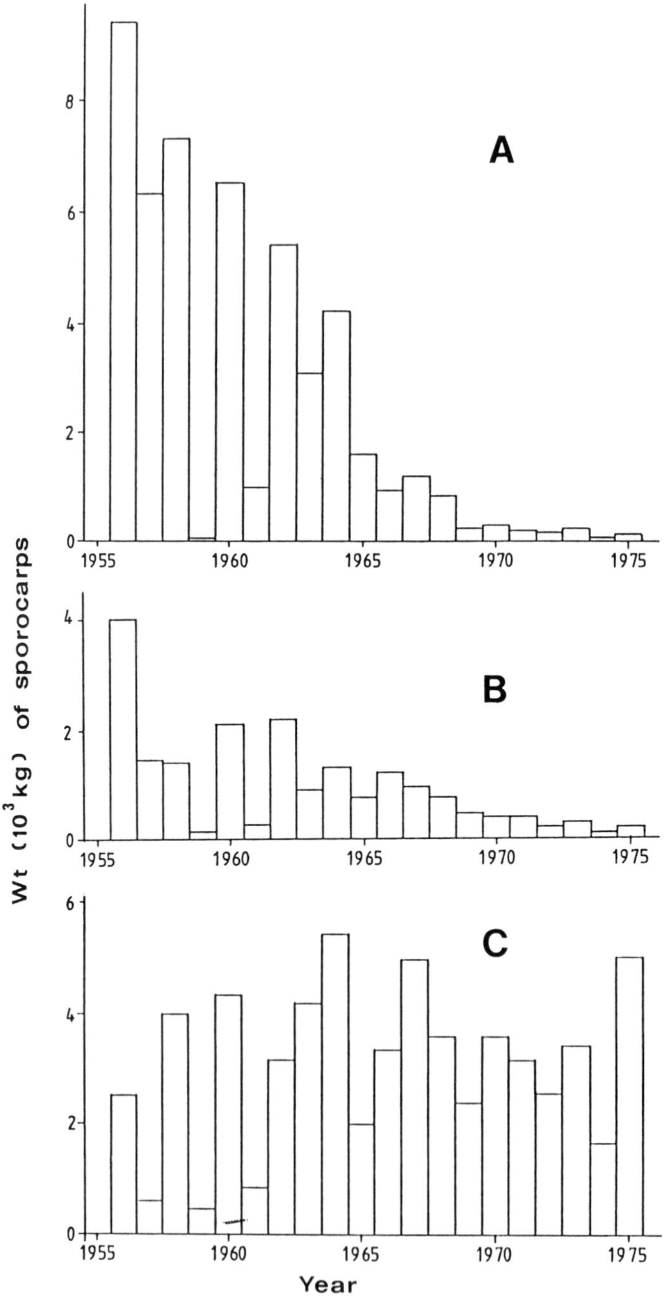

Figure 12.4. Weights of sporocarps of three edible fungi collected in forests in the Saarland (Germany) and supplied to the market in Saarbrücken between 1956 and 1975. (A) The mycorrhizal *Cantharellus cibarius*; (B) various species of boletes; and (C) the wood-inhabiting *Armillaria mellea* (from Derbsch and Schmitt, 1987).

(Jansen and van Dobben, 1987). A decrease of many hydnoid fungi was also documented in Germany on the basis of distribution maps from different periods, especially of those species associated with coniferous trees (Otto, 1990). A strong decline of *C. cibarius* was demonstrated in Saarland where Derbsch and Schmitt (1987) analyzed data on some edible fungi supplied to the Saarbrücken market between 1956 and 1975. The average weight of *C. cibarius* reaching the market between 1973 and 1975 was only 2% of that between 1956 and 1958; for boletes there was a 91% decline, whereas for the wood-inhabiting *Armillaria mellea* s.lat. there was an increase of 43% (Fig. 12.4). The same author estimated that the sporocarp production of *Russula rosacea* (i.e. *R. lepida*), a mycorrhizal associate of *Fagus sylvatica*, decreased in Saarland by 98% between 1967 and 1985.

Similar data have been published for Czechoslovakia. Fellner (1983) reported an 80% reduction in the variety of ectomycorrhizal fungi between 1958 and 1961, and two decades later when observing subalpine and montane forests in the Giant mountains. He also demonstrated a sharp decrease in the foray frequency of *Russula mustelina* between 1974 and 1985 in montane spruce forests in the Cheb district of western Czechoslovakia (Fellner, 1989).

The above studies, and other data, make it evident that many mycorrhizal fungi are decreasing markedly in large areas of Europe, in particular in densely populated or industrialized regions. This decline seems to be mainly attributable to indirect effects of air pollution, in particular to increases in the amount of available nitrogen (possibly in combination with acidification) and (or) to decreased tree vitality with consequent reductions in the transport of assimilates to roots and mycorrhizas (Termorshuizen and Schaffers, 1987; Arnolds, 1990). The decrease of sporocarps is often correlated with a decrease of mycorrhizal root frequency (e.g. Agerer, 1985; Schlechte, 1986; Jansen and de Vries, 1988).

The reduction of mycorrhizal fungi may lead to an increasing instability of forest ecosystems, among other things by increasing the water and nutrient stress in trees (Meyer, 1984).

Wood-inhabiting fungi

Wood is an extremely important substrate for macrofungi, especially for the Aphyllophorales and Heterobasidiomycetes which make up more than 75% of the lignicolous species. Bas (1976) estimated that 12% of the European agarics grow predominantly on wood, whereas according to Arnolds and de Vries (1989) 26% (868 species) of all macrofungi in The Netherlands belong to this category. Moreover, wood-decomposing basidiomycetes are of outstanding ecological importance in view of their almost unique capacity for the efficient breakdown of lignin (e.g. Rayner and Boddy, 1988).

The data on the decrease and vulnerability of lignicolous fungi seem to be contradictory at first sight. The Red Lists (see below) of Sweden (Hallingbäck, 1988), Norway (Høiland, 1988) and Finland (Rassi and Väisänen, 1987) comprise high proportions of this group, 41%, 39%, and 37%, respectively. In the former German Federal Republic (Winterhoff, 1984) and The Netherlands (Arnolds, 1989*b*), these proportions are only 16% and 12%, respectively. In the latter country even an increase of the frequency of many wood-inhabiting fungi was noticed (Arnolds, 1985, 1988*a*).

These contrasting data can be understood in the light of historical differences in forest practice. In the Nordic countries the majority of threatened lignicolous fungi are restricted to relics of virgin forest, very old trees, damp forests with constantly high air humidity, and some forest types with a restricted distribution. Most of these species probably became extinct long ago in densely populated and largely deforestated countries, such as The Netherlands. The increase of many species in these areas is mainly attributed to the ageing of many plantation forests, established during this century, offering larger quantities of suitable substrates, mainly branches and stumps (Arnolds, 1985).

Although many authors stress the importance of very old and virgin forests as habitats for wood-inhabiting fungi (e.g. Philippi, 1981; Winterhoff, 1989), surprisingly few exact data are available on the influence of different forestry measures on the diversity and floristic composition of this group of fungi. This is a promising area for future research.

Red Lists of fungi

The information on threatened fungi can be summarized in Red Lists, which contain the names of fungi considered as threatened or extinct in a certain area, often accompanied by details of distribution, ecology, and the reasons of threat. The first Red List to include fungi was published in the USSR in 1980 (Anon., 1980). Up to 1990 sixteen other national or regional lists on fungi have been published or are available in concept (Fig. 12.5; see Arnolds, 1989*b*). Most Red Lists are restricted to macrofungi, that for Finland (including Ustilaginales and Uredinales) being a notable exception (Rassi and Väisänen, 1987). The fungi enumerated are usually assigned to different categories of threat, ranging from extinct to potentially threatened. A fragment of the Red List of The Netherlands (one of the two lists published in English), including these categories, is given in Table 12.2.

The goals of these lists are not identical in all cases. In the USSR the species on the list are at the same time protected by law, and therefore only 20 species could be included. In all other countries the number of listed

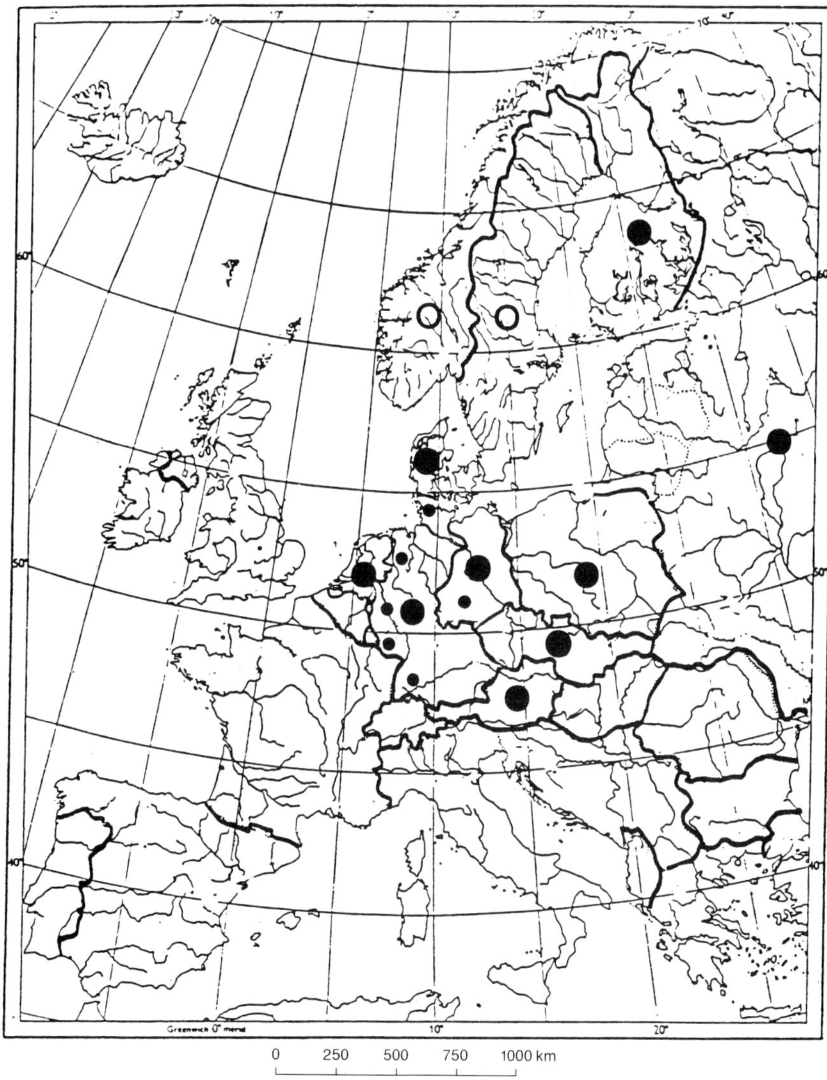

Figure 12.5. Distibution of Red Lists of macrofungi in Europe available in 1990. Large black dot = national list. Open dot = national list in concept. Small black dot = regional list.

species is much larger, varying from 132 (Norway) to 1208 (Saarland). The wider aims of these lists are:

1. To provide information for professional and amateur mycologists on the status of macrofungal species in order to draw their attention to threatened species and to areas where many such species are present.

Table 12.2. Extract from the Red List of macrofungi in The Netherlands (after Arnolds, 1989).

Names according to Arnolds (1984)

C = Categories of threatened species:

0: Probably extinct: species which were not recorded from The Netherlands since 1970, and which have not likely been overlooked

1: Threatened with extinction: very rare species, restricted to acutely threatened habitats or localities and rare species which have very strongly decreased in this century

2. Strongly threatened: Rare species, exclusively or predominantly growing in strongly threatened habitats, or which have strongly decreased in this century

3: Rare to scattered species, exclusively or mainly growing in threatened habitats and uncommon to scattered species which have distinctly decreased in this century

4: Potentially threatened: Rare species without distinct tendency to decrease and growing in habitats which seem not to be threatened at present.

Hab: Habitat code according to Arnolds (1984).
Sub: Substrate code according to Arnolds (1984).
Org: Code of associated organisms, according to Arnolds (1984).
E: Exploitation of habitat: M = mycorrhizal species, P = parasitic species, S = saprobic species.
F1: Frequency before 1970: CCC = very common, CC = common, C = rather common, R = rather rare, RR = rare, RRR = very rare, - = absent, ? = unknown.
F2: Frequency since 1970: see F1.

Lists: Records from other Red Data Lists (see Arnolds, 1989b).

Name	C	Hab	Sub	Org	E	F1	F2	Lists
Agaricus augustus	4	4.6	1.2	–	S	R	RR	
Agaricus bernardii	4	8.3	1.5	–	S	RR	RR	2,4,5,9
Agaricus campester	3	7.2	1.0	–	S	CCC	C	
Agaricus cupreobrunneus	3	7.7	1.4	–	S	C	R	1,2,4,5,9,11
Agaricus elvensis ss. Cooke	4	1.4	1.5	–	S	?	RRR	
Agaricus excellens	4	1.4	1.2	–	S	RR	RR	2,3,4,5,9
Agaricus geesterani	1	1.4	1.5	–	S	-	RRR	
Agaricus lanipes	4	1.4	1.5	–	S	RRR	RRR	4,11
Agaricus niveolutescens	4	1.6	1.2	–	S	RRR	RRR	4
Agaricus phaeolepidotus	4	1.5	1.2	–	S	RR	RR	2,3,4
Agaricus porphyrocephalus	2	7.7	1.4	–	S	RR	RRR	2,9
Agrocybe firma	4	1.5	3.0	1.0	S	RR	RR	1,2,3,4,6,9
Agrocybe paludosa	3	7.4	1.2	–	S	C	R	1,3,6,7,9
Agrocybe pusiola	3	7.7	1.4	–	S	RR	RR	2,3,4,9
Agrocybe vervacti	4	7.7	1.2	–	S	RRR	RRR	2,4,5,9,11

Table 12.2. *Continued.*

Name	C	Hab	Sub	Org	E	F1	F2	Lists
Aleurodiscus amorphus	0	3.0	3.5	6.3	S	RRR	-	1
Aleurodiscus disciformis	0	1.0	3.1	4.2	S	RRR	-	1,5,6,7,13
Amanita aspera	1	4.6	1.5	1.0	M	R	RR	2,3,4,6,7,9
Amanita crocea	2	1.5	1.2	1.0	M	RRR	-	4,11
Amanita eliae	0	1.7	1.5	1.0	M	RRR	-	3,4,5
Amanita friabilis	1	1.2	1.3	1.3	M	?	RRR	2,13

2. To provide information for nature conservationists and environmental planners to facilitate the interpretation of mycological data for planning regulation and protection of certain areas, or the acquisition of nature reserves, as well as for the evaluation of management practices.
3. To provide information for decision makers and politicians to enable them to estimate the threats to macrofungi and to develop measures and laws to prevent further declines or to improve the situation.
4. To provide basic data for the selection of species for monitoring programmes.
5. To provide basic data for the selection of species for possible protection by law.
6. To enable different lists to be compared in order to estimate the international status of the included species.

Fungi and nature management

Nature management is understood here as the variety of measures available for the maintenance, restoration, or increase of the biological values of particular areas, especially nature reserves (e.g. Bakker, 1989). Measures may be implemented to maintain the variety of ecosystems or to benefit the populations of certain organisms.

Several authors have stressed the fundamental differences between external management, directed to the minimalization of unwanted influences from outside a protected area, and internal management directed to create an optimal situation within a reserve by conscious measures (e.g. van Leeuwen, 1966). The possibilities of preventing or excluding unwanted influences from outside are increasingly becoming reduced, due to the increased importance of large-scale environmental changes, for instance the world-wide changes in climate and atmospheric chemistry, transnational air pollution and deposition of acidifying substances and nitrogen oxides, and regional lowering of the groundwater table and water pollution. Effective measures can only be taken against regional and more local influences, for

Table 12.3. The effects of various management systems and measures in average temperate forests on the species diversity of different fuctional groups of macrofungi.

Measure	Ectomycorrhizal fungi	Soil inhabiting saprobic fungi	Parasites on trees	Wood inhabiting saprobic fungi	Dung inhabiting fungi
selective tree cutting	0	0	0	0	0
no human influence at all	–	+	++	++	±
clear cutting	—	–	—	—	±
grazing by cattle	+?	–?	±	±	++
removal of litter	++	—	±	–	±
application of lime	–	+/–	+	±	+
application of N-fertilizers	—	–	±?	±	+

++ strong increase, + weak increase, ± neither increase nor decrease, – weak decrease, — strong decrease. All effects with regard to the selected standard treatment: selective cutting refers to mature stands of uniform age.

instance by the creation of buffer zones between a reserve and adjacent agricultural land and by the hydrological isolation of an area.

Internal management practices may range from the conscious termination of any human influence ("doing nothing"), to the continuation of traditional use or even the creation of new nature areas by digging, sod cutting, inundation, etc.

The mycoflora is only exceptionally considered in the planning of management in forests and nature reserves, and little detailed information is available on the effects of various practices on the mycoflora. As a logical sequel of the above, the effects of several forms of human interference on the mycoflora in temperate forests and seminatural grasslands are tentatively treated here as examples (Tables 12.3, 12.4). One important generalization is that there are no types of management which equally benefit all ecological groups of macrofungi; sometimes a measure useful for some threatened species can be harmful to others.

In the case of forests the cessation of management practices is often advocated by nature conservationists in order to restore natural ecosystems so that they include uneven age classes of trees and dead trunks in all stages of decomposition. It is true that virgin forests have become very rare in most regions, and that they are of outstanding importance for fungi, particularly wood-inhabiting species. However, the diversity of mycorrhizal fungi and maintainance of some species (e.g. *Cantharellus cibarius*, hydnoid fungi) is in some instances better served by the continuation of

Table 12.4. The effects of various management systems and measures in temperate, seminatural grassland with scattered trees on poor soils, on the species diversity of different functional groups of macrofungi.

Measure	Saprobic fungi of poor substrates	Saprobic fungi of rich substrates	Ectomycor-rhizal fungi	Dung inhabiting fungi
extensive grazing	0	0	0	0
mowing and removal of the sward	±	–	±	—
mowing, no removal of the sward	—	+	—	–
intensive grazing	–	+	–	+
application of additional dung	—	++	—	+
application of artificial fertilizers	—	+	—	±
removal of top soil	—	—	±	±
ploughing and resowing	—	—	—	±

++ strong increase, + weak increase, ± neither increase nor decrease, – weak decrease, — strong decrease. All effects with regard to the selected standard treatment: extensive grazing by herbivores without additional fertilizer application.

long-standing human practices which prevent the accumulation of thick litter layers, such as the removal of litter or the regular coppicing of trees (e.g. Grosse-Brauckmann and Grosse-Brauckmann, 1978; Jansen and van Dobben, 1987; Kuyper, 1989). It is to be expected that soil-inhabiting saprobes will be strongly reduced by such practices. The application of lime and fertilizers is under most circumstances unfavourable to the majority of fungi, but liming may have a neutral or weakly positive effect in soils with a low humus content (e.g. Kuyper, 1989). It is evident that the prevalent forestry practice of clear-cutting is unfavourable for most fungi, but even this measure promotes a restricted group; after replanting with young trees the abundance of early-stage mycorrhizal fungi is usually enhanced (e.g. Dighton and Mason, 1985).

Old, semi-natural, poor grasslands are amongst the most strongly threatened habitats. Grazing in low densities and mowing, followed by removal of the sward, are both practices with positive, although rather different, effects on the mycoflora (Arnolds, 1980; Nitare, 1988). The results of mowing strongly depend on the period of action; mowing just before the main fruiting season (in Western Europe late August or early September) is the best, but from a botanical point of view an earlier mowing is often preferred (e.g. Bakker, 1989). The application of organic manure has strongly negative effects on the characteristic macrofungi of poor grasslands, but may encourage coprobic fungi. However, even this is not universally true since some rare and endangered dung-inhabiting fungi are almost restricted to poor grasslands, such as *Poronia punctata.*

Special attention should be paid to old trees in poor grasslands since they may be accompanied by a rich mycorrhizal flora, including many rare species. In The Netherlands this situation is especially frequent in old, regularly mown roadsides on both sandy and clay soils. Field experiments in this habitat have indicated that the flora of both mycorrhizal and saprobic fungi was optimally developed when the sward was cut and removed in August each year (P.J. Keizer, pers. comm.).

Future research on the conservation of fungi

Research on fungi in relation to nature conservation is still in its infancy. In view of the dramatic changes in the mycoflora in large parts of Europe, including the strong decrease of important ecological groups, efforts in this field should be drastically increased in the near future. This also implies an extension of such studies to other groups of fungi, and to other parts of the world, in particular areas with a long standing mycological tradition, notably southern Europe, North America, and Japan.

There is a major need for more and accurate data on the geography and ecology of macrofungi. The successful gathering of this kind of data necessitates a high degree of cooperation between professional and amateur mycologists. The information collected should be presented in an easily accessible, conveniently arranged form, for instance as well-documented Red Lists and surveys of species of value as ecological indicators, analogous to those published on phanerogams (e.g. Ellenberg, 1974; Arnolds, 1989*c*). A presentation of results in a semi-popular form is necessary to reach workers in the fields of nature conservation, environmental politics, etc.

In turn, reliable studies on geography and ecology are dependent on a sound taxonomic and nomenclatural basis. Unfortunately, traditional floristic and monographic work on macrofungi is itself endangered in most countries. The time is probably now ripe for the coordination of such studies at an international level and to propose a separation of tasks, particularly between different taxonomic centres in Europe. International cooperation is necessary in all respects, for example in the development of uniform methods for mapping programmes, the initiation of an international monitoring programme in vulnerable habitats, and the compilation of supranational Red Lists. One important initiative in this respect was the founding of the European Council for Conservation of Fungi during the IX Congress of European Mycologists in Oslo in 1985; this will increasingly play a coordinating role in these activities.

Our knowledge of the relation between the mycoflora and different management practices in forests and nature reserves is still very scanty. Problems of this type offer a most promising field of research in applied ecology. Finally, the consequences of changes in the mycoflora for the functioning of ecosystems are still poorly understood. Increased insight into these relationships will without doubt contribute to a better appraisal of macrofungi as essential components of ecosystems, and of mycology as an essential branch of science.

Acknowledgement

Many thanks are due to my colleague Dr Th.W. Kuyper (Wijster), who gave me many valuable suggestions to improve the text.

Note

This paper is Communication No. 434 of the Biological Station, Wijster.

References

Agerer, R. (1985) Zur Ökologie der Mykorrhizapilze. *Bibliotheca mycologica* 97, 1–160.

Anon. (1980) *Contributions to the Red Data Book of the Tadzhik S.S.R.* Dushabe, Donish.

Arnolds, E. (1980) De oecologie en sociologie van Wasplaten. *Natura* 77, 17–44.

Arnolds, E. (1981) Ecology and coenology of macrofungi in grasslands and moist heathlands in Drenthe, The Netherlands. Vol. 1. *Bibliotheca mycologica* 83, 1–407.

Arnolds, E. (1982) Ecology and coenology of macrofungi in grasslands and moist heathlands in Drenthe, The Netherlands. Part 2. *Bibliotheca mycologica* 90, 1–501.

Arnolds, E. (1984) Standaardlijst van Nederlandse macrofungi. *Coolia* 26 (*Supplement*), 1-363.

Arnolds, E. [ed.] (1985) Veranderingen in de paddestoelenflora (mycoflora). *Wetenschappelijke Mededelingen van de Koninklijke Natuurhistorische Vereniging* 167, 1–101.

Arnolds, E. (1988*a*) The changing macromycete flora in the Netherlands. *Transactions of the British Mycological Society* 90, 391–406.

Arnolds, E. (1988*b*) Dynamics of macrofungi in two moist heathlands in Drenthe, The Netherlands. *Acta botanica Neerlandica* 37, 291–305.

Arnolds, E. (1988*c*) The Netherlands as an environment for agarics and boleti. In:

Bas, C., Kuyper, Th.W., Noordeloos, M.E. and Vellinga, E.C. (eds.), *Flora Agaricina Neerlandica* 1, 6–29. A.A. Balkema, Rotterdam.

Arnolds, E. (1989*a*) Former and present distribution of stipitate hydnaceous fungi (Basidiomycetes) in the Netherlands. *Nova Hedwigia* 48, 107–142.

Arnolds, E. (1989*b*) A preliminary Red Data List of macrofungi in the Netherlands. *Persoonia* 14, 77–125.

Arnolds, E. (1989*c*) The influence of increased fertilization on the macrofungi of a sheep meadow in Drenthe, The Netherlands. *Opera Botanica* 100, 7–21.

Arnolds, E. (1990) Decline of ectomycorrhizal fungi in Europe. *Agriculture, Ecosystems and Environment* 29, 209–244.

Arnolds, E. and de Vries, B. (1989) Oecologische statistiek van de Nederlandse macrofungi. *Coolia* 32, 76–86.

Bakker, J.P. (1989) *Nature Management by Grazing and Cutting.* Kluwer Academic Publishing, Dordrecht.

Bas, C. (1976) Een macro-oecologisch spectrum van de Europese Agaricales. *Coolia* 19, 86–93.

Boertmann, D. (1985) Vokshatte på overdrev i Vendsyssel. *Svampe* 12, 41–49.

Bumphus, J.A. and Aust, S.D. (1987) Biodegradation of environmental pollutants by the white rot fungus *Phanerochaete chrysosporium*: involvement of the lignin degrading system. *Bioessays* 6, 166–170.

Derbsch, H. and Schmitt, J.A. (1984) Atlas der Pilze des Saarlandes, Teil 1: Verbreitung und Gefährdung. In: Minister für Umwelt des Saarlandes (ed.), *Aus Natur und Landschaft im Saarland* 2, 1–536.

Derbsch, H. and Schmitt, J.A. (1987) Atlas der Pilze des Saarlandes, Teil 2: Nachweise, Ökologie, Vorkommen und Beschreibungen. In: Minister für Umwelt des Saarlandes (ed.), *Aus Natur und Landschaft im Saarland* 3, 1–816. Saarbrücken.

Dighton, J. and Mason, P.A. (1985) Mycorrhizal dynamics during forest tree development. In: Moore, D., Casselton, L.A., Woods, D.A. and Frankland, J.C. (eds), *Developmental Biology of Higher Fungi.* Cambridge University Press, Cambridge, pp. 117–139.

Ellenberg, H. (1974) Zeigerwerte der Gefässpflanzen Mitteleuropas. *Scripta Geobotanica* 9, 1–97.

Fellner, R. (1983) Mycorrhizae-forming fungi in climax communities at the timberline in Giant Mountains. *Ceská Mykologie* 37, 109.

Fellner, R. (1988) Effects of acid depositions on the ectotrophic stability of mountain forest ecosystems in Central Europe (Czechoslovakia). In: Jansen, A.E., Dighton, D. and Bresser, A.H.M. (eds), *Ectomycorrhizas and Acid Rain.* Berg en Dal, Bilthoven, pp. 116–121.

Fellner, R. (1989) Mycorrhiza-forming fungi as bioindicators of air pollution. *Agriculture, Ecosystems and Environment* 28, 115–120.

Fogel, R. (1976) Ecological studies of hypogeous fungi II. Sporocarp phenology in a western Oregon Douglas Fir stand. *Canadian Journal of Botany* 54, 1152–1162.

Grosse-Brauckmann, H. and Grosse-Brauckmann, G. (1978) Zur Pilzflora der Umgebung von Darmstadt vor 50 Jahren und heute (Ein Vergleich der floristischen Befunde Franz Kallenbachs aus der Zeit von 1918 bis 1942 mit dem gegenwärtigen Vorkommen der Arten). *Zeitschrift für Mykologie* 44, 257–269.

Hallingbäck, T. (1988) *A Preliminary List of Threatened Fungi in Sweden.* Report of the Department of Ecology and Environmental Research, University of Uppsala, Uppsala.

Hille, M. and Scholz, P. (1988) Der mykologische und lichenologische Durchforschungsgrad der Naturschutzgebiete der D.D.R. *Archiv für Naturschutz und Landschaftsforschung, Berlin* 28, 139–151.

Høiland, K. (1988) *A Preliminary List of Threatened Macrofungi in Norway.* Botanical Garden and Museum, Oslo (unpubl.).

Jansen, A.E. (1984) Vegetation and macrofungi of acid oakwoods in the northeast of the Netherlands. *Pudoc, Agricultural Research Reports* 923, 162.

Jansen, A.E. and de Vries, F.W. (1988) *Qualitative and quantitative research on the relation between ectomycorrhiza of* Pseudotsuga menziesii, *vitality of host and acid rain. Report 1985–1987.* Landbouwuniversiteit, Wageningen.

Jansen, E. and van Dobben, H.F. (1987) Is decline of *Cantharellus cibarius* in the Netherlands due to air pollution? *Ambio* 16, 211–213.

Kreisel, H. (1960) Pilze in Naturschutzgebieten. *Naturschutz-Arbeit und naturkundliche Heimatforschung im Bezirk Rostock-Schwerin-Neubrandenburg* 7, 36–38.

Krieglsteiner, G.J. (1977) *Die Makromyzeten der Tannen-Mischwälder des Inneren Schwäbisch-Frankischen Waldes (Ostwürttemberg) mit besonderer Berücksichtigung des Welzheimer Waldes.* Lempp Verlag, Schwäb Gmünd.

Kuyper, Th.W. (1989) Auswirkungen der Walddüngung auf die Mykoflora. *Beiträge zur Kenntnis der Pilze Mitteleuropas* 5, 5–20.

Kuyper, Th.W. and de Vries, B. (1990) Effects of fertilisation on the mycoflora of a pine forest. In: Oldeman, R.A.A., Schmidt, P., Arnolds, E.J.M. and Staudt, F. (eds), *Forest Ecosystems and their Components.* Wageningen, Agricultural University, Wageningen, pp. 102–111.

van Leeuwen, C.G. (1966) Het botanisch beheer van natuurreservaten op struktuuroecologische grondslag. *Gorteria* 3, 16–28.

Meyer, F.H. (1984) Mykologische Beobachtungen zum Baumsterben. *Allgemeines Forst Zeitschrift* 39, 212–228.

Nitare, J. (1988) Jordtungor, en svampgrupp på tillbakegång i naturliga fodermarker. *Svensk Botanisk Tidskrift* 82, 341–368.

Otto, P. (1990) *Die terrestrischen Stachelpilze der D.D.R. – Taxonomie, Ökologie, Verbreitung und Rückgang.* Dissertation, Martin-Luther Universität Halle-Wittenberg.

Parent, G.H. and Thoen, D. (1986) Etat actuel de l'extension de l'aire de *Clathrus archeri* (Berkeley) Drung (syn.: *Anthurus archeri* (Berk.) Ed. Fischer) en Europe et particulièrement en France et au Benelux. *Bulletin trimestriel de la Société mycologique de France* 102, 237–272.

Philippi, G. (1981) Die Bedeutung von Altholzbeständen aus botanischer Sicht. *Beihefte zu den Veröffentlichen für Naturschutz und Landschaftspflege in Baden-Württemberg* 20, 19–22.

Rald, E. (1985) Vokshatte som indikatorarten for mykologisk vaerdifulde overdrevslokaliteter. *Svampe* 11, 1–9.

Rald, E. (1986) Vokshattelokaliteter på Sjaelland. *Svampe* 13, 1–10.

Rassi, P. and Väisänen, R. (1987) *Threatened Animals and Plants in Finland.* Ministry of Environment, Helsinki.

Rayner, A.D.M. and Boddy, L. (1988) *Fungal Decomposition of Wood, its biology and ecology.* John Wiley and Sons, Chichester.

Richardson, M.J. (1970) Studies on *Russula emetica* and other agarics in a Scots pine plantation. *Transactions of the British Mycological Society* 55, 217–229.

Rücker, T. and Peer, T. (1988) Die Pilzflora des Hellbrunner Berges: Ein historischer Vergleich. *Berichte der Naturwissenschaftlich-medischen Vereins, Salzburg* 9, 147–161.

Sammler, P. (1988) Die Pilzflora sandiger Kiefernforste in der Beelitzer und Fresdorfer Heide. *Gleditschia* 16, 223–240.

Schlechte, G. (1986) Zur Mykorrhizapilzflora in geschädigten Forstbeständen. *Zeitschrift für Mykologie* 52, 225–232.

Senn-Irlet, B.I. (1986) *Ökologie, Soziologie und Taxonomie alpiner Makromyzeten (Agaricales, Basidiomycetes) der Schweizer Zentralalpen.* Dissertation, Universität Bern.

Termorshuizen, A. and Schaffers, A.P. (1987) Occurrence of carpophores of mycorrhizal fungi in selected stands of *Pinus sylvestris* in the Netherlands in relation to stand vitality and air pollution. *Plant and Soil* 104, 209–217.

Thoen, D. (1976) Facteurs physiques et fructification des champignons supérieurs dans quelques pessières d'Ardenne Meridionale (Belgique). *Bulletin mensuel de la Societé Linnéenne de Lyon* 45(8), 269–284.

Winterhoff, W. (1978) Gefährdung und Schutz von Pilzen. *Beihefte zu den Veröffentlichungen für Naturschutz und Landschaftspflege in Baden-Württemberg* 11, 161–167.

Winterhoff, W. (1984) Vorläufige Rote Liste der Grosspilze (Makromyzeten). In, Blab, J., Nowak, E., Trautmann, W. and Sukopp, H. (eds), *Rote Liste der gefährdeten Tiere und Pflanzen in der Bundesrepublik Deutschland.* 4. Aufl. Kilda, Greven, pp. 162–184.

Winterhoff, W. (1989) Die Bedeutung der baden-württembergischen Bannwälder für den Pilzartenschutz. *Mitteilungen der Forstlichen Versuchs und Forschungsanstalt Baden-Württemberg, Waldschutzgebiete* 4, 183–190.

Winterhoff, W. and Krieglsteiner, G.H. (1984) Gefährdete Pilze in Baden-Württemberg. *Beihefte zu den Veröffentlichungen für Naturschutz und Landschaftspflege in Baden-Württemberg* 40, 1–120.

Wöldecke, K. and Wöldecke, K. (1988) Erhaltet die Lisei! – Ein Laubmischwald als Refugium gefährdeter Grosspilze und Gefässpflanzen im Lemgow (Landkreis Lüchow-Dannenberg). *Jahrbuch der Naturwissenschaftlichen Vereins des Fürstentums Lüneburg* 38, 131–156.

Wöldecke, K. and Wöldecke, K. (1990) Zur Schutzwürdigkeit eines Cladonio-Pinetums mit zahlreichen gefährdeten Grosspilzen auf der Langendorfer Geest-Insel (Landkreis Lüchow-Dannenberg). *Beiträge zur Naturkunde Niedersachsens* 43, 62–83.

13

The Teaching of Mycology

J. Webster, *Department of Biological Sciences, Hatherly Laboratories, University of Exeter, Prince of Wales Road, Exeter EX4 4PS, UK.*

ABSTRACT Some ways in which mycology teaching can be made lively are discussed. Topics include attendance at fungus forays, the relevance of mycology, fungal classification, what to teach, when to teach, the use of text books, visual aids, practical classes, provision of living material, cultivation of plant pathogens, the use of cultures, recipe books, technicians, the training of research assistants, demonstrators and academic staff, the role of the student, drawings and projects.

Introduction

It is over 40 years since I began my present occupation as a university teacher. There are two main parts to this job: research and teaching. The results of research find their outlet in the form of papers read at conferences or published in journals and books. In contrast, very few conferences are devoted to the teaching of mycology, and the published literature on teaching techniques is relatively scanty. In an attempt to redress this balance I should like to discuss some of the ways in which the teaching of mycology can be made more lively.

Fungus forays

Fungus forays attract a wide variety of people, including naturalists, gastronomists, and professional mycologists. It is very instructive for students to join fungus forays and to see fungi at work and at home, that is in the field. There is no substitute for first-hand experience. There are so many things which can be learnt, for example the importance of collecting the *entire* basidiome, and not to leave the volva behind, the value of using

Frontiers in Mycology. Honorary and General Lectures from the Fourth International Mycological Congress, Regensburg 1990. Edited by D.L. Hawksworth.

the senses of touch (texture), taste, and smell. The relationship between mycorrhizal fungi and their tree hosts can be readily appreciated. It is important to scrape away the surface litter to demonstrate mycorrhizal roots, mycelium, and mycelial cords. The difference between white- and brown-rot wood decay can be demonstrated. Although it is possible to identify some fungi in the field, it is best to bring back the specimens to a work room or laboratory and then attempt to identify the finds. Fortunately there is a wide choice of illustrated books which help identification. At first many students are content to match their specimens to the illustrations, but this has its drawbacks. Keys using macroscopic characters will be adequate at first to lead to recognition of genera, and for many students this is sufficient, for example to be able to recognize an *Amanita* or to distinguish between a *Lactarius* and a *Russula.* I always encourage students to come to grips with the problems of identification, and to attempt to name their own collections. Whilst they are doing this I put out a named display of easily recognizable specimens, which they can use to check their own identifications. I usually end the excursion with a slide show in the form of a competition. It is often salutary to the students to realize that many amateur naturalists know far more than they do. However, it is important not to discourage students, some of whom are bewildered and put off by the wealth of field material. I try to overcome this by pinning up coloured wall charts with illustrations, for example, of poisonous or edible fungi. I encourage the braver ones to have confidence in their identification by eating such well-known favourites as *Boletus edulis, Hydnum repandum,* or *Lepiota procera.* I also like to bring in a few striking specimens each weekend and display them.

Those who teach mycology in universities and similar institutions can do much to help foster an understanding of larger fungi by leading forays and encouraging children to learn about fungi and how to deal with them. Books such as Kendrick's (1986) *A Young Person's Guide to the Fungi,* or Dobbs' (1962) *Fungi for Fun* show what is possible. Teaching mycology can be done at many levels.

For the serious student longer forays are possible, such as the spring or autumn forays organized by the various mycological and natural history societies lasting several days. Some have travel grants or subsidized rates to encourage students to attend.

The relevance of mycology

Mycology is taught within the framework of several possible disciplines, for example, biology, botany, microbiology, agriculture, forestry, and medicine. My own background is that of a botanist. With the change in outlook away from the traditional morphological and phylogenetical approach towards a more functional one, students are questioning the relevance of traditional mycology. Fortunately there are several excellent articles and books which set out some of the reasons why fungi are important. They include the stimulating book by Large (1958) *The Advance of the Fungi*, which traces the history of plant diseases caused by fungi and their control. Christensen's (1975) *Molds, Mushrooms and Mycotoxins* covers a wide range of topics from poisonous and hallucinogenic mushrooms to ergot and ergotism, fungus spores and respiratory allergy, and fungi pathogenic to man and animals. Another book with more emphasis on industrial mycology is Gray's (1959) *The Relation of Fungi to Human Affairs*. There are many other readable books covering more specialized topics such as the role of fungi in biotechnology, food spoilage, timber decay, mycorrhizas, plant pathology, human mycoses, etc. Attractive video-tapes are available which illustrate the relevance of mycology; one which I have used is *The Rotten World About Us* produced by the British Broadcasting Corporation (BBC) in 1978. There is thus ample justification for including courses in mycology for every student of biology.

Classification

Most students need a framework of classification within which to order the organisms they study. Classification has two functions. Ideally it should reflect phylogeny and relationships. On a more practical level it is valuable as a cataloguing system, that is as a basis for information retrieval. Students should understand that classification is subjective, relying on the judgement of individual taxonomists who may assess the relative importance of the available criteria differently. It should not then come as a surprise to find that there is no one agreed system of classification. I prefer not to discuss in detail the relative merits of different systems of classification, but instead, to take one of them as a framework. Others may take a different view. Techniques such as fine structural similarities, common biochemical pathways, similarities in chemical structure and DNA homology are proving powerful tools in indicating relationships, and probably in the future we shall have firmer grounds on which to base a phylogenetically related classification. However, the classification in such groups as the ascomycetes is changing so rapidly that it would be difficult for an undergraduate to keep up with it.

What to include

The choice of material for inclusion in undergraduate mycology courses will depend on the stage at which they are taught, the discipline they are acquiring, the time available, etc. For me an important criterion is the availability of suitable material *in the living state*, which can be readily collected or cultivated in some form. Much useful information is available in the form of practical guide books, media recipes, sources of material, etc., in which the accumulated wisdom and knowledge of other teachers and research workers can be tapped. These practical guides will be referred to later.

If mycology courses are taught in more than one year, some progression in content and aim is essential. I believe that the introductory course should attempt to introduce the structure, reproduction, and life-cycles of representatives of the major fungal groups, and provide a vehicle for learning the essential vocabulary of technical terms as a foundation for later work. At the same time it is possible to choose examples illustrating the relevance of fungi, for example in relation to plant pathology, enzymology, genetics, and mycorrhizal symbiosis. In later years more detailed consideration can be given to particular groups or to topics such as fungal physiology, differentiation, or ecology. Instead of the more formal, set practical work, more open-ended research-type projects can be very popular. These often excite the better students and set them on the road to research. I will return to projects later.

When to teach mycology

If a choice of time for teaching is open, in my opinion the autumn is the best in temperate regions. This is because so many of the larger fungi are fruiting then and fruit-bodies can be collected for laboratory study before they are killed by frosts. Many plant pathogens can also be found at this time. It is, of course possible to teach at other times, especially if the emphasis is more on laboratory work using cultures than on field-based material. As I shall show later, it is possible by the use of a deep-freeze, a refrigerator, cloches, and glasshouses to extend the period over which living material can be provided. For certain groups of fungi a later time of the year may be preferable; for example, the aquatic hyphomycetes whose spores are abundant in foam reach a seasonal peak of abundance after the fall of deciduous tree leaves, so that in the Northern Hemisphere November and December are ideal months to study them.

Text-books

Practice varies in the recommendation of suitable text-books. In North American universities the use of a single text for a course is more common than in Europe. I prefer to recommend a choice of books from the several available. Mycology is changing rapidly in emphasis and expanding in content, so it is valuable to see a succession of new ideas and approaches. Although fungi are ubiquitous and cosmopolitan, which means that many fungi are common the world-over, there is a need for more "regional" text-books based on locally abundant material. Nowhere is this more true than in the tropics. I would like to encourage teachers from such countries to write books suitable for their region, based on plentiful local material, or to get together to write a multi-authored text suitable for use throughout the moist tropics.

Visual aids

Mycology is a very visual subject, with aesthetically pleasing material available from the fine structural to the macroscopic, which lends itself to illustrations in several media. Most laboratories possess suitable photo-micrographic or macroscopic cameras to take good illustrations suitable for lectures and demonstrations. Students find it helpful to study poster-type displays of striking photographs of fungi. I pin up a corridor display of photographs of fungi close to the laboratory where students are working on fungi. These photographs supplement the practical material and can provide a valuable focus for discussion.

The ease with which transparencies can be copied means that slide exchange is easily possible, and the Mycological Society of America, through its newsletter, *MSA News*, is active in promoting such exchanges. The British Mycological Society owns an excellent set of colour slides of mainly larger fungi available on loan to members. Sets of transparencies are also available from commercial sources. Mention has already been made of video-tapes on fungal subjects. The equipment is readily available to prepare video-tapes of microscopic subjects, and I recommend this strongly for moving subjects such as the release of zoospores, behaviour of motile gametes, and discharge of ascospores, aeciospores and ballistospores.

There are many commercially available cine-films on fungal subjects for hire. I can particularly recommend those produced by the Institut für Wissenschaftlichen Film (IWF; Nonnenstieg 72, 3400 Göttingen). An interesting development which they are pioneering is the use of video-discs which combine many of the features of a cine-film and a video-recording and can store images from many thousands of transparencies. Edited excerpts of their films are now being assembled on a mycological video-disc, in which it is possible to follow frame-by-frame, developmental sequences taken from time-lapse films; for example of stages in nuclear division in the plasmodium of a slime mould, the development of the sporangium of *Pilobolus*, the zygospore of *Rhizopus*, or in real time the escape of gametes and mating in *Allomyces*, and the release of zoospores in *Saprolegnia, Pythium* or *Albugo.*

It is important for us, as teachers, not only to take advantage of the films, video-tapes, and video-discs available, but, where we have special expertise, to become involved in the preparation of new material. The making of a film can be a valuable educational exercise in itself.

Programmed learning in mycology has not been used as widely as it might. The technology exists for mounting a collection of diagrams, photomicrographs and macroscopic pictures of fungi on a tape-slide projector with an accompanying text and programmed slide changes. If studied along with a suitable text, this may provide useful for distance-learning or remedial teaching for slow learners or absentees. The amount of time needed to prepare such presentations is high, but if multiple copies were available for distribution there might be compensations. I have used such a system to supplement practical classes, but generally prefer to supply living material and for the students to have first-hand experience.

Practical classes

The best opportunity to get to know a student, and his or her capabilities, is in demonstrating in practical classes. Even when the classes are large, it is highly desirable for the person giving the course, however senior, to demonstrate, and not to delegate this entirely to colleagues. It is best for the person in charge to be responsible, for checking personally any material, culture, or demonstration that is provided.

Working demonstrations can enliven practicals. It is simple to demonstrate splash dispersal of basidiospores of *Lycoperdon*, peridioles of *Cyathus stercoreus*, or the conidia of *Nectria cinnabarina.* The discharge of the glebal mass of *Sphaerobolus* over a distance of several metres excites interest, as does the functioning of the falling closed ascomata (cleistocarps) of *Phyllactinia guttata*, the accuracy of shooting of *Pilobolus* sporangia towards a target, or the homing reaction of a monokaryotic hypha of *Coprinus cinereus* towards a compatible conidium.

Living material

The provision of fresh material is of the utmost importance if mycology is to retain a lively image. This may involve locating material some days or weeks beforehand, and arranging to collect it shortly before the practical class. In some cases, for example fungi growing on woody substrata (*Xylaria*, polypores, jelly fungi), such material may be collected well in advance and stored in a suitable shady moist environment before use. I collect the eggs of *Phallus impudicus* 2–3 days before they are needed and incubate them in moist peat in a covered plant propagator to ensure the availability of expanded basidiomes.

One of the problems in providing fresh material is that the availability of good specimens may not coincide with the time of the class, or a particular stage is not available when needed. With a little ingenuity, a deep-freeze, and a refrigerator, some material can be demonstrated in good condition months out of season. For example, I need to demonstrate asci of *Rhytisma acerinum* in October, although they are normally in this state in April. Overwintered infected leaves collected from the ground in March, before the hymenium is fully developed, will keep for months in a deep-freeze. Within two weeks of incubation in moist chambers such material produces plentiful asci. Similar techniques can be adopted with sclerotial fungi, such as *Sclerotinia curreyana* a parasite of *Juncus effusus.* We collect the sclerotia, easily found at the base of bleached *Juncus* culms in late winter, and store them in a refrigerator. In nature such sclerotia would form apothecia in April, but apothecia readily develop in November when sclerotia from the refrigerator are provided with moisture, light, and warmth. *Claviceps purpurea* sclerotia can be treated in the same way, and here the chilling is a stimulus to subsequent germination.

Cultivation of plant pathogens

Providing living material of diseased plants showing reproductive stages of their fungal pathogens at the appropriate time for mycology classes can be made much easier if they are cultivated in plant pathology gardens. Webster (1985) has given examples of small-plot techniques for growing necrotrophic and biotrophic fungal pathogens for class use. When used in conjunction with frames, cloches, and glasshouses, the growing season can be extended so that material can be available virtually throughout the year. It is especially valuable to use hosts infected with systemic pathogens. In some cases inoculum can be preserved dry or frozen, or freeze-dried so that the continued cultivation of infected host material is unnecessary. Instead of using agriculturally dangerous pathogens such as *Synchytrium endobiotricum* on *Solanum tuberosum*, substitution of unrelated species on "weed hosts" such as *Synchytrium taraxaci* on *Taraxacum* species enables living material to be used. Other examples of substitution are the cultivation of *Albugo* and *Peronospora* on the "weed host" *Capsella*, or powdery mildews such as *Erysiphe polygoni* on *Polygonum aviculare* or *E. heraclei* on *Heracleum sphondylium.* Although these examples relate to British host-pathogen combinations, the principle is a valuable one and could certainly find application elsewhere.

The use of cultures in teaching

Cultures are indispensible in teaching. There is much useful literature on suitable culture material, media recipes, and for preparing culture material for use at the precise time and date that it is required. I would particularly like to commend an article by Emerson (1958) in which he set out protocols for cultivating zoosporic fungi. Emerson's ideas have been expanded with many fine examples in Fuller & Jaworski's (1987) *Zoosporic Fungi in Teaching and Research.* Details are provided of sources of cultures, of recipes for media, times of subculture, special treatments, etc. In my experience information of this sort is very helpful. It enabled me to pin-point accurately the timing of meiospore release in *Allomyces macrogynus* which was useful for class, and also when making a film.

The zoosporic fungi above all others can be guaranteed to excite students. It is an unforgettable experience to watch the release of the zoospores of *Pythium* or *Saprolegnia,* and with a little experience the timing of these events can be predicted within a few minutes.

A less detailed account of the cultivation of zygomycetes (O'Donnell, 1979) is nevertheless useful. More general techniques including the use of cultures have been assembled in Stevens (1974), Esser (1985), Booth (1971), and Gams *et al.* (1987).

I would like to advocate more publications of this sort, not necessarily in book form. Many of us have pet organisms which we have used in our research and which lend themselves to teaching. For example, I have used *Itersonilia perplexans* in research on the mechanism of ballistopore discharge. It is an excellent object for demonstrating this to students, and very easily grown in culture (Ingold, 1990). I hope to see one of the mycological societies or journals sponsoring a loose-leaf file system with brief authoritative articles by experts. It would have the merit that it could be added to and updated. A modern development is the use of a computer program on techniques for cultivating fungi by Motta (1990), compatible with an IBM personal computer.

There are, of course, several excellent fungus culture collections. Some of these, for example the Centraalbureau voor Schimmelcultures (CBS) in CBS Baarn and the International Mycological Institute (IMI) at Kew run their own teaching courses on mycology, drawing on their culture collections for class material. CBS publish a text-book with outlines of techniques and media recipes (Gams *et al.*, 1987). However, it has to be said that the authenticated cultures from these sources can be expensive, and many mycologists prefer to isolate their own. Most universities keep a small collection of cultures used in teaching and research. University teachers are generally willing to provide or exchange cultures with colleagues in other universities.

The maintenance of a teaching collection need not be very expensive in labour. Many common moulds and even urediniospores of rusts survive freeze drying. Many fungi survive preservation under mineral oil. In some other cases resistant structures such as resting sporangia can be dried, for example in *Allomyces* and *Blastocladiella emersonii* on strips of sterile filter paper. The ascospores of *Pyronema* will remain viable for several months in sterile dry soil, whilst the glebal masses of *Sphaerobolus* or the peridoles of *Cyathus stercoreus* will survive in the dry state for several years and can be used to start off fresh cultures. The sclerotia of *Physarum polycephalum* revive readily after wetting, whilst those of *Claviceps purpurea* can be used, after surface sterilization, to start fresh cultures.

I have spent a good deal of time extolling the merits of living material in preference to the dead. Emerson (1964) wrote "... the prepared slide, the dried specimen and the pickle jar. All these are dependable, such as they are. They do not require advance planning, environmental control, knowledge of habitats or habits, visits to the laboratory on a Sunday midnight, or other ulcer engendering activities. If, however, dead specimens exert no strain on the physiology of the instructor, they are equally unlikely to stir either the hormones or the imagination of the student".

Support staff

The provision of living material is labour-intensive and needs skilled technician support. Support is needed in preparing media and sterile vessels, maintaining the culture collection, and putting up the cultures at the right time, under the right conditions on the appropriate medium. These skills are not acquired quickly. We, as university teachers, must be prepared to help train our technicians, by encouraging them to attend our own classes, relevant courses elsewhere, and by direct training. Reliable, observant, intelligent technicians are priceless, and should be cherished. We can also encourage them by acknowledging their help in our research publications or, better, by giving them joint authorship of our papers.

Most university teachers serve their apprenticeship as research students, post-doctoral fellows, or as research demonstrators or teaching assistants. Although only a low proportion of these eventually become teachers of mycology, their training is clearly important. Whilst some of the most distinguished mycological teachers have acquired their skills the hard way, and have taught themselves, there is much to be said for a young mycologist to attach himself to an experienced colleague. There are some fascinating genealogies of intellectual descent showing that many of the well-known names in mycology can trace their origins in the subject to distinguished forebears. The recently published short history of Constantine Alexopoulos (Blackwell, 1988) shows that his scientific pedigree can be traced through W.G. Farlow back to Anton de Bary. Alexopoulos in turn has begat (in the scientific sense) a series of distinguished research students, some of whom have also been very fertile.

Many other such examples could be given. Ralph Emerson acknowledges his own debt and that of several others to W.H. Weston. Many British mycologists, including myself, have reason to be thankful to Charles G. Chesters and to C. Terence Ingold, both of whom in their teaching and writings have had an enormous influence in awakening the mycological awareness of younger colleagues.

Young mycology teachers not only learn by example, but also learn on the job. It is very important to get them involved in every aspect of teaching, including collecting material, growing cultures, setting up demonstrations, leading tutorial discussions and seminars, and giving a few lectures in topics close to their research interests. They should participate in regular *post-mortem* sessions attended also by the technician responsible at which any shortcomings in the provision of materials, that is other aspects of the classwork are discussed with a view to improvement the next time. They should be encouraged to attend conferences and workshops and to present the results of their own work in the form of papers, posters and demonstrations.

The mycological societies around the world can help by mounting and encouraging workshops on particular groups of fungi and on new techniques. Whilst the conventional paper-reading sessions and poster demonstrations are useful, it is even better to have classes where the participants can gain first-hand practical experience.

Active encouragement for excellence in the teaching of mycology is given by the Mycological Society of America in its William H. Weston award. This has been awarded since 1980 to nine senior mycologists. It is an idea which other societies could take up, possibly with emphasis on younger teachers.

The role of the student

What does the good mycology teacher expect of his students, apart from the obvious virtues of intelligent interest and dedicated hard work? I believe that the lectures should excite the student to read round the subject for himself, not merely from text-books, but from the original scientific literature. This means providing references to relevant articles. The students should be required to write essays and to learn to express themselves clearly. In small tutorial groups they should be encouraged to open up discussion and be critical of what they have read. In the laboratory the emphasis should be on handling material for themselves, the making of temporary preparations, hand-cut sections, and epidermal strips. This is laborious and time-consuming, but much is learnt in the process. The experienced demonstrator will know how to get the best out of a particular specimen or culture, for example to look for a particular stage of development, or which part of the culture to look at to find the material of interest. Passing on this "know-how" can increase the students' understanding of the fungus.

I believe that it is important for students to prove things to their own satisfaction. For example, I ask students to project basidiospores from the telia of *Puccinia graminis* on a grass or cereal host onto leaves of potted *Berberis* seedlings, and ten days later they can find the pycnia, thus demonstrating heteroecism. They can spermatize the haploid pustules by Pasteur pipette and obtain aecia. I ask them to project the ascospores of *Pleospora herbarum* onto agar and see the *Stemphylium* anamorph one week later, thus proving pleomorphy. They inoculate their own cultures with compatible strains of *Phycomyces*, so demonstrating heterothallism.

Drawings

I am a firm believer in the value of drawing, not merely as a means of recording what a student sees, but as a means of testing his powers of observation and his ability to interpret it. If asked to draw a basidium of the cultivated mushroom *Agaricus bisporus*, many students show them with four spores rather than two. The appendages surrounding the zygospore of *Phycomyces* are often portrayed as arising from the zygospore itself, whilst examination of developing zygospores shows that they arise from the suspensor. The conidiogenous cells of *Aspergillus oryzae* are often shown as arising directly from the vesicle. It is important to correct such errors.

My own practice is not only to check the quality and accuracy of drawings as I demonstrate, but to collect up, correct and annotate the practical books at the end of each practical, so that the student can have rapid feedback on quality control, and can improve.

Projects

A good way of capturing students' interest and to test their capacity for original thinking and to do research is to see how they tackle open-ended investigations in the form of projects. They are probably best undertaken towards the end of a three year course. The ideas for projects can come from the students themselves or from the teacher, as a spin-off from his own research, reading or teaching. Projects can be devised as pilot investigations to test out some of one's wilder ideas before deciding whether or not to embark on a fuller investigation. It does not matter too much if the project does not yield positive results, so long as the student tackles the project in a sensible way, using sound techniques, proper controls, and appropriate statistical analysis. Sometimes if the project goes well the work merits publication as it stands, or encourages further work to be done. In Exeter our students spend ten weeks at the end of the second year working on projects, and they are expected to have their reports completed by the end of the long summer vacation. I should like to give examples of project work which has led directly to publication, or to further work which has been published.

It has long been suspected that nematophagous fungi such as *Arthrobotrys* attract their prey, and that capture is not merely the result of chance or idle curiosity. Traps are formed in response to the presence of nematodes and can also be stimulated to develop by adding horse serum. The idea that the traps themselves are attractive to nematodes was tested by an undergraduate, Jenny Field, in a very simple set-up (Field and Webster, 1977). She showed that mycelial discs of several predacious fungi, when stimulated to form traps by the addition of nematode extract or horse serum were much more attractive to nematodes then unstimulated traps. The differences were highly significant, and were not due to the extract or horse serum. We now use this method as a standard teaching technique in practicals on predacious fungi. James Harper, in his experimental work on the coprophilous fungus succession, showed that *Coprinus heptemerus* was very combative and inhibited the growth and fruiting of other coprophilous fungi (Harper and Webster, 1964). A clue to the mechanism of inhibition was discovered by an undergraduate project student Terry Burns who showed that if *Pilobolus* was inoculated into rabbit pellets with *Coprinus* the *Pilobolus* failed to fruit as soon as hyphal contact occurred. This was followed up by Ikediugwu who showed, using *Ascobolus*, that contact with *Coprinus* lead to cell death and loss of turgor in *Ascobolus*. This phenomenon, termed hyphal interference, has since been shown to be a very effective mechanism of competition in a wide range of basidiomycetes (Ikediugwu and Webster, 1970*a*, *b*; Ikediugwu *et al.*, 1970; Rayner and Webber, 1984). Undergraduate project work on zoospore discharge in *Pythium middletonii*, which is an excellent subject for teaching, has also led to publications (Webster and Dennis, 1967*a*, *b*), and a project by Henry Clarke on the effects of aeration in Ingoldian aquatic hyphomycetes in turn led to fuller investigations and publication (Towfik and Webster, 1973; Webster, 1975; Sanders and Webster, 1980). Isolation of *Olpidiopsis gracilis*, a mycoparasite of *Pythium*, was followed by a study of its host range, which extended to certain species of *Phytophthora* (Pemberton *et al.*, 1990). These examples show that just as the findings of research can have their effects on teaching, so project work, a form of teaching, can also have its effects on research.

Conclusion

I have tried to show that mycology need *not* be a dull subject, indeed that by relatively simple means it can be really lively, rewarding, and fun to learn. It can also be rewarding and fun to teach.

References

Blackwell, M. (1988) C.J. Alexopoulos. A short history. *Transactions of the British Mycological Society* 90, 153–158.

Booth, C. [ed.] (1971) *Methods in Microbiology.* Vol. 4. Academic Press, London and New York.

Christensen, C.M. (1975) *Molds, Mushrooms and Mycotoxins.* University of Minnesota Press, Minneapolis.

Dobbs, E. (1962) *Fungi for Fun.* Basil Blackwell, Oxford.

Emerson, R. (1958) Mycological organisation. *Mycologia* 50, 589–621.

Emerson, R. (1964) Performing fungi. *The American Biology Teacher* 6, 90–100.

Esser, K. (1985) *Kryptogamen. Cyanobacterien, Algen, Pilze, Flechten. Prakticum und Lehrbuch.* Springer Verlag, Berlin.

Field, J.I. and Webster, J. (1977) Traps of predacious fungi attract nematodes. *Transactions of the British Mycological Society* 68, 467–469.

Fuller, M.S. and Jaworski, A. [eds] (1987) *Zoosporic Fungi in Teaching and Research.* Southeastern Publishing Company, Athens, Georgia.

Gams, W., van der Aa, H.A., van der Plaats-Niterink, A.J., Samson, R.A. and Stalpers, J.A. (1987) *CBS Course of Mycology.* 3rd edn. Centraalbureau voor Schimmelcultures, Baarn.

Gray, W.D. (1959) *The Relation of Fungi to Human Affairs.* Henry Holt, New York.

Harper, J.E. and Webster, J. (1964) An experimental analysis of the coprophilous succession. *Transactions of the British Mycological Society* 47, 511–530.

Ikediugwu, F.E.O. and Webster, J. (1970*a*) Antagonism between *Coprinus heptemerus* and other coprophilous fungi. *Transactions of the British Mycological Society* 54, 181–204.

Ikediugwu, F.E.O. and Webster, J. (1970*b*) Hyphal interference in a range of coprophilous fungi. *Transactions of the British Mycological Society* 54, 205–210.

Ikediugwu, F.E.O., Dennis, C. and Webster, J. (1970) Hyphal interference by *Peniophora gigantea* against *Heterobasidion annosum. Transactions of the British Mycological Society* 54, 307–309.

Ingold, C.T. (1990) The ballistospore. *The Mycologist* 4, 36–37.

Kendrick, W.B. (1986) *A Young Person's Guide to the Fungi.* Mycologue Publications, Waterloo, Ontario.

Large, E.C. (1958) *The Advance of the Fungi.* Jonathan Cape, London.

Motta, J. (1990) *A Programmed Guide for the Cultivation of Fungi.* [A relational database program for the Mycology Laboratory; version 1.0.] Mycotechnology Services, Silver Spring, Maryland.

O'Donnell, K.L. (1979) *Zygomycetes in Culture.* Department of Botany, University of Athens, Georgia.

Pemberton, C.M., Davey, R.A., Webster, J., Dick, M.W. and Clark, G. (1990) Infection of *Pythium* and *Phytophthora* species by *Olpidiopsis gracilis* (Oomycetes). *Mycological Research* 94, 1081–1085.

Rayner, A.D.M. and Webber, J.F. (1984) Interspecific mycelial interactions – an overview. In: Jennings, D.H. and Rayner, A.D.M. (eds), *The Ecology and Physiology of the Fungal Mycelium.* Cambridge University Press, Cambridge, pp. 383–417.

Sanders, P.F. and Webster, J. (1980) Sporulation responses of some aquatic hyphomycetes in flowing water. *Transactions of the British Mycological Society* 74, 601–605.

Stevens, R.B. [ed.] (1974) *Mycology Guidebook.* University of Seattle Press, Seattle and London.

Towfik, S.H. and Webster, J. (1973) Sporulation of aquatic hyphomycetes in relation to aeration. *Transactions of the British Mycological Society* 59, 353–364.

Webster, J. (1975) Further studies of sporulation of aquatic hyphomycetes in relation to aeration. *Transactions of the British Mycological Society* 64, 109–127.

Webster, J. (1985) Plant pathology plots. *Sydowia* 38, 358–368.

Webster, J. and Dennis, C. (1967*a*) A technique for obtaining zoospores of *Pythium middletonii. Transactions of the British Mycological Society* 50, 329–331.

Webster, J. and Dennis, C. (1967*b*) The mechanism of sporangial discharge in *Pythium middletonii. New Phytologist* 66, 307–313.

Index